Lecture Notes in Electrical Engineering 1361

The book series *Lecture Notes in Electrical Engineering* (LNEE) publishes the latest developments in Electrical Engineering—quickly, informally and in high quality. While original research reported in proceedings and monographs has traditionally formed the core of LNEE, we also encourage authors to submit books devoted to supporting student education and professional training in the various fields and applications areas of electrical engineering. The series cover classical and emerging topics concerning:

- Communication Engineering, Information Theory and Networks
- Electronics Engineering and Microelectronics
- Signal, Image and Speech Processing
- Wireless and Mobile Communication
- Circuits and Systems
- Energy Systems, Power Electronics and Electrical Machines
- Electro-optical Engineering
- Instrumentation Engineering
- Avionics Engineering
- Control Systems
- Internet-of-Things and Cybersecurity
- Biomedical Devices, MEMS and NEMS

For general information about this book series, comments or suggestions, please contact leontina.dicecco@springer.com.

To submit a proposal or request further information, please contact the Publishing Editor in your country:

China

Jasmine Dou, Editor (jasmine.dou@springer.com)

India, Japan, Rest of Asia

Swati Meherishi, Editorial Director (Swati.Meherishi@springer.com)

Southeast Asia, Australia, New Zealand

Ramesh Nath Premnath, Editor (ramesh.premnath@springernature.com)

USA, Canada

Michael Luby, Senior Editor (michael.luby@springer.com)

All other Countries

Leontina Di Cecco, Senior Editor (leontina.dicecco@springer.com)

** This series is indexed by EI Compendex and Scopus databases. **

Patrick Siarry · M. A. Jabbar ·
Simon King Sing Cheung · Xiaolong Li
Editors

Proceedings of the 2023 International Conference on Wireless Communications, Networking and Applications

 Springer

Editors
Patrick Siarry
Laboratory of Images, Signals and Intelligent
Systems
Paris-Est Créteil University
Vitry-sur-Seine, France

Simon King Sing Cheung
Info Technology Office
Hong Kong Metropolitan University
Kowloon, Hong Kong

M. A. Jabbar
Department of CSE
Vardhaman College of Engineering
Hyderabad, Telangana, India

Xiaolong Li
Electronics and Computer Engineering
Technology
Indiana State University
Terre Haute, IN, USA

ISSN 1876-1100   ISSN 1876-1119 (electronic)
Lecture Notes in Electrical Engineering
ISBN 978-981-96-2408-9   ISBN 978-981-96-2409-6 (eBook)
https://doi.org/10.1007/978-981-96-2409-6

This Springer imprint is published by the registered company Springer Nature Singapore Pte Ltd.
The registered company address is: 152 Beach Road, #21-01/04 Gateway East, Singapore 189721, Singapore

If disposing of this product, please recycle the paper.

Preface

The 7th Annual 2023 International Conference on Wireless Communications, Networking and Applications (WCNA2023) was held on December 29 to 31, 2023 at Shenzhen, Guangdong, China.

It aims to bring researchers, engineers, and students to the areas of wireless communications, networking, and applications. The conference will provide a forum for sharing experiences and original research contributions on those topics. Researchers and practitioners are invited to submit their contributions to WCNA2023.

WCNA2023 proceeding tends to collect the most up-to-date, comprehensive, and worldwide state-of-the-art knowledge on wireless communications, networking, and applications. All the accepted papers have been submitted to strict peer-review by 2–4 expert referees and selected based on originality, significance, and clarity for the purpose of the conference. The conference program is extremely rich, profound, and featuring high-impact presentations of selected papers and additional late-breaking contributions. We sincerely hope that the conference will not only show the participants a broad overview of the latest research results on related fields but also provide them with a significant platform for academic connection and exchange.

The Technical program committee members have been working very hard to meet the deadline of review. The final program consists of 38 papers. There are four editors. All the editors are internationally recognized leading experts in their research fields, who have demonstrated outstanding proficiency and have achieved distinction in their profession. The proceedings would be published on Springer Book Series Lecture Notes in Electrical Engineering as a volume quickly, informally, and in high quality.

We would like to express our sincere gratitude to all the members of the international program committee and organizers for their hard work, precious time, and endeavor in preparing for the conference. Our deepest thanks also goes to the volunteers and staffs for their long-hours work and generosity they have given to the conference. Last but not least, we would like to thank each and every of the authors, speakers, and participants for their great contributions to the success of WCNA2023.

WCNA2023 Organizing Committee

Committees

General Chair

Siarry Patrick — Universite Paris-Est Creteil, France

Editors

Siarry Patrick	Universite Paris-Est Creteil, France
M. A. Jabbar	Vardhaman College of Engineering, India
Simon K. S. Cheung	Hong Kong Metropolitan University, China
Xiaolong Li	Indiana State University, USA

Technical Program Committee

Guillermo Escrivá-Escrivá	Universitat Politècnica de València, Spain
Surinder Singh	Sant Longowal Institute of Engineering and Technology, India
Pejman Goudarzi	Iran Telecom Research Center (ITRC), Iran
Antonio Muñoz	University of Malaga, Spain
K. Somasundaram	Amrita Vishwa Vidyapeetham, India
Daniela Litan	Oracle Developer, Romania
Artis Mednis	University of Latvia, Latvia
Hari Mohan Srivastava	University of Victoria, Canada
Sumit Kushwaha	Kamla Nehru Institute of Technology, India
Petko Hristov Petkov	Technical University of Sofia, Bulgaria
Pankaj Bhambri	I.K.G. Punjab Technical University, India
Marek Blok	Gdańsk University of Technology, Poland
Phongsak Phakamach	Rajamangala University of Technology Rattanakosin, Thailand
Mohammed Rashad Baker	Imam Ja'afar Al-Sadiq University, Iraq
Ahmad Fakharian	Islamic Azad University, Iran
Nikhil Marriwala	Kurukshetra University, India
Marco Listanti	University of Roma "La Sapienza," Italy

Jibendu Sekhar Roy	KIIT University, India
Sivaradje Gopalakrishnan	Puducherry Technological University, India
A. K. Verma	Thapar Institute of Engg & Technology, India
Kamran Arshad	Ajman University, UAE
Gyu Myoung Lee	Liverpool John Moores University, UK
Fathollah Bistouni	Islamic Azad University, Iran
Sachin Kumar	Kyungpook National University, South Korea
Hoang Trong Minh	Telecommunications Engineering, Vietnamese
Hari Shankar Singh	Electronics and Communication Engineering, India
Hooman Hematkhah	Chamran University (SCU), Iran
Alexandros-Apostolos A. Boulogeorgos	University of Piraeus, Greece
Muge Erel-Ozcevık	Manisa Celal Bayar University, Turkey
Lamri Sayad	University of Msila, Algeria
Ly-Minh-Duy Le	Ho Chi Minh University of Technology and Education, Vietnam
Anand Nayyar	Duy Tan University, Vietnam
Ahlem Ben Younes	ENSIT, University of Tunis, Tunisia
Chi-Wai Chow	National Yang Ming Chiao Tung University, Taiwan
Alexei G. Shishkin	Moscow State University, Russia
Loc Nguyen	Loc Nguyen's Academic Network, India
Fatiha Benkouider	Amar Telidji University of Laghouat, Algeria
Valentina Emilia Balas	Aurel Vlaicu University of Arad, Romania
Gheorghe Grigoras	"Gheorghe Asachi" Technical University of Iasi, Romania
Noor Zaman	Taylor's University, Malaysia
Qiang (Shawn) Cheng	University of Kentucky, USA
Yilun Shang	Northumbria University, UK
Suman Kr. Dey	National Institute of Technology Rourkela, India
Ahishek Shukla	A.P.J. Abdul Kalam Technical University, India
Bentaieb Samia	University Belhadj Bouchaib Aiin Temouchent, Algeria
Hyunsung Kim	Kyungil University, South Korea
A. Manikandan	Amrita School of Engineering, India
Jun Tao	Jianghan University, China
Najib Fadlallah	Lebanese University, Lebanon
Ning Zhang	Beijing Union University, China
S. Jeyadevi	Kamaraj College of Engineering and Technology, India
Dimitris Kanellopoulos	University of Patras, Greece
Lu Leng	Nanchang Hangkong University, China

Aizaz U. Chaudhry	Carleton University, Canada
Fateh Mebarek-Oudina	University of 20 Août 1955-Skikda, Algeria
Farrukh Arslan	Purdue University, USA
Hamid Reza Naji	Graduate University of Advanced Technology
Mehdi Salimi	St. Francis Xavier University, Canada
Khalid A. AlAfandy	Abdelmalek Essaadi University, Morocco
Ramya Tekumalla	Georgia State University, USA
Shancheng Zhao	Jinan University, China
Fu Tak Chung	Technological and Higher Education Institute of Hong Kong, China
Almetwally Mostafa	College of Computers Science & Information Systems King Saud University, KSA
Anand Nayyar	Duy Tan University, Vietnam
Phongsak Phakamach	Rajamangala University of Technology Rattanakosin, Thailand
Ntapat Worapongpat	Bangkok Thonburi University (BTU), Thailand
Zhen Chen	South China University of Technology, China
Atanaska Dimitrova Bosakova-Ardenska	University of Food Technologies, Bulgaria
Mário F. S. Ferreira	University of Aveiro, Portugal
Mohammad Mehdi Rashidi	University of Electronic Science and Technology of China, China
Pasura Aungkulanon	King Mongkut's University of Technology North Bangkok, Thailand
Gang Zhang	Harbin Institute of Technology, China
Hua-Yi Lin	China University of Technology, China
Ong Thian Song	Multimedia University Malaysia, Malaysia
Yiannis Koumpouros	University of West Attica, Greece
Enrique Guzmán Ramírez	Technological University of the Mixteca, México
Essalih Mohamed	Cadi Ayyad University, Morocco
Faiza Mekhalfa	Center for Development of Advanced Technologies
Setti Ahmed Soraya	Mascara University, Algeria
Shigeaki Sakurai	Toshiba Corporation, Japan
Aouad Siham, Ensias	Mohammed V University of Rabat, Morocco
Patrick Lanusse	Maître de Conférences HC-Ex, HDR, France
Daniela Litan	Hyperion University, Romania
Namoune Abdelhadi	University Relizane, Algeria
Tavarov Saidjon	South Ural State University (National Research University), Russian Federation
Juntao Fei	Hohai University, China
Constantin Klimov	Moscow Aviation Institute, Russia
Teh Sin Yin	Universiti Sains Malaysia, Malaysia

Dariush Akbarian	Islamic Azad University, Iran
Andrew Fish	University of Brighton, UK
Seyed Hamed Moosavirad	Shahid Bahonar University of Kerman, Iran
S. Muthukumar	Indian Institiute of Information Technology, India
Zaid Ameen AbdulJabbar Al-Selmi	University of Basrah, Iraqi
Tao Yu	China Academy of Management Science, China
Ali Fadhil Naser	Al-Furat Al-Awsat Technical University, Iraq
Klimis Ntalianis	University of West Attica, Greece
A. S. M. Sanwar Hosen	Jeonbuk National University, South Korea
Ningzhi Wang	Hainan University, China
Zeeshan Ahmad	Ningbo University of Technology, China
Yinglei Song	Jiangsu University of Science and Technology, China
Vilson Luiz Dalle Mole	Federal Technological University of Paraná, Brazil
Elisabeta Mihaela Ciortea	University of Alba Iulia, Romania
Cheng Siong Chin	Newcastle University, Singapore
Shadi Abudalfa	University College of Applied Sciences, Palestine
Xiaoping Zhou	Beijing University of Civil Engineering & Architecture, China
Zhiyu Jiang	University of Chinese Academy of Sciences, China
Xinxing Wu	University of Kentucky, USA
Yihong Chen	China West Normal Nuniversity, China
Patrick Siarry	Université Paris-Est Créteil Val de Marne, France
T. Velmurugan	D.G. Vaishnav College, Arumbakkam, Chennai, India
Janmenjoy Nayak	Maharaja Sriram Chandra Bhanja Deo (MSCBD) University, India
Abdel-Badeeh M. Salem	Ain Shams University, Egypt
V. Jayalakshmi	Vels Institute of Science Technology and Advanced Studies, India
Elżbieta Macioszek	Silesian University of Technology, Poland
Kefeng Guo	Army Engineering University of PLA, China
Zhe Chen	Dalian University of Technology, China
Zhixiao Wang	Xi'an University of Technology, China
Shadi Abudalfa	University College of Applied Sciences, Palestine
Huan Yu	Chengdu University of Technology, China
Joydev Ghosh	The New Horizons Institute of Technology, India
Abdelber Bendaoud	Djilali Liabès University of Sidi Bel-Abbès, Algerian
Chi-Hua Chen	Fuzhou University, China

Wenfeng Wang	Shanghai Institute of Technology, China
Juma Mary Atieno	South Eastern Kenya University (SEKU), China
Karim El Moutaouakil	Sidi Mohamed Ben Abdellah University, Morocco
Samir Ladaci	National Polytechnic School, Algeria
Habil Corneliu Doroftei	Alexandru Ioan Cuza University of Iasi, Romania
Ranjith Kumar Gatla	Institute of Aeronautical Engineering, India

Editors for WCNA2023

Prof. Patrick Siarry
Universite Paris-Est Creteil, France

Biography: Patrick Siarry was born in France in 1952. He received the PhD degree from the University Paris 6, in 1986, and the Doctorate of Sciences (Habilitation) from the University Paris 11, in 1994. He was first involved in the development of analog and digital models of nuclear power plants at Electricité de France (E.D.F.). Since 1995, he has been a professor in automatics and informatics. His main research interests are computer-aided design of electronic circuits and the applications of new stochastic global optimization heuristics to various engineering fields. He is also interested in the fitting of process models to experimental data, the learning of fuzzy rule bases, and neural networks.

Prof. M. A. Jabbar
Vardhman College of Engineering, India

Biography: Dr. M.A. Jabbar is a distinguished professional in the field of Computer Science and Engineering, with a specialization in Artificial Intelligence (AI) and Machine Learning (ML). He serves as the Professor and Head of the Department of Computer Science and Engineering (AI&ML) at Vardhaman College of Engineering, Hyderabad, Telangana, India. His work primarily focuses on advancing AI and ML, and his dedication and expertise have earned him numerous accolades. These include the Best Faculty Researcher Award from the CSI Mumbai Chapter, the Best HOD of the Year Award from CSI, the Outstanding Leadership Award from IEEE Hyderabad Section, and the Distinguished Contributor Award from the IEEE CS Chapter in 2021.

His remarkable research contributions span artificial intelligence, machine learning, and their interdisciplinary applications, solidifying his reputation as a leader in these fields. In addition to his research, Dr. M.A. Jabbar holds senior memberships in globally recognized professional organizations such as the Association for Computing Machinery (ACM) and the Institute of Electrical and Electronics Engineers (IEEE). He is also a lifetime member of the Computer Society of India (CSI) and the Indian Science Congress Association (ISCA). His diverse roles in editorial and research capacities reflect his unwavering commitment to advancing computer science and its transformative applications.

Prof. Simon King Sing Cheung
Hong Kong Metropolitan University, China

Biography: Dr. Cheung is currently the Director of IT at the Hong Kong Metropolitan University. He received his BSc and PhD in Computer Science from the City University of Hong Kong, and was admitted as IET fellow, IMA fellow, BCS fellow, HKIE fellow,

HKCS fellow. He is active in research, with near 200 publications in the form of books, book chapters, journal articles and conference papers in two distinct areas, namely, innovation and technology in education, and software and system engineering. He won the Outstanding Research Publication Award from the Open University of Hong Kong in 2016, the 1st class Achievement in Computer and IT from Shenzhen Science and Technology Association in 2016, the 1st class Outstanding CIO Award from the Hong Kong IT Joint Council in 2015, and an Honoree for IT Excellence from the CIO Asia in 2015. He has been listed in the Who's Who in Science and Engineering since 2007.

Prof. Dr. Xiaolong Li
Bailey College of Engineering & Technology, Electronic & Computer Engineering Technology, Indiana State University, USA

Biography: Xiaolong Li obtained a Bachelor's degree in Electronic and Information Engineering from Huazhong University of Science and Technology in 1999, followed by a Master's degree in Electronic and Information Engineering from the same university in 2002, and a PhD in Electronic and Computer Engineering from the University of Cincinnati (2006). Since 2008, he has been an assistant professor in the Department of Electronic and Computer Engineering Technology at Indiana State University.

Main Research Interests: Wireless and mobile networks, wireless ad hoc networks, wireless Internet, modeling and performance analysis, QoS in communication networks, wireless mesh networks, microcontroller based applications, security in wireless networks.

Contents

Unveiling Challenges in Migrating from 5G to 6G: Insights from Wireless Communication Networks

Shaista Farhat[1], Sumaiya Shaikh[2], M. A. Jabbar[1]([⊠]), and Shagufta Farhat[3]

[1] Department of Computer Science and Engineering (AI and ML), Vardhaman College of Engineering, Hyderabad, India
`jabbar.meerja@gmail.com`
[2] Department of Information Technology, Vardhaman College of Engineering, Hyderabad, India
[3] Department of Computer Science and Engineering (Data Science), Lords Institute of Engineering and Technology, Hyderabad, India

Abstract. The evolution from 5G to 6G wireless communication networks signals the dawn of a transformative era in technological innovation and connectivity. However, this evolution is fraught with multifaceted challenges that extend beyond mere technological upgrades. This paper unveils the hidden obstacles that must be navigated to achieve a seamless migration to 6G. Key challenges include the integration of advanced technologies such as terahertz communication, quantum computing, and artificial intelligence, which necessitate unprecedented levels of research and development. Additionally, the requirement for ultra-low latency, massive connectivity, and enhanced security introduces complexities in network architecture and design. Regulatory and standardization issues, alongside the need for substantial infrastructural investments, further complicate the migration process. Socio-economic factors, including digital divide concerns and the environmental impact of new infrastructure, also play critical roles. By discussing these challenges, this paper provides a comprehensive understanding of the prerequisites for successful 6G deployment, laying the groundwork for future research and policy development aimed at overcoming these obstacles.

Keywords: 6G Wireless Communication · 5G to 6G Migration · Terahertz Communication · Quantum Computing · Artificial Intelligence

1   Introduction

Wireless communication systems have become the modern-day equivalent of Eureka moments, thanks to rapid technological advancements and the evolution of symmetric technologies for the Internet of Things over recent decades. Currently, there are five generations of mobile wireless cellular communication systems, with the latest being the fifth generation (5G) wireless network. Since 1980, a new generation of wireless cellular communication has emerged roughly every decade. This progression includes the first generation of analog FM cellular systems in 1981, the second generation in 1992,

P. Siarry et al. (Eds.): WCNA 2023, LNEE 1361, pp. 1–13, 2025.
https://doi.org/10.1007/978-981-96-2409-6_1

the third generation (3G) in 2001, and the fourth generation (4G), also known as Long-Term Evolution (LTE), in 2011. In 2020, the standardization of 5G communications was finalized, and the system is now being deployed globally [1] (Fig. 1).

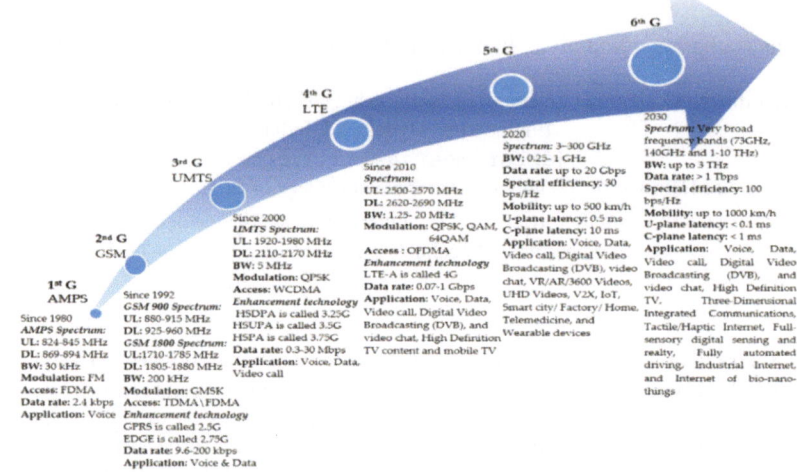

Fig. 1. Notable achievements for different generations of communication [2]

5G, the fifth generation of mobile networks, succeeds 1G, 2G, 3G, and 4G as the latest global wireless standard. It establishes a new type of network aimed at connecting virtually everyone and everything, including machines, objects, and devices. Engineered to provide higher multi-Gbps peak data speeds, ultra-low latency, enhanced reliability, massive network capacity, increased availability, and a more consistent user experience for a greater number of users, 5G wireless technology offers superior performance and efficiency. This advancement not only enhances user experiences but also enables the connection and transformation of various industries.

6G, or sixth-generation wireless, represents the next leap in cellular technology following 5G. Operating at higher frequencies than 5G, 6G networks will offer significantly increased capacity and much lower latency. Commercial launch of 6G internet is anticipated in 2030, with a key goal of achieving communication latency of one microsecond. This would make it 1,000 times faster—or have 1/1000th the latency—of the one millisecond throughput seen with 5G (Fig. 2).

The transition from 5G to 6G is motivated by the need to meet increasing demands for higher performance, enhanced connectivity, and new technological advancements as shown in the above Fig. 3. Here are several key reasons for this shift:

Higher Data Rates and Capacity: 6G will provide significantly higher data rates and greater network capacity than 5G, essential for supporting more data-intensive applications and a larger number of connected devices.

Enhanced Connectivity: With the growing number of devices and sensors connecting to the internet, 6G will offer the necessary infrastructure to support massive Internet of Things (IoT) deployments, ensuring reliable and efficient communication.

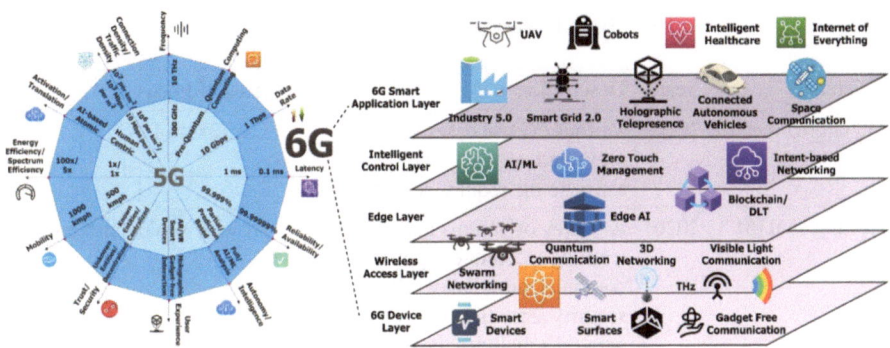

Fig. 2. Migration from 5G to 6G network [3]

Improved Spectrum Efficiency: Utilizing higher frequency bands, including the terahertz spectrum, 6G will offer more bandwidth and improved spectrum efficiency, helping to manage the increasing data traffic.

Advanced Applications and Services: 6G will enable new applications and services that 5G cannot support, such as advanced robotics, sophisticated AI-driven services, and immersive holographic communications.

Global Connectivity: 6G aims to extend high-speed connectivity to remote and underserved areas, helping bridge the digital divide and ensure more inclusive access to technology.

Enhanced Network Intelligence: Incorporating advanced AI and machine learning techniques, 6G networks will optimize performance, dynamically manage resources, and enhance the overall user experience.

Sustainability: Focusing on energy efficiency and sustainability, 6G aims to reduce the environmental impact of mobile networks and support green technologies.

In summary [4], migrating to 6G is essential to keep up with technological advancements, overcome the limitations of 5G, and meet future consumer and industry needs.

2   Literature Survey

The evolutionary history of wireless mobile communications was documented by author Asghar et al. [9], who also looked at the exciting prospects of the "beyond 5G era." The goal of this period is to improve the relationship between communication technology and mankind, bringing connectivity to a wider audience and helping the marginalized and poor segments of society. They talked about the technological elements enabling 6G and went over the main technical roadblocks once again. Realizing free, pervasive, ultra-high availability internet around-the-clock is still a challenging objective with several potential research avenues. Even with the advancements brought about by 5g, cyber-physical fusion—the convergence of terrestrial, aerial, and space communications—that is affordable, fast, and low-latency internet is still a dream. They investigated the major problems and difficulties associated with 5G, highlighting major roadblocks and important research issues for 6G (Table 1).

Table 1. Description of existing work

Reference	Year	Description	Area of Focus
M. Giordani et al [4]	2020	Various 6G use cases were presented, along with efforts to estimate the requirements for 6G	Use cases of 6G
F. Tang et al.[5]	2020	A survey of various machine learning techniques applied to communication, networking, and security aspects in 6G vehicular networks includes the evolution of intelligent radio, intelligentization, and self-learning with proactive exploration	AI with Vehicular Network
B. Mao et al. [6]	2020	For 6G IoT networks, an AI-based adaptive security specification technique is suggested, allowing IoT devices to connect to the cellular network over a range of frequency bands, including terahertz and millimetre waves	AI based adaptive security
W. Jiang[7]	2021	The author provides a thorough analysis of the state-of-the-art in 6G technology, offers insights into ongoing developments, and identifies areas requiring further exploration	6G Technology
Haiquan Lu[8]	2024	XL-MIMO is expected to significantly boost network capacity, coverage, connection density, sensing capabilities, and localization due to its exceptionally high spectral efficiency and spatial resolution. Specifically, XL-MIMO can alleviate network congestion and enhance the quality of service for users in densely populated urban areas, while also extending network coverage and delivering high-speed connectivity to previously underserved remote communities	XL-MIMO

While 5G is still in its early phases and has only recently begun to roll out in a few nations and areas, researchers and developers are already looking into and testing 6G, with an emphasis on novel communication methods and technologies. The sixth generation mobile system, anticipated to be released in 2030, is expected to bring significant changes to future mobile wireless networks [10]. The authors Alraih et al.in 2022 suggested that the "5G triangle" will transform into the "6G hexagon," introducing new dimensions to enable industrial use cases. 6G is anticipated to provide data rates of up to 10 Tbps and ultra-low latency ranging from 10 to 100 microseconds. Moreover, it will greatly improve spectrum efficiency and connection density, potentially delivering performance levels 10 to 100 times higher than 5G.

Rohit Singh et al. in 2024 [11] offered the comprehensive article, detailed overview of the evolution of wireless communication technologies, tracing its journey from foundational concepts to cutting-edge developments. Moreover, the article delves into the intricate landscape of implementing 6G, addressing three pivotal challenges while also shedding light on the concerted global efforts towards standardization. Additionally, it presents a groundbreaking proof of concept wherein the transmission of Terahertz (THz) signals is demonstrated using Orbital Angular Momentum (OAM) multiplexing, showcasing its potential for THz transmission. They have not only encapsulated existing research but also illuminates promising research opportunities and future visions inspired by IMT-2030 recommendations, aiming to catalyze further exploration and innovation in the field of wireless communication.

3 Overview of 6G Network

The International Telecommunication Union (ITU) standardizes wireless generations every decade. Subsequent generations named as next G utilizes more sophisticated way to denote the gap in the "air interface" signifies the next generation as "6G". 6G is the sixth – generation wireless communication mobile network standard widely used in cellular data networks in telecommunications [10]. It is the forthcoming generation of wireless communication technology that is anticipated to follow and surpass 5G networks. This advanced network aims to provide significantly higher data transfer speeds, extremely low latency, and the capacity to connect an even greater number of devices simultaneously. According to the report titled "6G The Next Hyper – Connected Experience for All", the ITU start working in 2021 to create 6G. Around 2030, the deployment will be close to ubiquitous. 6G network takes the advantage of the present 5G network infrastructure, which promises to provide diverse connectivity at microsecond speed.

The "AI native" design philosophy for the 6G architecture will allow intelligence to make the network intelligent, flexible, and able to learn from and adjust to shifting network conditions. As it develops, it will become a "network of sub-networks," enabling more flexible and effective updates. Intelligent radio and the separation of hardware and algorithms will serve as the foundation for this new framework, which will provide heterogeneous and upgradeable hardware capabilities. As will be further demonstrated in the ensuing subsections, each of these features will make use of AI approaches.

Some of the characteristics of 6G network are listed below:

  i. Terahertz (THz) frequencies
 ii. Extremely High Data rates
iii. Ultra – Low Latency

iv. Massive Connectivity
v. Enhanced Security
vi. Integration with Emerging Technologies
vii. Improved Energy Efficiency

Terahertz (THz) Frequencies: It utilizes higher frequency bands, particularly in the terahertz range (100 GHz to 10 THz), allowing for much higher bandwidth and data rates.

Extremely High Data rates: this is capable of achieving data rates up to 1 terabit per second (Tbps), which is significantly faster than 5G's peak rates of up to 20 gigabits per second (Gbps).

Ultra-Low Latency: The latency is designed to offer end-to-end latency as low as 1 ms or even lower, facilitating real-time communication and applications that require immediate feedback.

Massive Connectivity: It can support up to 10 million connected devices per square kilometer, which is ten times the capability of 5G networks. This is crucial for the proliferation if the Internet of Things (IoT) and smart cities.

Enhanced Security: The security measures incorporate advanced security protocols, potentially leveraging quantum cryptography to ensure secure communication channels.

Integration with Emerging Technologies: Many technologies are integrated like: Artificial Intelligence – AI and ML will be integral to managing network resources, optimizing performance, and ensuring security; Quantum computing – Utilizes quantum computing for complex data processing and encryption tasks.

Improved Energy Efficiency: Among the SDG goals, sustainability is designed with energy – efficient protocols and technologies to reduce the overall energy consumption of the network infrastructure (Table 2).

3.1 Meeting the Requirements of New Scenario

The "Typical Scenarios and Key Capabilities of 6G" whitepaper [12], released by the IMT-2030 (6G) Promotion Group, outlines five typical scenarios: further enhanced mobile broadband, ubiquitous massive connection, superlative ultra-reliable and low latency communication, quality guaranteed network AI service, and integrated communication and sensing. Similarly, The International Telecommunication Union - Radio Communication Sector (ITU-R) published "Framework and Overall Objectives of the Future Development of IMT for 2030 and Beyond" [13].

To address the demands of new scenarios, 6G network architecture must incorporate sensing services, AI services, and computing services, alongside enhanced communication services. The whitepaper [4] classifies the new services required by these scenarios into three main categories. The first category is super communication services, aimed at enhancing 5G communications. The second category is basic information services, which encompass positioning, sensing, network exposure information, and standard industry information. The third category is converged computing services, which integrate AI capabilities. Additionally, new scenarios present specific requirements for 6G networks, such as ultra-low power or even zero-power IoT device access, which can be

Table 2. Advancement and improvement from 5G to 6G

Feature	5G	6G
Frequency range	Up to 100 GHz	100 Gz to 10 THz
Maximum Data Rate	Up to 20 Gbps	Up to 1 Tbps
Latency	1 – 10 ms	< 1 ms
Device Density	1 million devices/km^2	10 million devices/km^2
Bandwidth	1 GHz	Up to 100 GHz
Spectral Efficiency	Moderate	Very High
Energy Efficiency	Improved over 4G	Significantly enhanced
AI Integration	Limited	Extensive
Security	Enhanced	Advanced (Quantum Cryptography)
Network Architecture	Centralized/ Distributed	Decentralized with Edge Computing
Edge Computing	Emerging	Integral
Application Examples	Enhanced mobile broadbrand, IoT, autonomous vehicles	Real time holography, advanced AR/VR, remote surgery, smart cities
Reliability	High	Ultra – High
Connection Density	Up to 1 million/km^2	Up to 10 million/km^2
Mobility Support	Up to 500 km/h	Up to 1000 km/h

achieved through simplified air interface technologies and network architectures. Ubiquitous connectivity needs to extend beyond 5G coverage through the seamless integration of terrestrial and non-terrestrial networks. Therefore, new functions and procedures must be developed to support these advanced 6G scenarios and services. The requirements of introducing new technologies have to meet the requirements of the technology. Among them a few are listed below:

1. AI Technology
2. Wireless Sensing Technology
3. Passive Internet of Things Technology
4. Satellite mobile Communication technology

AI Technology: AI encompasses a range of technologies grounded in machine learning and deep learning, which facilitate analysis, prediction, object categorization, natural language processing, recommendations, and intelligent data retrieval. Presently, AI has effectively addressed numerous challenges that were previously difficult to manage. These include image recognition and natural language processing in computer science, as well as motion control and trajectory planning in robotics.

Many issues in communication systems cannot be accurately modeled, such as the variation patterns of wireless channels and the nonlinear effects of power amplifiers.

However, by integrating AI with mobile networks, it is possible to extract implicit relationships, features, and knowledge from vast amounts of wireless communication data. Consequently, AI capabilities can enable more precise modeling of complex problems. Through a data-driven approach, AI can map the relationship between input information and potential solutions. AI is also applicable to many problems in communication systems for which closed-form solutions are not easily available or do not exist. AI-based approaches can provide direct or approximate solutions to these problems. By using AI, a neural network can treat multiple correlated functional modules as a joint optimization problem within a communication system. This approach transforms a complex multi-module correlation issue into a straightforward data fitting or regression problem, yielding a near-optimal solution.

Despite being one of the ten main topics of study for 6G, artificial intelligence is not a stand-alone discipline. Rather, it is integrated with all other study fields, such as intelligent reflecting surfaces, full-duplex communication, and cell-free massive MIMO. In 6G networks, data-driven, trained systems can improve each of these sectors' performance, simultaneously enhancing sustainability and energy efficiency. The air interface will be further optimized by using taught machine learning models for signal processing tasks including channel estimation, equalization, and demapping, beyond the capabilities of present 4G LTE and 5G NR networks.

In addition to working together on projects like the 6G-Access, Network of Networks, Automation & Sim-plification (6G-ANNA) lighthouse project, Rohde & Schwarz supports research programs in Europe, Asia, and the US. This project seeks to create a 6G design featuring end-to-end architecture, aiming to simplify interactions between humans, technology, and the environment through the use of new sensors and algorithms that detect human movements.

3.2 Challenges of AI in 6G Network

The integration of AI in 6G networks presents several significant challenges that need to be addressed to realize the full potential of these advanced systems. Here are some of the key challenges:

1. Data Privacy and Security
2. Computational Complexity and Energy Efficiency
3. Real-Time Processing and Latency
4. Interoperability and Standardization
5. Data Quality and Availability
6. Model Training and Adaptation
7. Scalability
8. Ethical and Regulatory Concerns

Data Privacy and Security: AI handles sensitive data which utilizes more data to train and operate effectively. Ensuring the privacy and security of this data, particularly sensitive user information is a major challenge. Securing data transmission will protect the data from interception and tampering during transmission across the network is crucial.

Computational Complexity and Energy Efficiency: In High Computational demand, Deep learning models in particular are computationally demanding and require a large amount of processing power, which can be difficult to supply at the network's edge. The other is Energy Consumption, the high energy consumption of AI operations can impact the overall energy efficiency of 6G networks. Efficient AI models and hardware are needed to mitigate this issue.

Real – Time Processing and Latency: The Low – Latency requirements, many 6G applications, such as autonomous driving and remote surgery, require real – time processing. Ensuring that AI systems can process data and make decisions with minimal latency is a critical challenge. The Edge AI integration will distribute AI processing to edge nodes to reduce latency while maintaining performance and accuracy is complex.

Interoperability and Standardization: 6G networks will involve a wide variety of devices and systems. Ensuring that AI solutions are interoperable across different platforms and standards is essential. With the lack of Standards, the rapid evolution of AI technologies and 6G networks means that standardization is still in progress. Establishing common standards for AI integration in 6G is necessary for widespread adoption.

Data Quality and Availability: AI systems can only process data of the same quality as the training set. It can be difficult to guarantee that broad, high-quality data sets are available for AI model training. In case of Data Scarcity there are some cases obtaining sufficient data, especially for new or niche applications, can be difficult.

Model training and Adaptation: In model training, training the AI models for 6G applications can be complex and time consuming, requiring specialized knowledge and resources. AI models must be adaptable to changing network conditions and evolving user requirements, necessitating continuous learning and updating.

Scalability: While implementing, network scalability provide AI solutions that can scale with the growing size and complexity of 6G networks is a significant challenge. Developing AI systems that can operate efficiently in a distributed network environment, with decentralized data processing and decision – making is complex.

Ethical and Regulatory Concerns: For Bias and Fairness, ensuring that AI systems are fair and unbiased in their decision – making processes is a critical ethical challenge. The regulatory compliance of navigating the regulatory landscape for AI and ensuring compliance with various legal and ethical standards is essential.

3.3 Terahertz (THz) Communication

Although 5G currently operates at extremely high millimeter-wave (mmWave) frequencies, 6G seeks to outperform 5G NR in terms of latency reduction and data transmission rates by operating at even higher frequencies, beyond 100 GHz. Academic research is now investigating frequencies as high as 330 GHz for communication applications.

Among the 6G technology components are reconfigurable intelligent surfaces (RIS), AI and ML, (ISAC) or (JCAS), and THz communication. It is quite likely to be essential for achieving the desired levels of throughput and very low latencies. Additionally, fascinating new applications like holographic communication are made possible by THz communication.

The terahertz and sub-terahertz bands, which provide a lot of bandwidth to overcome spectrum scarcity issues, are essential for the development of 6G wireless networks. In addition to 5G, 6G technology offers ultra-high-capacity wireless backhaul connections, high-definition holography, and wireless networks on chips made possible by tiny terahertz transceivers. 6G terahertz communication deployment has special difficulties, such as high data rates and scarce commercial components, which call for creative solutions to get beyond.

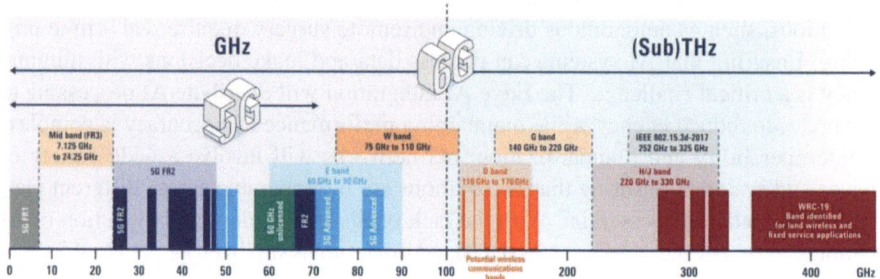

Fig. 3. Terahertz Communication [13]

The telecom industry is actively researching terahertz technology, and research projects at 6G firms, academic institutions, and research centers in the US, Europe, and Asia are funded by Rohde & Schwarz. One of the goals of Rohde & Schwarz, an active collaborator in the 6G-TERAKOM project, is to investigate and create a viable wireless system operating in the terahertz spectrum with integrated antennas. Currently, one of the most promising frequency bands for 6G development is the D band, which spans 110 GHz to 170 GHz.

With a significant bandwidth available, terahertz transmission for 6G can efficiently alleviate the spectrum scarcity issue that wireless networks are now facing. Exciting new opportunities for wireless terabit-per-second (Tbps) connections result from this. Continue reading as we explore 6G technologies and Terahertz communication (Fig. 4).

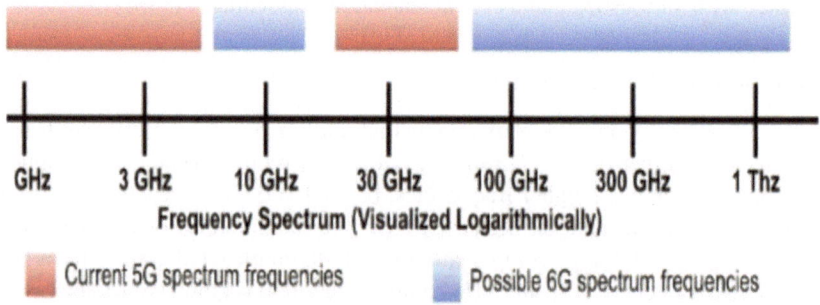

Fig. 4. Spectrum of Frequencies for 5G and Upcoming 6G Applications [14]

3.4 Future Uses of Terahertz Band in 6G

In the realm of 6G technology, a multitude of advancements will collaborate seamlessly to deliver the various capabilities mentioned earlier. Significant progress in research and development is anticipated across several ground-breaking discoveries within terahertz-related fields. Device Advancements: Innovations in electronic, photonic, and plasmonic devices for terahertz (THz) communication and sensing.

- Innovations in smart antenna systems, such as meta surfaces, lenses, and arrays, allow for flexible THz transmission.
- Signal processing: Terabit-per-second (Tbps) lines in THz communication are supported by ultra-broadband signal processing techniques.
- Propagation modelling is the process of building models to support system design for various THz scenarios, such as indoor and space environments.
- Methods for improving near-field THz communication through wave front optimization.
- The design of waveforms and modulation methods for ultra-broadband THz communication and sensing is the focus of this article.
- MIMO Technology: Investigating THz systems using huge multiple-input multiple-output (MIMO) technology.
- Integrated Sensing: Combining THz technology's communication and sensing features.
- Radar and Communication: integrating data communication in THz systems with radar capabilities.
- Networking Protocols: Creating innovative protocols for THz communication networks that are ultra-directional.
- Energy Efficiency: Examining THz communication systems' energy consumption and effectiveness.

3.5 Issues in Deploying Terahertz (THz) in 6G

- Deployment of sixth-generation (6G) terahertz (THz) technology differs significantly from 5G frequency bands.
- Key distinctions between sub-THz and THz systems impact fine-grained design, including difficult propagation circumstances, extreme data rate requirements exceeding 100 Gbit/s, scarcity of commercial radio frequency components, and the need to function in relatively small coverage areas.
- Innovative approaches are necessary to tackle the unique challenges of terahertz frequencies.
- One approach involves using parallel processing to effectively manage extreme data rates and other method involves pre-distorting the signal to eliminate the necessity for traditional mixers.
- Typically, communication radios follow a signal generation path using a mixer, but at terahertz frequencies, traditional mixers are impractical due to significantly higher power demands.
- Instead, the conventional mixer approach is bypassed by directly feeding information into the signal source itself.

4 Conclusion

The migration from 5G to 6G wireless communication networks presents numerous hidden Obstacles that must be tackled to fully unlock the potential of next-generation connectivity. Key areas of focus include the integration of terahertz communication for ultra-high-speed data transfer and the incorporation of artificial intelligence to enhance network efficiency, optimize resource allocation, and support advanced applications. By tackling these challenges, the transition to 6G can facilitate unprecedented advancements in communication technology, enabling smarter, more agile, and highly adaptive networks that cater to evolving user demands and emerging technological landscapes.

The transition from 5G to 6G wireless communication networks promises to unlock new possibilities and revolutionize connectivity. Future research will focus on overcoming challenges related to terahertz communication to achieve ultra-high-speed data transfer and developing advanced artificial intelligence algorithms to optimize network performance and resource management. Innovations in network architecture, including the integration of edge computing and intelligent reflecting surfaces, will enhance efficiency and flexibility. As these technologies mature, they will enable transformative applications such as real-time holography, advanced augmented reality, and seamless connectivity across terrestrial and non-terrestrial networks, paving the way for a hyper-connected, intelligent world.

References

1. Chowdhury, M.Z., Shahjalal, M., Ahmed, S., Jang, Y.M.: 6G wireless communication systems: applications, requirements, technologies, challenges, and research directions. IEEE Open J. Communications Society **1**, 957–975 (2020). https://doi.org/10.1109/OJCOMS.2020.3010270
2. Alsharif, M.H., Kelechi, A.H., Albreem, M.A., Chaudhry, S.A., Zia, M.S., Kim, S.: Sixth generation (6G) wireless networks: vision, research activities, challenges and potential solutions. Symmetry **12**, 676 (2020). https://doi.org/10.3390/sym12040676
3. De Alwis, C., et al.: Towards 6G; Key Technological Directions. ICT Express, **9**(4) (2023). ISSN – 2405 – 9595. Accessed 20 May 2024
4. Giordani, M., Polese, M., Mezzavilla, M., Rangan, S., Zorzi, M.: Toward 6G networks: use cases and technologies. IEEE Commun. Mag. **58**(3), 55–61 (2020)
5. Tang, F., Kawamoto, Y., Kato, N., Liu, J.: Future intelligent and secure vehicular network toward 6G: machine-learning approaches. Proc. IEEE **108**(2), 292–307 (2020)
6. Mao, B., Kawamoto, Y., Kato, N.: AI-based joint optimization of QoS and security for 6G energy harvesting internet of things. IEEE Internet Things J. Early Access, (2020). 7.1109/JIOT.2020.2982417
7. Jiang, W., Han, B., Habibi, M.A., Schotten, H.D.: The road towards 6G: a comprehensive survey. IEEE Open J. Communications Society **2**, 334–366 (2021). https://doi.org/10.1109/OJCOMS.2021.3057679
8. Lu, H., et al.: A tutorial on near-field XL-MIMO communications towards 6G. IEEE Communications Surveys & Tutorials (2024)
9. Asghar, M.Z., Memon, S.A., Hämäläinen, J.: Evolution of wireless communication to 6G: potential applications and research directions. Sustainability **14**, 6356 (2022). https://doi.org/10.3390/su14106356

10. Alraih, S., et al.: Revolution or evolution? technical requirements and considerations towards 6G mobile communications. Sensors **22**, 762 (2022). https://doi.org/10.3390/s22030762
11. Singh, R., et al.: Towards 6G Evolution: Three Enhancements, Three Innovations, and Three Major Challenges. Cornell University, 16 Feb 2024
12. Letaief, K.B., Chein, W., Shi, Y., Zhang, Y.A.: The Roadmap to 6G – AI Empowered Wireless Networks, Researchgate (2019)
13. https://www.rohde-schwarz.com/us/solutions/wireless-communications-testing/wireless-standards/6g/thz-communication/thz-communication_257042.html#gallery-5. Accessed 20 May 2024
14. https://www.linkedin.com/pulse/japans-breakthrough-6g-technology-unveiling-worlds-first-borghare-ltnjc/. Accessed 20 May 2024

Design of Location and Assistive Service Software Based on Android System

Wenbin Wang, Kezhen He$^{(\boxtimes)}$, and Yuchun Ma

Yazhou Bay Innovation Institute, Hainan Tropical Ocean University, Sanya 572022, China
kezhen.he@qq.com

Abstract. In the process of using smartphones, elderly individuals often unintentionally set their phones to silent mode or turn off Wi-Fi, resulting in disruptions to answering calls and receiving WeChat video messages. This article implements a whitelist feature that automatically restores the phone to normal mode, monitors Wi-Fi status and reminds users to enable it. Additionally, it can discreetly check the geographical location and send low balance notifications to whitelisted phones, assisting the elderly in using smartphones without obstacles.

Keywords: Smartphone · Silent Mode · Location Inquiry · Assistive Services

1 Implementation of Main Form Functions and Services

The primary functions of the location and assistive service software are completed in the background, enabling monitoring of the phone. The foreground is mainly managed by a main form, facilitating the addition, deletion, backup, and restoration of the whitelist. When completing the background functions, it is necessary to *review* and compare these whitelists to make decisions: only calls from whitelisted numbers can automatically switch the phone from silent or vibration mode to normal mode and set the volume to the maximum. Location inquiries from whitelisted SMS messages will receive a response. The whitelist is represented only by phone numbers, displayed through a custom ListView component that allows manual adjustment of the order. When receiving a low balance SMS from the service provider, this information will be sent to the first phone number in the whitelist [1]. Data in the whitelist is written to an external text file using the FileOutputStream object through multithreading technology, and it can be copied to external storage for backup. It can also be restored by calling the FileInputStream object from external storage, facilitating device replacement operations.

When system resources decrease, Android forcefully terminates a s ervice to ensure the normal operation of applications with user focus. If a service is bound to an Activity with user focus, it generally will not be forcefully terminated. If a service is declared as a "foreground service," it is unlikely to be terminated. Otherwise, if a service has been started and running for a long time, the system will gradually lower its priority in the background task list over time, and such services are likely to be forcefully terminated [2]. Starting the service can be done in the main form or in the BroadcastReceiver

P. Siarry et al. (Eds.): WCNA 2023, LNEE 1361, pp. 14–21, 2025.
https://doi.org/10.1007/978-981-96-2409-6_2

completed during boot. In the onCreate method of the service code, the timestamp strDate is obtained, then the system method startForeground is called to display the service notification bar icon, with the timestamp as the icon content, as shown in the following code [3].

```
public void onCreate() {
super.onCreate();
    Calendar cl = Calendar.getInstance();
    Date dt = cl.getTime();
    String strDate = dt.toString();
    startForeground(FOREGROUND_ID,
    buildForegroundNotification(strDate));
}
```

The custom function buildForegroundNotification is used to construct the notification bar icon, returning a Notification object named notify. The PendingIntent object pi is directed to the main form MainActivity. By calling setContentIntent on notify and passing this pi object, clicking the notification icon can open the main form. The method setOngoing is set to true, indicating an ongoing background task. The method setPriority is passed the parameter Notification.PRIORITY_MAX, ensuring that the icon is placed at the forefront.

```
private Notification buildForegroundNotification(String strContent) {
    Intent intent = new Intent(this, MainActivity.class);
    intent.setFlags(Intent.FLAG_ACTIVITY_NEW_TASK        |        Intent.FLAG_ACTIVITY_CLEAR_TASK);
    PendingIntent  pi  =  PendingIntent.getActivity(this,  0,  intent,  PendingIntent.FLAG_UPDATE_CURRENT);
    Notification notify = new Notification.Buil der(this)
    .setOngoing(true)
    .setPriority(Notification.PRIORITY_HIGH)
    .setContentIntent(pi)
    .setContentTitle("Address Server")
    .setContentText(strContent)
    .setSmallIcon(R.drawable.ic_launcher)
    .build();
    return notify;
}
```

2   Monitoring Changes in Wi-Fi Settings

To monitor the Wi-Fi status, it is necessary to declare the ACCESS_NETWORK_STATE and ACCESS_WIFI _STATE permissions in the configuration file and utilize a custom broadcast receiver, WiFiChangeReceiver, for monitoring [4]. The WifiManager object is obtained through the system method getSystemService, allowing the invocation of the getWifiState method to retrieve the Wi-Fi status. If the status is

WIFI_STATE_DISABLED, the setWifiEnabled (true) method of the WifiManager object is called to enable the Wi-Fi switch. On lower versions of the Android system, it can be directly enabled, while on higher versions, user authorization is still required. The key source code for the WiFiChangeReceiver broadcast receiver, monitoring Wi-Fi settings, is as follows:

```
still required. The key source code for the WiFiChangeReceiver broadcast receiver,
monitoring Wi-Fi settings, is as follows:
    public class WiFiChangeReceiver extends Broadcast Receiver {
        @Override
        public void onReceive(Context context, Intent intent) {
            WifiManager wifiManager = (WifiManager)
context.getSystemService(Context.WIFI_SERVICE);
            int status =wifiManager.getWifiState();
            switch(status){
            case WifiManager.WIFI_STATE_DISABLED:
                Toast.makeText(context,         "WIFI_STATE_        DISABLED",
Toast.LENGTH_SHORT). show();
                wifiManager.setWifiEnabled(true);
                break;
            default:
                break;
                }
            }
        }
```

3   Call Status Monitoring

Declare the READ_PHONE_STATE permission in the configuration file and implement a custom broadcast receiver, PhoneStateReceiver. When the state content is EXTRA_STATE_RINGING, indicating an incoming call, retrieve the phone number and store it in inPhone. If the phone number is an empty value (anonymous call), return directly [5]. If the phone number is not in the whitelist collection strPhoneSet or the whitelist collection starts with "000000," no action is taken. Otherwise, set the silent mode to RINGER_MODE_NORMAL and the ringtone volume to the maximum. In this context, the whitelist phone number "000000" has a special function. When this number is moved to the front in the main form, it indicates that the software does not automatically switch to normal mode for whitelisted incoming calls [6]. At the same time, it sets the volume to the maximum.

```
public class PhoneStateReceiver extends Broadcast Receiver {
    String strPhoneSet;
    private String inPhone="";
public void onReceive(Context context, Intent intent) {
        String state = intent.getStringExtra(Telephony
Manager.EXTRA_STATE);
        if(TextUtils.isEmpty(state)) return;
        if (!state.equals(TelephonyManager.EXTRA_
STATE_RINGING)) return;
        strPhoneSet = MainActivity.readStorage();
        inPhone = intent.getStringExtra(Telephony
Manager.EXTRA_INCOMING_NUMBER);
        if(TextUtils.isEmpty(inPhone)) return;
        if((!strPhoneSet.contains(inPhone)) ||
          strPhoneSet.startsWith("000000")) return;
        Sound.setSoundMode(context, AudioManager.
RINGER_MODE_NORMAL);
        Sound.setMaxInCallVolume(context);// Setting
Maximum Volume
    }
}
```

4   Acquisition and Forwarding of Low Balance SMS

When receiving a text message from a mobile service provider starting with "100," extract the text content of the message. Use this as a parameter to call the tipRemainder function, extracting the floating-point number between "余额" (balance) and "元" (yuan). If this value is greater than the given minimum value (such as 20 yuan), return. Otherwise, convert the whitelist collection strPhoneSet into a string array and compare them one by one, sending the low balance SMS to the first whitelisted phone number [7]. The definition of the custom function tipRemainder is as follows:

```
private void tipRemainder(String body){
    String strValue = StringProcess.getDecimalString
(body, "余额","元');
    float fVal = Float.parseFloat(strValue);
    if(fVal >= MONEY_MIN) return;
    String strPhoneSet = MainActivity.readStorage();
    String strArray[] = strPhoneSet.split("/");
    for(int i=0; i<strArray.length; i++){
        if(PhoneBook.isMobileNumber(strArray[i])){
        // mobile number?
            String content = " balance: " +
Float.toString(fVal) + " yuan ";
            SmsProcess.sendMessage(content,
strArray[i]); // Send SMS
            break;
        }
    }
}
```

The first line of code in the tipRemainder function calls the getDecimalString function with three parameters. The first parameter is the text content of the message, the second parameter is "balance", and the third parameter is "yuan". If the message text contains "balance" and "yuan," a regular expression is ultimately used to remove characters other than numbers and decimal points. Finally, it returns the decimal string [8].

```
public static String getDecimalString(String strData,
String strHead, String strTail){
    //" The current balance is: 17.32 yuan. ",
"balance", "yuan" --> 17.32
    if(TextUtils.isEmpty(strData)) return "";
    if(TextUtils.isEmpty(strHead)) return "";
    if(TextUtils.isEmpty(strTail)) return "";
    if((!strData.contains(strHead)) ||
(!strData.contains(strTail))) return "";
    int nStart = strData.indexOf(strHead);
    String strRet = strData.substring(nStart + 2);
    int nEnd = strRet.indexOf(strTail);
    strRet = strRet.substring(0, nEnd);
    strRet = strRet.replaceAll("[^0-9?!\\.]", "");
    return strRet;
}
```

5 Location Inquiry and Response

The query for geographical location is carried out through Baidu Maps. Involving the development of applications using Baidu Maps, the initial steps include applying for a Baidu account and then requesting an SDK development key from the Baidu Maps console [9]. Input information such as "Application Name," "Application Type," and development tool fingerprint, submit the request to obtain an "Application AK." This "Application AK" will be used in the project configuration file, AndroidManifest.xml.

Upon receiving a SMS command to query the geographical location, the assistive service software first checks if the number sending the SMS is in the whitelist. If it is, the Baidu Maps API function is called to initiate the location query. In the onReceiveLocation callback function of the custom class MyLocationListenner, the latitude and longitude data is retrieved. The data is then prefixed with "lld," and the string of latitude and longitude data is encrypted using BASE64. Finally, this encrypted data is sent back to the querying phone [10]. After decrypting the SMS, the querying phone marks the latitude and longitude data on Baidu Maps.

During the process of handling location query SMS, the phone remains in silent mode. The sent and received SMS messages are automatically deleted, ensuring it doesn't disrupt the phone owner's work and life. The key code of the custom class MyLocationListenner is as follows:

```
public class MyLocationListenner implements
BDLocationListener {
    @Override
    public void onReceiveLocation(BDLocation
location) {
        if (location == null) return;
        // Location data, with accuracy measured in
meters
        String strHead = "lld/";
        String strAddress = Double.toString(location.
getLongitude()) + "/";
        strAddress += Double.toString(location.
getLatitude());
        SmsProcess.sendMessage(strHead +
            ByteProcess.enStringToBase64(strAddress), inPhone);
        mLocClient.unRegisterLocationListene(myListener);
        mLocClient.stop();
        myListener = null;
    }
}
```

6 Software Testing

The operational effect of the Location and Assistive Service software is shown in Fig. 1. The main interface primarily utilizes the "Add" button to add whitelist numbers, the "Remove" button to delete whitelist numbers, and arrows to move data up and down. The whitelist entry "000000" has a special function: if it is at the top, incoming calls from whitelisted numbers will not automatically switch to normal mode and set the maximum volume. Otherwise, incoming calls from whitelisted numbers, regardless of the current mode, will automatically be set to the maximum volume. Closing the main interface can be done by clicking the notification icon, which will automatically reopen the main interface. When turning off Wi-Fi, the permission request interface shown in Fig. 1 will appear automatically. Upon receiving a low balance SMS from the service provider, it will automatically forward to the first phone number as shown in Fig. 1. For location query SMS from the whitelist, the obtained latitude and longitude data will be encrypted and returned.

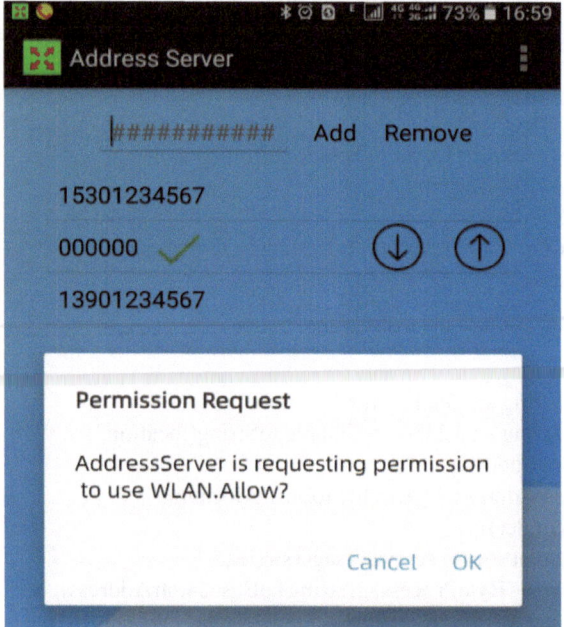

Fig. 1. Software Performance

7 Conclusion

Smartphones bring convenience to users, but for the elderly population with poor eyesight or low literacy, there are many inconveniences in the usage process. The developed Location and Assistive Service software in this article operates in the background as a persistent service, monitoring changes in Wi-Fi settings and silent mode. This ensures that the phone can better answer whitelisted calls and WeChat video messages. Additionally, for low balance notifications from service providers, it can automatically forward to the first phone number in the whitelist, prompting the recipient to recharge. It also facilitates location queries, handling all these messages in silent mode without disrupting the normal life of the phone owner.

References

1. Guo, F., Wu, W., et al.: Realization and precision analysis of real-time precise point positioning with android smartphones. Geomatics and Information Science of Wuhan University **46**(7), 1053–1062 (2021). https://doi.org/10.13203/j.whugis20200527
2. Alazab, M., Alazab, M., Shalaginov, A., et al.: Intelligent mobile malware detection using permission requests and API calls. Futur. Gener. Comput. Syst. **107**, 509–521 (2020)
3. Yang, F., Chan, K.C., Fang, Y., et al.: Vision-based indoor corridor localization via smartphone using relative distance perception and deviation compensation. SN Computer Science **5**(3) (2024)
4. Zhou, Q., Xing, J, Yang, Q., et al.: Measuring intrinsic human activity information using WiFi-based attention model. Measurement **195** (2022)
5. Letizia, L., Miran, M., Andrej, G., et al.: Design and evaluation of personalized services to foster active aging: the experience of technology pre-validation in italian pilots. Sensors **23**(2), 797 (2023)
6. Xia, Z., Chong, S.: WiFi-based indoor passive fall detection for medical internet of things. Computers and Electrical Engineering **109**, 108763 (2023)
7. Zhang, M., Jia, J., Chen, J., et al.: Indoor localization fusing WiFi with smartphone inertial sensors using LSTM networks. IEEE Internet Things J. **8**(17), 13608–13623 (2021)
8. Wu, L., Wei, X., Meng, L., et al.: Privacy-preserving location-based traffic density monitoring. Connect. Sci. **34**(1), 874–894 (2022)
9. Ji, X.: Application of optical motion capture device based on android intelligent platform in sports field auxiliary recognition system. Optical and Quantum Electronics **56** (2023)
10. Yang, B., Yang, Z., Chen, X., et al.: Analysis of consistency between sensitive behavior and privacy policy of android applications. J. Computer Appl. **44**(3), 788–796 (2024)

EffiNet: An Efficient, Low-Complexity Neural Network for Monocular Depth Estimation on Embedded Devices

Xiaohan Tu[1]([✉]), Chuanhao Zhang[1], Haiyan Zhuang[1], Zhen Sang[1], and Mengran Liu[2]

[1] Zhengzhou Police University, Zhengzhou 450053, China
{tuxiaohan,zhangchuanhao,zhuanghaiyan,sangzhen}@rpc.edu.cn
[2] Tianjin Public Security Division, Beijing Railway Public Security Bureau, Tianjin 300142, China
1912231206@mail.sit.edu.cn

Abstract. Monocular Depth Estimation (MDE) plays an important role in many fields. However, current MDE requires complex models that are computationally intensive and unsuitable for real-time applications. Some MDE methods also have a high memory access frequency (MAF), leading to substantial data transfer demands. This paper presents EffiNet, a novel network structure designed for MDE. We emphasize high performance and lightweight computing. EffiNet significantly reduces computational complexity while maintaining accurate depth estimation by utilizing MobileNetV2 as the backbone for the encoder. The newly designed decoding algorithm optimizes the model by implementing block processing strategies, efficient upsampling, and point convolution operations. The decoder enhances performance under various scene conditions while keeping computational costs low. This approach also minimizes MAF, thereby reducing data transmission requirements and improving execution speed and energy efficiency. Experimental results demonstrate that EffiNet achieves a reduction in inference time by at least 18.18% on Jetson Nano and 24.39% on Jetson TX2, and a maximum 96.07% reduction in computational complexity (FLOPs), compared to existing methods. Additionally, the accuracy index δ_1 shows an improvement of 0.44% to 65.70%. These results validate the superior performance and suitability of resource-constrained mobile and edge computing devices. These advancements position EffiNet as an innovative solution in the field of MDE, capable of delivering high efficiency and reducing resource consumption.

Keywords: convolutional neural networks (CNNs) · deep learning · embedded devices · monocular depth estimation (MDE) · memory access frequency MAF

© The Author(s) 2025
P. Siarry et al. (Eds.): WCNA 2023, LNEE 1361, pp. 22–35, 2025.
https://doi.org/10.1007/978-981-96-2409-6_3

1   Introduction

With the rapid development of machine vision and autonomous driving technology, the importance of monocular depth estimation (MDE) has become increasingly prominent. MDE refers to predicting depth in three-dimensional (3D) from two-dimensional (2D) images. Traditional MDE mainly depends on specialized hardware devices, including stereo cameras and LiDAR systems. They directly measure the depth information. However, these approaches are often expensive. They have limitations in equipment size or energy consumption. In contrast, MDE only needs an ordinary camera. It significantly reduces the cost and the complexity of the equipment.

Recently, researchers have used deep learning models like convolutional neural networks (CNNs) to learn depth information from depth-labeled data [1–9]. These models can learn local features like edges and textures to infer depth. Recent MDE research has significantly improved accuracy. For instance, relying on CNNs and attention mechanisms, researchers can better capture image details and context, enhancing MDE accuracy [10,11]. Self-supervised and unsupervised learning strategies are also been employed in MDE to reduce the reliance on large amounts of labeled data [12–17]. Despite these improvements, MDE still encounters several challenges. First, high-precision MDE often requires complex model structures, resulting in a large amount of calculation and difficulty in meeting the needs of real-time applications. Secondly, many existing MDE methods have high MAF [18]. This results in high data transfer requirements. Finally, current methods run inefficiently on resource-constrained devices, limiting their application potential on edge computing devices.

In response to the above problems, this paper proposes a new efficient network structure EffiNet. It is specially designed for MDE and has the characteristics of lightweight and high-performance computing. Specifically, to construct EffiNet, we use MobileNetV2 [19] as the backbone network of the encoder. This significantly reduces the computational complexity of the model through optimized decoding algorithms while maintaining high estimation accuracy. Secondly, we designed a new decoding algorithm that significantly improves the model performance and efficiency under different scene conditions based on [20]. In the algorithm, we adopt a block processing strategy. By dividing the channels of input features and processing them separately, we effectively reduce the computational complexity and unnecessary computational burden. Then, we employ efficient upsampling and point convolution operations after each block. This operation not only improves the spatial resolution of features, but also keeps the computational cost low and adapts to the performance limitations of edge computing devices. Finally, we optimized the MAF in each algorithm operation, reducing data transmission requirements and further improving execution speed and energy efficiency.

The proposed method in this paper shows significant improvements on multiple key performance indicators. Compared with existing MDE methods, our inference time is reduced by at least 18.18% on Jetson Nano and 24.39% on Jetson TX2. These results show that we effectively reduce the running load on edge devices. At the same time, EffiNet achieved a maximum 96.07% reduction in computational complexity (FLOPs). These results demonstrate that we significantly alleviate the computational demands of MDE. In the accuracy index δ_1, this method improves by 0.44% to 65.70% compared with other methods, showing its superior performance. These results highlight the innovation of our method in improving computing efficiency and reducing resource consumption, and is particularly suitable for application in resource-constrained mobile and edge computing devices.

2   Related Work

CNNs were first applied to monocular depth estimation (MDE) by the study [1]. Subsequent research utilized advanced deep learning structures such as ResNets [3,4,6]. Laina et al. [3] introduced full CNNs, highlighting efficient up-sampling for enhanced resolution. Hu et al. [6] explored pixel relevance in MDE, while the approach by Ma et al. [4] integrated color images with depth data for depth estimation. Lightweight, fast MDE strategies were later developed using encoder-decoder networks with up-sampling, balancing efficiency and accuracy [5,8]. Multi-scale strategies were also employed [2,7,9,21]. Probabilistic models were proposed to handle depth as a distribution to improve estimation accuracy and confidence [10,11]. Additionally, attention mechanisms were integrated to enhance depth precision, highlighting the use of attention for non-linear representation and long-range correlations [22–24]. Fang et al. [25] reviewed MDE models and loss functions, noting the application of relative depth and the shift towards self-supervised methods to reduce reliance on labeled data [26,27].

Other studies use self-supervised and unsupervised methods for monocular depth estimation (MDE). For example, Cheng et al. [17] developed a fusion approach to better aggregate different levels of feature information. Klingner et al. [14] and Tosi et al. [15] introduced self-supervised techniques using semantic guidance and knowledge distillation respectively. Yin et al. [12] leveraged 3D scene geometry in their framework to jointly learn MDE, optical flow, and ego-motion. Similarly, Gur et al. [13] employed depth from focus cues to improve unsupervised MDE accuracy. Unlike these prior studies, we introduce a novel decoding algorithm, designed to reduce computational load and enhance resolution efficiently. We also specifically utilizes split operations and dual convolution processes that manage complexity and memory demands better than standard convolutional layers. By effectively balancing computational efficiency with precision, EffiNet provides a robust solution for real-time MDE applications, ensuring high performance even on devices with limited computational resources.

3   Proposed Methods

We propose the efficient network EffiNet for MDE. EffiNet is designed with a focus on lightweight and high-performance computing. As depicted in Fig. 1, it utilizes MobileNetV2 [19] as the backbone for its encoder. MobileNetV2 is

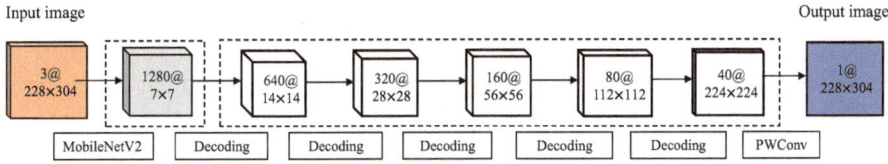

Fig. 1. EffiNet.

a popular choice as an encoder in efficient neural networks, even if it is not the latest model. 'N@WxL' denotes the dimensions of the features, depicted as cubical representations. EffiNet still offers strong performance on a variety of tasks, especially mobile-centric vision applications. Especially, it strikes an optimal equilibrium between precision and computational cost. As we focus on depth feature extraction, the classification layer of MobileNetV2 is omitted. Subsequently, we introduce a decoder to reconstruct depth while maintaining low computational demand. Our decoder leads to efficient and effective reconstruction of high-resolution outputs from encoded inputs. The decoder includes Algorithm 1 and upsampling operations.

Algorithm 1. Decoding Algorithm

Require: a feature F
Ensure: a new feature F'
1: $b \leftarrow a/2$
2: Split F into F_1 and F_2 by channels; the features of the first b channels are denoted by F_1; the features of the remaining channels are presented by F_2.
3: Apply the first convolution to F_1 and obtain F_3
4: Concatenate F_2 with F_3 and obtain F_4
5: Apply the second convolution to F_4 and obtain F'

Algorithm 1 shows the decoding process relying on [20]. Assume the channel of the input feature F is a. We calculate half of the a channel and obtain b in Step 1. In Step 2, the feature F is divided equally into two separate components based on its channels. The first b channels of F are assigned to F_1, and the remaining channels are represented as F_2. This splitting is essential for the distinct operations that follow for each feature subset. In Step 3, we apply the first convolution operation to F_1 and produce an intermediate feature F_3. The first convolution operation corresponds to Conv1 in Fig. 2. Conv1 applies a 3×3 convolutional filter to each input channel, producing half as many output channels. The stride of Conv1 is one, and the padding is one. We add no bias to the convolution results. Conv1 is responsible for transforming the initial feature subset into a form that is further processed. Here, we selectively perform convolution operations on the features from some of these channels, because features exhibit a high degree of similarity across various channels. In this way, we reduce the MAF and enhance the computational efficiency while also conserving the memory resources. This strategy is especially beneficial in scenarios where the available memory is a limiting factor, such as in embedded systems with restricted hardware capabilities. Following Conv1, F_3 is merged with F_2 in Step 4. The combined feature F_4 is obtained. This step integrates the transformed subset of features with the untouched ones. By doing so, a unified feature set is created for further manipulation. Lastly, in Step 5, the second convolution is applied to the concatenated feature F_4. The second convolution operation corresponds to Conv2 in Fig. 2. Conv2 has one 1×1 kernel. The stride is one, and the padding is zero. It compacts the channel information without changing the height and width of F_4. The layer performs a linear transformation to the channel vector of each pixel across F_4. Conv2 refines F_4 into the output feature F'.

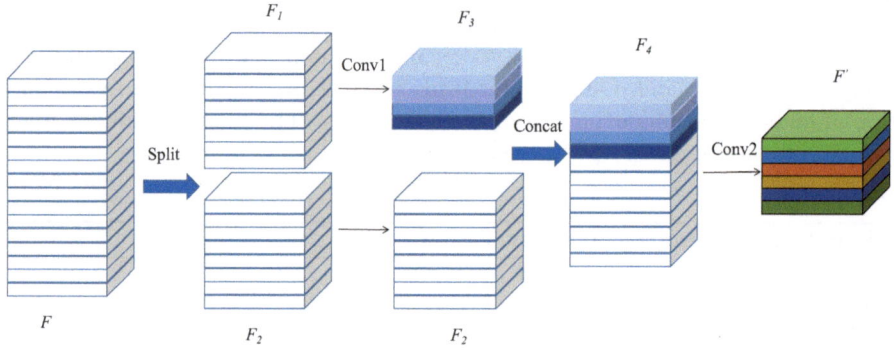

Fig. 2. The proposed decoding algorithm.

The decoding flow of Algorithm 1 is illustrated in Fig. 2. Our approach targets select input channels, aiming to lower computational demands. This ensures improved efficiency. The input feature F is split into two distinct features F_1 and F_2. The feature F_1 is then processed through a convolution layer Conv1. Conv1 produces a new feature F_3. F_3 is subsequently concatenated with F_2 to form F_4. This combined feature F_4 is finally passed through a second convolution layer Conv2. Conv2 outputs the feature F'. Conv1 and Conv2 are designed to operate solely on the portion of channels, thereby significantly reducing computational burden. The diagram demonstrates a decoding mechanism. We introduce a strategic modification to the traditional convolution process. Importantly, our method effectively extracts spatial features.

Then, we calculate the computational complexity of our decoding algorithm and regular convolutions. The FLOPs are a precise measure of the total number of floating-point operations that must be performed, making them an accurate indicator of computational complexity. For an ordinary convolution, the formula is expressed as C to estimate its computational complexity:

$$C = \epsilon^2 \times \mu \times \phi \times \iota \times \omega, \tag{1}$$

where ϵ^2 is the dimensions of the convolutional filter; μ and ϕ are respectively the quantity of channels after and before the convolution; ω and ι are respectively the horizontal and vertical measures of the resulting features.

The computational complexity of Conv1 is:

$$
\begin{aligned}
C_1 &= \epsilon^2 \times \mu/4 \times \phi/2 \times \iota \times \omega \\
&= (\epsilon^2 \times \mu \times \phi \times \iota \times \omega)/8,
\end{aligned}
\tag{2}
$$

where Eq. 2 denotes that the computational complexity of Conv1 is eight times smaller than that of an ordinary convolution.

The computational complexity of Conv2 is:

$$
\begin{aligned}
C_2 &= \epsilon^2 \times \mu/2 \times (\phi \times 2/3) \times \iota \times \omega \\
&= (\epsilon^2 \times \mu \times \phi \times \iota \times \omega)/3,
\end{aligned}
\tag{3}
$$

where Eq. 3 indicates that Conv2 operates with a computational demand reduced to one-third that of a standard convolutional process.

Next, we calculate the MAF. It equals the sum of the weights of the convolutional filter (parameters), the amount of input data that the operation processes (input activation), and the amount of data it produces as output (output activation). The parameters are the weights of the convolutional filter, which are calculated as $\epsilon^2 \times \mu \times \phi$. The input activation is the amount of data that needs to be loaded from memory for a convolution, which is $\iota \times \omega \times \phi$. The output activation is the data that are written to memory as the output of a convolution, which is $\iota \times \omega \times \mu$.

For an ordinary convolution, its MAF is:

$$
\begin{aligned}
M &= \iota \times \omega \times \mu + \iota \times \omega \times \phi + \epsilon^2 \times \mu \times \phi \\
&= \iota \times \omega \times (\phi + \mu) + \epsilon^2 \times \mu \times \phi.
\end{aligned}
\tag{4}
$$

The MAF of Conv1 is:

$$
\begin{aligned}
M_1 &= \epsilon^2 \times \mu/4 \times \phi/2 + \iota \times \omega \times \phi/2 + \iota \times \omega \times \mu/4 \\
&= (\epsilon^2 \times \mu \times \phi)/8 + \iota \times \omega \times (\phi + \mu/2)/2.
\end{aligned}
\tag{5}
$$

From Eq. 5, we know that the MAF in Conv1 is at least half smaller than that of the ordinary convolution.

The MAF of Conv2 is:

$$
\begin{aligned}
M_2 &= \epsilon^2 \times \mu/2 \times (\phi \times 2/3) + \iota \times \omega \times \phi \times 2/3 + \iota \times \omega \times \mu/2 \\
&= (\epsilon^2 \times \mu \times \phi)/8 + \iota \times \omega \times (\phi \times 2/3 + \mu/2).
\end{aligned}
\tag{6}
$$

In Eq. 6, we see that the MAF of Conv2 is reduced by at least $1/3$ to half in each item. Therefore, the overall MAF of Conv1 and Conv2 is smaller than that of an ordinary convolution. Consequently, Conv1 and Conv2 are designed to be fast by lowering computational complexity and MAF compared to regular convolutions. Algorithm 1 adopts Conv1 and Conv2 to extract spatial features, reducing unnecessary memory usage and computation. Relying on Algorithm 1, we propose a decoder that is lightweight while maintaining high accuracy.

The decoder is shown in Algorithm 2. The algorithm requires an input M, which is the output from our encoder. The algorithm ensures an output M', which is the final result of the decoder process. The algorithm begins by initializing a variable feature M_j with the value of M (line 2). This step sets up the initial state for the subsequent operations in the decoder. From lines 3 to 7, there is a loop that iterates five times, representing five stages of upsampling. In each iteration, two main operations are performed on M_i and M_j: Algorithm 1 and bilinear interpolation. Algorithm 1 is applied to M_j. This operation transforms M_j into M_i (line 4). Next, a bilinear operation, denoted as δ, is applied to M_i (line 5). This operation increases the spatial resolution of features. The result of this operation is assigned to M_j, preparing for the next iteration or the final step. After the loop, the pointwise convolution is employed on M_j (line 7). The kernel size of the pointwise convolution is ϵ^2 ($\epsilon=1$). The stride is one, as we mix channel information without spatial aggregation. The padding is zero because there is no need to pad the input for spatial reasons. The pointwise convolution performs channel-wise transformations without altering the spatial dimensions of the input. This convolution serves to refine and adjust the feature map, resulting in the final output M'.

Algorithm 2. Decoder

Require: the output from the encoder M
Ensure: the output from the decoder M'
 1: The bilinear operation is denoted as δ.
 2: Initialize M_j; $M_j \leftarrow M$
 3: **for** $i = 1, 5$ **do**
 4:     Apply Algorithm 1 to M_i and obtain M_i
 5:     Apply δ to M_i and obtain M_j
 6: **end for**
 7: Apply the pointwise convolution and obtain M'

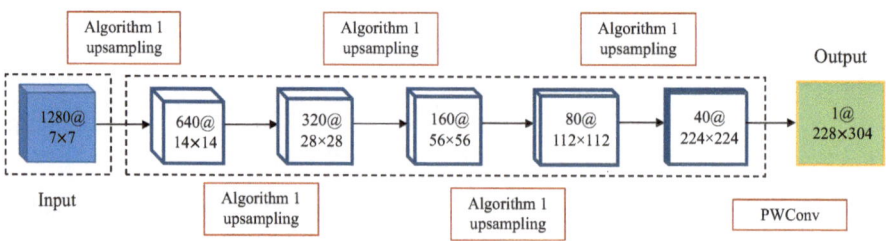

Fig. 3. The model size output by each module of the decoder.

We present the flowchart of how the feature size changes from the decoder (Fig. 3). The encoder produces the feature of size 7×7 with 1280 channels (1280@7×7). The number on the left of the @ symbol is the quantity of channels. The number on the right is the feature size. This feature of 1280@7×7 is then passed through a sequence of algorithms and upsampling steps. Algorithm 1 focuses on the center position of features. It halves the number of channels. It captures the importance of positions within the receptive field more effectively than uniformly applied regular convolutions. The upsampling employs the bilinear method to double the spatial sizes of features. After Algorithm 1 and the initial upsampling step, the feature size increases from 1280@7×7 to 640@14×14. Following the second application of Algorithm 1 and subsequent upsampling, the feature expands to 320@28×28. Continuing with Algorithm 1 and the third round of upsampling, the feature size grows to 160@56×56. The fourth iteration of Algorithm 1 followed by upsampling transforms the feature to 80@112×112. Finally, the last application of Algorithm 1 and the fifth upsampling enlarges the feature to 40@224×224. After the last upsampling step, a pointwise convolution (PWConv) is applied, resulting in the output feature with a single channel of size 1@228×304. The outputs are fine-grained.

In summary, the decoder involves applications of Algorithm 1 and bilinear interpolation, subsequently undergoing a pointwise convolution to generate a final output. In the decoder, feature maps are progressively refined and upsampled to reconstruct higher-resolution outputs. The decoder performs better in the following aspects:

- Diminishing latency: Our decoder addresses the issue of high latency caused by high MAF in traditional convolutions like DWConv. By reducing computational redundancy and MAF, the decoder presents itself as a competitive alternative in MDE.
- Selective application on input channels: The decoder applies convolution filters only on a part of the input channels. This approach results in lower FLOPs than regular convolutions. The selective application enhances the exploitation of on-device computational capacity.
- Efficiency in upsampling: The bilinear upsampling after each convolutional operation allows for a gradual increase in the spatial dimensions of features. The approach is particularly effective for generating higher-resolution outputs from lower-resolution inputs. We offer a good balance between performance and output quality. This is advantageous in applications where computational resources are a concern.

The decoder adopts a structured approach for feature extraction and upsampling. It offers a blend of efficiency, speed, and effective feature extraction, making it a valuable advancement in the design and performance optimization of neural networks. EffiNet uses the decoder to reconstruct depths from the encoded features efficiently. It utilizes the strengths of MobileNetV2 while introducing an efficient decoder. This design allows for high-quality MDE in a lightweight, computationally efficient package. This makes EffiNet an attractive solution for a wide range of applications, especially those requiring mobile or edge sensors.

4   Experiments

4.1   Experiment Detail

We use the KITTI [28] and NYU-Depth-v2 [29] dataset for training and evaluation. The experiment platforms include NVIDIA TX2 and Nano GPU. When training EffiNet, the loss function is the mean absolute error. The batch size is eight. For inference, we set the batch size to one, ensuring a precise and detailed evaluation of the model performance on individual samples. To evaluate MDE, we adopt δ_t ($t = 1$, 2, and 3), RMSE, computational complexity, number of parameters, and inference time. These metrics are commonly utilized by various methods [1–4, 21, 30, 31].

4.2   Results and Analysis

Here, our methods are compared with different strategies. Particularly, the ablation study for our proposed modules is detailed in Sect. 4.2. Section 4.2 presents that our techniques are contrasted with the most advanced methods in terms of precision, inference time, computational complexity, and model complexity.

Table 1. Ablation Study

Encoder	Decoder	FLOPs	GPU	δ_1	RMSE	Parameters	Input size
Encoder [5]	Decoder [8]	1.3	47	75.9	0.589	7.08	224 × 224
Encoder [5]	Decoder[21]	0.8	132	78.1	0.593	4.01	224 × 224
Encoder [8]	Decoder[21]	0.9	141	78.2	0.557	3.44	228 × 304
Encoder [2]	Decoder[21]	0.9	938	72.3	0.726	**2.54**	228 × 304
Ours	Decoder[21]	**0.6**	123	77.2	0.579	3.44	228 × 304
Encoder [3]	Decoder [8]	10.6	226	80.1	0.533	38.93	228 × 304
Encoder [3]	UpConv [3]	22.9	3,010	78.9	0.604	43.02	228 × 304
Encoder [2]	Decoder [8]	21.3	321	69.1	0.658	15.69	228 × 304
Ours	Decoder [8]	0.9	55	64.3	0.723	3.95	228 × 304
Encoder [8]	UpProj [4]	14.8	753	78.2	0.609	17.87	228 × 304
Encoder [8]	Decoder [2]	10.1	1,329	71.9	0.724	134.94	228 × 304
Encoder [2]	Decoder [2]	10.5	1,564	69.7	0.753	132.24	228 × 304
Ours	Ours	1.4	**45**	**91.6**	**0.432**	5.50	228 × 304

Ablation Studies. To demonstrate the benefits of our modules, we perform tests using the Nano GPU board on NYU-Depth-v2. The outcomes of these tests are summarized in Table 1. When various methods employ our encoder or decoder, there is a balance achieved between computational efficiency and accuracy. This balance is crucial in optimizing the performance of MDE models, ensuring they are both fast and accurate. Specifically, it shows a at least 25.00% reduction in FLOPs when our encoder is employed, as listed in lines 2–6. Compared to the FLOPs observed (22.9 GFLOPs for 'Encoder [3] & UpConv [3]'), our encoder (Decoder [8]) has a reduction of 96.07%. When the encoder employs our encoder, compared with that of other decoders, the GPU inference time (in milliseconds) of our decoder is reduced by 18.18%; the δ_1 of our decoder goes up 18.65%; the RMSE of our decoder decreases by 25.39%. In summary, our method demonstrates a significant reduction in FLOPs and GPU processing time compared to the most computationally intensive methods. We achieved a marked improvement in accuracy (as indicated by δ_1 and RMSE metrics) over other methods. This highlights that our encoder and decoder are efficient in balancing computational demands with high accuracy in MDE.

Quantitative Results on Different Datasets. On NYU-Depth-v2: We conducted executions of various MDE methods across different edge devices on NYU-Depth-v2. We choose Nano and TX2 GPU, as they are commonly used in IoT environments. We re-implement the most recent and classic methods on edge devices for comparison in Table 2. The '-' symbol refers to the model [10] is complex and cannot run on edge devices, making it impossible to obtain their inference time in Nano and TX2. This symbol reflects the limitations of the model in the computational demands, which exceed the processing capabilities of Nano and TX2. Our methods (EffiNet) show an improvement in inference time, FLOPs, δ_1, and parameters. For example, in the metric of inference time on Jetson TX2, EffiNet is 24.39% lower compared with others [2,3,7–11,17,21–23,26,27]. On average, EffiNet shows a reduction of approximately 71.24% in inference time. Relative to others, EffiNet decreases at least by 18.18% in the metric of inference time on Jetson Nano. In comparison to the recent method, such as [9], EffiNet exhibits an improvement in δ_1 by 0.44%. The RMSE metric of our method also surpasses that of most other methods.

Table 2. Comparison Results on NYU-Depth-v2

on NYU-Depth-v2	TX2	Nano	FLOPs	δ_1	RMSE	Input size	Parameters
Eigen *et al.* [1]	23	872	2.1	61.1	0.907	228 × 304	32,735,923
Laina *et al.* [3] (Up)	237	3,010	22.9	78.9	0.604	228 × 304	43,017,728
Laina *et al.* [3] (UpProj)	319	3,403	42.7	81.1	0.573	228 × 304	63,563,008
Liu *et al.* [8]	41	55	0.9	78.6	0.548	228 × 304	3,948,314
Eigen and Fergus [2]	195	2,664	23.4	76.9	0.641	228 × 304	226,271,170
Eigen and Fergus [2]	96	1,564	10.5	69.7	0.753	228 × 304	132,237,250
Lee *et al.* [7]	790	4,733	89.5	88.5	0.392	416 × 544	47,000,688
Chen *et al.* [9]	195	1,987	35.4	91.2	**0.334**	576 × 448	60,000,000
Tu *et al.* [21]	73	122	0.5	77.6	0.568	228 × 304	1,627,970
Mertan *et al.* [11]	531	1,391	74.6	71.8	0.700	384 × 384	115,226,816
Xia *et al.* [10]	-	-	311.8	86.1	0.512	257 × 353	76,981,505
Liu *et al.* [22]	201	3,458	13	83.3	0.495	228 × 304	31,940,623
Zhao *et al.* [23]	564	13,182	45.6	71.2	0.710	240 × 320	54,564,675
Lee and Kim [26]	256	3,105	31.3	83.7	0.538	224 × 224	356,570,988
Xian *et al.* [27]	283	3,266	61.8	78.1	0.660	384 × 384	110,508,482
Cheng *et al.* [17]	126	1,783	29.0	83.8	0.508	288 × 384	31,940,000
EffiNet	31	45	1.4	**91.6**	0.432	228 × 304	5,498,754

Table 3. Comparison Results on KITTI

on KITTI	TX2	Nano	FLOPs	δ_1	RMSE	Input size	Parameters
Eigen et al. [1]	99	171	5.0	69.2	7.156	228 × 912	95,298,395
Laina et al. [3] (Up)	801	2,659	67.0	58.9	6.301	228 × 912	43,017,728
Hu et al. [6]	2,085	26,431	321.0	-	5.437	228 × 912	67,569,473
Ma and Karaman [4]	846	2,764	124.4	59.1	6.266	228 × 912	63,563,008
Wofk et al. [5]	211	212	3.2	84.3	5.191	228 × 912	3,960,930
Liu et al. [8]	33	45	0.6	91.0	3.656	128 × 416	3,948,314
Tu et al. [21]	93	97	**0.3**	91.1	3.796	128 × 416	1,849,362
Liu et al. [22]	704	4,890	89.0	89.1	4.311	228 × 912	31,940,623
Zhao et al. [23]	564	13,182	45.6	79.8	4.679	228 × 912	54,564,675
Li et al. [24]	789	3,537	56.3	97.5	**2.143**	376 × 1,241	273,000,000
Fang et al. [25]	132	344	20.1	87.3	4.537	128 × 416	19,873,156
Klingner et al. [14]	206	532	11.8	87.9	4.693	192 × 640	16,302,792
Tosi et al. [15]	25,013	165,915	30.3	86.5	4.608	192 × 640	9,388,578
Yin and Shi [12]	251	245	10.6	80.2	5.737	128 × 416	50,222,512
Gur and Wolf [13]	349	1,670	25.1	84.6	5.187	128 × 416	46,382,113
EffiNet	29	41	1.0	**97.6**	3.322	128 × 416	5,498,754

On KITTI: For additional comparisons with other approaches, we execute an evaluation of KITTI in Table 3. Our approaches demonstrate efficacy in inference time, precision, and the complexity of computation and model. For example, EffiNet shows a considerable reduction in TX2 inference time (in milliseconds) compared to other methods. Compared to other methods, EffiNet achieves a reduction in inference time on TX2 ranging from 12.12% to 99.88%. In Nano inference time, the smallest reduction of EffiNet is against 'Liu et al. [8]', with 8.89%; the largest reduction of EffiNet is against 'Tosi et al. [15]', with 99.97%. In FLOPs, EffiNet demonstrates reductions compared to other methods. In δ_1, the minimum and maximum improvement of EffiNet is approximately 0.10% and 65.70%, respectively. In summary, EffiNet significantly improves inference time, FLOPs, accuracy, and parameters compared to other methods. This highlights the effectiveness of our methods in balancing different metrics, particularly in the context of hardware-constrained sensors like those utilizing Jetson Nano.

5   Conclusion

This paper presents EffiNet, a novel network structure tailored for MDE that addresses the challenges of high computational cost and efficiency in real-time applications, particularly on resource-constrained devices. The three primary innovations introduced by this method substantially enhance the performance and feasibility of MDE technologies in practical scenarios. First, the adoption of MobileNetV2 as the backbone for the encoder significantly reduces computational complexity. Second, the newly designed decoding algorithm optimizes performance across various scene conditions. The model reduces unnecessary computational load by implementing a block processing strategy that processes input feature channels separately. This, coupled with efficient upsampling and point convolution operations, enhances feature resolution while managing computational expenses. Lastly, the optimization of MAF within the algorithm operations minimizes data transmission requirements, thereby boosting execution speed and improving energy efficiency. The results demonstrate our significant improvements in computational efficiency and MAF. Future directions for this research will focus on further reducing the latency and enhancing the accuracy of EffiNet.

Acknowledgment. This work was supported by the the Natural Science Foundation of Henan Province (No. 242300420693), the science and technology plan of PRC Ministry of Public Security (No. 2023JSYJC28), the Key Scientific Research Project Plan of Colleges and Universities in Henan Province (No. 23A520042 and No. 23B520019), the Fundamental Research Funds for Central Universities of Zhengzhou Police University (No. 2023TJJBKY012 and No. 2023TJJBKY011), and the Zhengzhou Police College Teaching Reform Project (No. JY2024013 and No. JY2024Z09).

References

1. Eigen, D., Puhrsch, C., Fergus, R.: Depth map prediction from a single image using a multi-scale deep network. Proc. Adv. Neural Inf. Process. Syst. 2366–2374 (2014)
2. Eigen, D., Fergus, R.: Predicting depth, surface normals and semantic labels with a common multi-scale convolutional architecture. In: Proceedings of the IEEE/CVF International Conference on Computer Vision, pp. 2650–2658 (2015)
3. Laina, I., Rupprecht, C., Belagiannis, V., Tombari, F., Navab, N.: Deeper depth prediction with fully convolutional residual networks. In: Proceedings of IEEE Fourth International Conference on 3D Vision, pp. 239–248 (2016)
4. Ma, F., Karaman, S.: Sparse-to-dense: depth prediction from sparse depth samples and a single image. In: Proceedings of the IEEE International Conference on Robotics and Automation, pp. 4796–4803 (2018)
5. Wofk, D., Ma, F., Yang, T., Karaman, S., Sze, V.: Fastdepth: fast monocular depth estimation on embedded systems. In: Proceedings of the IEEE International Conference on Robotics and Automation, pp. 6101–6108 (2019)
6. Hu, J., Zhang, Y., Okatani, T.: Visualization of convolutional neural networks for monocular depth estimation. In: Proceedings of the IEEE/CVF International Conference on Computer Vision, pp. 3869–3878 (2019)
7. Lee, J.H., Han, M.-K., Ko, D.W., Suh, I.H.: From big to small: Multi-scale local planar guidance for monocular depth estimation, *CoRR*, vol. abs/1907.10326, July 2019
8. Liu, S., Yang, L.T., Tu, X., Li, R., Xu, C.: Lightweight monocular depth estimation on edge devices. IEEE Internet Things J. **9**(17), 16 168–16 180 (2022)
9. Lei, C., Zhengyou, L., Yu, S.: A monocular image depth estimation method based on weighted fusion and point-wise convolution. IET Comput. Vis. (2023)
10. Xia, Z., Sullivan, P., Chakrabarti, A.: Generating and exploiting probabilistic monocular depth estimates. In: Proceedings of IEEE/CVF Conference on Computer Vision and Pattern Recognition, pp. 62–71 (2020)
11. Mertan, A., Sahin, Y.H., Duff, D.J., Unal, G.: A new distributional ranking loss with uncertainty: illustrated in relative depth estimation. In: Proceedings of IEEE International Conference on 3D Vision, pp. 1079–1088 (2020)

12. Yin, Z., Shi, J.: Geonet: unsupervised learning of dense depth, optical flow and camera pose. In: Proceedings of IEEE/CVF Conference on Computer Vision and Pattern Recognition, pp. 1983–1992 (2018)
13. Gur, S., Wolf, L.: Single image depth estimation trained via depth from defocus cues. In: Proceedings of IEEE/CVF Conference on Computer Vision and Pattern Recognition, pp. 7683–7692 (2019)
14. Klingner, M., Termöhlen, J.-A., Mikolajczyk, J., Fingscheidt, T.: Self-supervised monocular depth estimation: solving the dynamic object problem by semantic guidance. In: Proceedings of European Conference Computer Vision, pp. 582–600 (2020)
15. Tosi, F., et al.: Distilled semantics for comprehensive scene understanding from videos. In: Proceedings of IEEE/CVF Conference on Computer Vision and Pattern Recognition, pp. 4653–4664 (2020)
16. Tu, X., et al.: Lidar point cloud recognition and visualization with deep learning for overhead contact inspection. Sensors **20**(21), 6387 (2020)
17. Cheng, D., Chen, J., Lv, C., Han, C., Jiang, H.: Using full-scale feature fusion for self-supervised indoor depth estimation. Multimedia Tools Appl. 1–19 (2023)
18. Tu, X., Zhang, C., Liu, S., Xu, C., Li, R.: Point cloud segmentation of overhead contact systems with deep learning in high-speed rails. J. Netw. Comput. Appl. **216**, 103671 (2023)
19. Sandler, M., Howard, A., Zhu, M., Zhmoginov, A., Chen, L.-C.: Mobilenetv2: inverted residuals and linear bottlenecks. In: Proceedings of the IEEE/CVF Conference on Computer Vision and Pattern Recognition, pp. 4510–4520 (2018)
20. K. Han, Y. Wang, Q. Tian, J. Guo, C. Xu, and C. Xu, "Ghostnet: More features from cheap operations," in *Proc. IEEE/CVF Conf. Comput. Vis. Pattern Recognit.*, 2020, pp. 1577–1586
21. Tu, X., Xu, C., Liu, S., Li, R., Xie, G., Huang, J., Yang, L.T.: Efficient monocular depth estimation for edge devices in internet of things. IEEE Trans. Ind. Informat. **17**(4), 2821–2832 (2021)
22. Liu, S., Tu, X., Xu, C., Li, R.: Deep neural networks with attention mechanism for monocular depth estimation on embedded devices. Futur. Gener. Comput. Syst. **131**, 137–150 (2022)
20. Zhao, Y., Kong, S., Shin, D., Fowlkes, C.: Domain decluttering: simplifying images to mitigate synthetic-real domain shift and improve depth estimation. In: Proceedings of the IEEE/CVF Conference on Computer Vision and Pattern Recognition, pp. 3327–3337 (2020)
24. Li, Z., Chen, Z., Liu, X., Jiang, J.: Depthformer: exploiting long-range correlation and local information for accurate monocular depth estimation. Mach. Intell. Res. **20**(6), 837–854 (2023)
25. Fang, Z., Chen, X., Chen, Y., Van Gool, L.: Towards good practice for CNN-based monocular depth estimation. In: Proceedings of the IEEE Winter Conference on Applications of Computer Vision, pp. 1080–1089 (2020)
26. Lee, J.-H., Kim, C.-S.: Monocular depth estimation using relative depth maps. In: Proceedings of the IEEE/CVF Conference on Computer Vision and Pattern Recognition, pp. 9729–9738 (2019)
27. Xian, K., et al.: Monocular relative depth perception with web stereo data supervision. In: Proceedings of the IEEE/CVF Conference on Computer Vision and Pattern Recognition, pp. 311–320 (2018)
28. Huang, G., Liu, Z., Maaten, L.V.D., Weinberger, K.Q.: Densely connected convolutional networks. In: Proceedings of the IEEE Conference on Computer Vision and Pattern Recognition, pp. 2261–2269 (2017)

29. Nathan Silberman, P.K., Hoiem, D., Fergus, R.: Indoor segmentation and support inference from rgbd images. In: Proceedings of European Conference on Computer Vision, pp. 746–760 (2012)
30. Tu, X., et al.: Learning depth for scene reconstruction using an encoder-decoder model. IEEE Access **8**, 89 300–89 317 (2020)
31. Tu, X., Yang, L.T., Liu, S., Li, R.: Accelerated feature extraction and refinement for improved aerial scene categorization. IEEE Trans. Geosci. Remote Sens. **1**(1), 1–1 (2024)

Adaptive Communication Spectrum Sensing Algorithm Based on Energy Detection

Weihong Shi[1,2,3], Tao Tang[1,2,3], Runhui Zhao[1,2,3(✉)], Hong Wen[1,2,3], and Xuewei Feng[1,2,3]

[1] School of Aeronautics and Astronautics, University of Electronic Science and Technology of China, Chengdu 611731, People's Republic of China
{shiweihong,202221100402}@std.uestc.edu.cn, 1528129899@163.com, uestcrunhui@gmail.com, sunlike@uestc.edu.cn

[2] Aircraft Swarm Intelligent Sensing and Cooperative Control Key Laboratory of Sichuan Province, Chengdu 611731, China

[3] Sichuan Intelligent loT Communication Technology Engineering Research Center, UESTC, Chengdu, China

Abstract. Based on the traditional energy detection algorithm, an adaptive communication spectrum sensing algorithm based on energy detection is proposed for signals whose noise power is difficult to be estimated accurately. The algorithm takes into account the problem of large error in calculating the energy detection judgment threshold by noise power when the estimated value of noise power is inaccurate, and is able to perform adaptive communication spectrum sensing for signals with less a priori information and noise power that cannot be estimated accurately. Simulation results show that the adaptive communication spectrum sensing algorithm based on energy detection can effectively improve the performance of spectrum sensing for signals, and the algorithm can effectively detect and sense signals under the condition of reasonable setting of judgment coefficients.

Keywords: Cognitive radio · Spectrum sensing · Energy detection · Adaptive

1 Introduction

With the rapid development of today's wireless field, spectrum resources have gradually become an extremely valuable and scarce resource. Spectrum sensing techniques can rationally use spectrum resources and enable devices to dynamically monitor and understand the surrounding spectrum usage [1–3]. Mainstream spectrum sensing techniques include three broad categories: energy detection, matched filter detection, and cyclic smooth feature detection [4, 5]. L. Yang, J. Fang, H. Duan and H. Li developed a computationally efficient power spectrum reconstruction method based on spectrum sensing [6]. R. Sarikhani and F. Keynia combined cooperative spectrum sensing with machine learning for signaling reduction in SU networks [7]. A. Gao, C. Du, S. X. Ng and W. Liang proposed a cooperative spectrum sensing approach for cognitive radio networks based on multi-intelligent reinforcement learning [8]. P. C. Sofotasios, E. Rebeiz, L.

© The Author(s) 2025
P. Siarry et al. (Eds.): WCNA 2023, LNEE 1361, pp. 36–44, 2025.
https://doi.org/10.1007/978-981-96-2409-6_4

Zhang, T. A. Tsiftsis, D. Cabric and S. Freear investigated the performance of an energy detector over generalized $k - \mu$ and $k - \mu$ extreme fading channels, which have been shown to provide remarkably accurate fading characterization [9]. I. Sobron, P. S. R. Diniz, W. A. Martins and M. Velez used a low-cost function to intern adaptive spectrum sensing [10].

The advantage of the energy detection algorithm is that it is simple to implement and can quickly detect the presence of signals that occupy the spectrum. The disadvantage of the energy detection algorithm is that it has low sensitivity, is only suitable for detecting high-power signals, is not sensitive to the detection of low-power signals, and is easily affected by noise [11]. In this paper, on the basis of the traditional energy detection algorithm, for the traditional energy detection algorithm is easily affected by noise and the defects of poor performance for low power signals, we propose an adaptive communication spectrum sensing algorithm based on the energy detection method, which is able to obtain a more accurate judgment threshold value in the case of uncertainty in the noise power, so as to carry out adaptive communication spectrum sensing for the signals with different noise power levels which improves the accuracy of signal spectrum sensing.

2 System Model

2.1 Traditional Energy Detection Model

Spectrum sensing is a key technique in the field of wireless communications that allows a system to identify salient features in external signals by receiving them and recognizing them based on a set of assumptions. The core purpose of this process is to ensure that the system is able to effectively distinguish between different types of signals, with a particular focus on detecting signals from Primary Users (PUs). Primary Users are those users who have the right to use the spectrum, and the spectrum sensing technique attempts to protect the rights of these users by preventing incoming Secondary Users (SU) from interfering with the signals of the Primary Users [12, 13].

In spectrum sensing, the judgment threshold of the energy detection method is determined by the difference between the signal and noise energies, and the presence or absence of a PU signal is usually obtained by comparing the detection statistics with the judgment threshold [14, 15]. The received signal has significant characteristics when under the hypothesis conditions and behaves in two states, the spectrum sensing process can be represented as the following binary hypothesis testing model:

$$\begin{cases} H_0 : Y \leq \sigma_w^2 \\ H_1 : Y \geq \sigma_w^2 \end{cases} \tag{1}$$

where $\begin{cases} H_0 : Y \leq \sigma_w^2 \\ H_1 : Y \geq \sigma_w^2 \end{cases}$ is the noise signal, assumed to be Gaussian white noise;

$\begin{cases} H_0 : Y \leq \sigma_w^2 \\ H_1 : Y \geq \sigma_w^2 \end{cases}$ is the PU signal; $\begin{cases} H_0 : Y \leq \sigma_w^2 \\ H_1 : Y \geq \sigma_w^2 \end{cases}$ is the signal detected by the receiver

at the cognitive receiver end; $\begin{cases} H_0 : Y \leq \sigma_w^2 \\ H_1 : Y \geq \sigma_w^2 \end{cases}$ is the number of sample points. The two states are: $\begin{cases} H_0 : Y \leq \sigma_w^2 \\ H_1 : Y \geq \sigma_w^2 \end{cases}$ indicates that the PU is not using the channel, $\begin{cases} H_0 : Y \leq \sigma_w^2 \\ H_1 : Y \geq \sigma_w^2 \end{cases}$ indicates that the PU is using the channel.

In the energy detection method, the detection statistic is defined as:

$$\begin{cases} H_0 : Y \leq \sigma_w^2 \\ H_1 : Y \geq \sigma_w^2 \end{cases} \tag{2}$$

If the noise energy is known, then the binary hypothesis model for energy detection can be expressed as:

$$\begin{cases} H_0 : Y \leq \sigma_w^2 \\ H_1 : Y \geq \sigma_w^2 \end{cases} \tag{3}$$

where σ_w^2 is the noise energy.

2.2 Conventional Energy Detection Threshold

According to the central limit theorem, for sufficiently large N, the test statistic Y satisfies under two assumptions the central The characteristics of the chi-square distribution can be expressed as:

$$\begin{cases} H_0 : Y \sim \chi(\sigma_w^2, \frac{2}{N}\sigma_w^4) \\ H_1 : Y \sim \chi((1+\alpha)\sigma_w^2, \frac{2}{N}(1+\alpha)^2\sigma_w^4) \end{cases} \tag{4}$$

where α denotes the signal-to-noise ratio value, and $Y \sim \chi()$ denotes obeying the central chi-square distribution.

From the above distributional characteristics it can be derived that for a given judgment threshold value, the false alarm probability and detection probability of the energy detection method are respectively:

$$P_f = Pr(Y > \lambda \mid H_0) = Q\left(\frac{\lambda - \sigma_w^2}{\sqrt{2/N}\sigma_w^2}\right) \tag{5}$$

$$P_d = Pr(Y > \lambda \mid H_1) = Q\left(\frac{\lambda - (1+\partial)\sigma_w^2}{\sqrt{2/N}(1+\partial)\sigma_w^2}\right) \tag{6}$$

where $Q()$ is the standard Gaussian complementary cumulative distribution function.

According to the Constant False Alarm Probability Criterion (CFAR) and Constant Detection Probability Criterion (CDR), when the false alarm probability or the detection probability is certain, the corresponding detection judgment threshold can be derived from the above two expressions.

When the false alarm probability is P_f, the judgment threshold λ_f is denoted as:

$$Q^{-1}(P_f) = \frac{\lambda_f - \sigma_w^2}{\sqrt{\frac{2}{N}}\sigma_w^2} \Rightarrow \lambda_f = \sigma_w^2(1 + \sqrt{\frac{2}{N}}Q^{-1}(P_f)) \tag{7}$$

When the probability of detection is P_d, the judgment threshold λ_d is denoted as:

$$Q^{-1}(P_d) = \frac{\lambda_d - (1 + \partial)\sigma_w^2}{\sqrt{\frac{2}{N}(1 + \partial)\sigma_w^2}} \Rightarrow \lambda_d = \sigma_w^2(1 + \partial)(1 + \sqrt{\frac{2}{N}}Q^{-1}(P_f)) \qquad (8)$$

3 Adaptive Spectrum Awareness Factor

Since traditional energy detection methods require accurate estimation of the noise energy value, an adaptive spectral perception coefficient based on segmented energy detection is proposed on this basis ξ_λ.

When the signal noise energy cannot be estimated accurately, the threshold value of energy detection is greatly affected by the noise and cannot be judged accurately. Consider dividing the overall frequency band of the signal into a number of sub-frequency bands, calculating the average signal energy and the energy of each sub-frequency band, respectively, and comparing the sub-frequency band energy with the average energy, the resulting adaptive communication spectrum perception coefficient ξ_λ can be expressed as.

$$\xi_i = \mu_i \frac{E_i}{E_s}; \quad i = 1,2, \cdots N \qquad (9)$$

where, μ_i is the compensation coefficient, which is used to compensate the adaptive spectrum perception coefficient, E_i denotes the energy value of each sub-band, E_s denotes the average energy value of the signal, and N denotes the number of sub-bands divided.

The selection of the compensation factor μ_i is related to the number of sub-band divisions and the magnitude of the energy value in each sub-band.

Compared to the threshold value of conventional energy detection, the adaptive communication spectrum sensing threshold value can be expressed as:

$$\lambda_{fi} = \xi_i \ \sigma_w^2(1 + \sqrt{\frac{2}{N}}Q^{-1}(P_f)) \qquad (10)$$

$$\lambda_{di} = \xi_i \ \sigma_w^2(1 + \partial)(1 + \sqrt{\frac{2}{N}}Q^{-1}(P_f)) \qquad (11)$$

Since the noise energy value cannot be estimated accurately, this adaptive coefficient is proposed to improve the deficiency of noise power estimation in traditional spectrum sensing to some extent. Compared with the traditional method, the calculation of this adaptive coefficient not only considers the average energy of the signal, but also pays more attention to the energy distribution of each frequency band. On the basis of the adaptive spectrum sensing coefficient, this paper proposes an adaptive communication spectrum sensing algorithm based on energy detection, which enables the system to intelligently adjust the threshold value of spectrum sensing, so as to more accurately sense the actual situation of the spectrum environment.

4 Adaptive Communication Spectrum Sensing Algorithm

The adaptive communication spectrum sensing algorithm is mainly divided into three parts: sub-band division of the signal, calculation of adaptive spectrum sensing coefficients and adaptive sensing of the signal.

Adaptive communication spectrum sensing algorithm based on energy detection first of all need to sub-banding of the signal frequency band, sub-banding based on its signal frequency domain envelope of the degree of ups and downs to be divided, the signal frequency domain envelope between two points of the degree of ups and downs is less than the set threshold, that is, between the two points are considered to belong to the same sub-frequency band, if it is greater than the threshold, it is considered to belong to a different sub-frequency band of the two points, the sub-frequency banding can be expressed as follows:

$$\begin{cases} Env(f)_{j+1} \in f_i : abs(Env(f)_{j+1} - Env(f)_j) \leq \eta \\ Env(f)_{j+1} \notin f_i : abs(Env(f)_{j+1} - Env(f)_j) > \eta \end{cases}; Env(f)_j \in f_i, j = 1,2,\cdots M \quad (12)$$

where $Env(f)_{j+1}$ is the next point where the envelope needs to be divided, $Env(f)_j$ is the point where the envelope has been divided, f_i is the sub-band to be divided, η is the set threshold for sub-band division, and M is the number of frequency domain points of the signal. A schematic diagram of sub-banding is shown in Fig. 1.

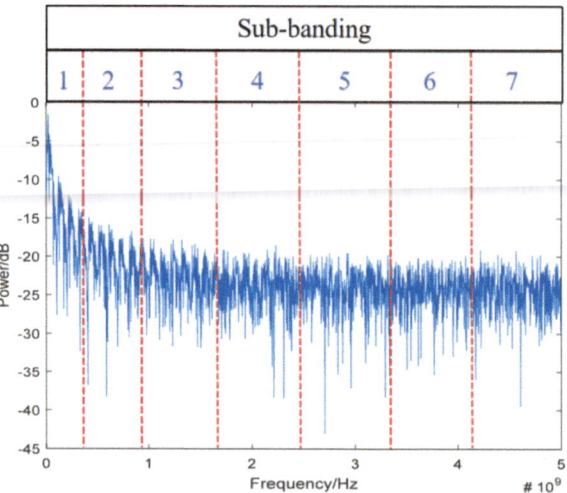

Fig. 1. Schematic diagram of sub frequency band division

Sub-banding based on the undulation of the frequency domain envelope ensures that the energy fluctuation of each neighboring sub-band is within a reasonable range, and the energy difference of each neighboring sub-band can be expressed as:

$$\begin{cases} abs(f_i - f_{i+1}) \leq \varphi_i \\ abs(f_{i-1} - f_i) \leq \varphi_i \end{cases} \quad (13)$$

where f_i is the current sub-band, f_{i+1} and f_{i-1} are the next and previous sub-bands, respectively, and φ_i is the set energy fluctuation range.

After the sub-band division is completed, the adaptive spectrum sensing coefficient ξ_i of each sub-band is calculated, and then the energy detection threshold value is corrected according to the adaptive spectrum sensing coefficient of each sub-band.

The sub-bands of the signal are computed to obtain the adaptive coefficients ξ_i corresponding to the sub-bands, and then individual spectrum sensing is performed for each sub-band, and the sensing threshold value of the sub-band can be expressed as:

$$\lambda_i = \xi_i \lambda \tag{14}$$

Compared with the traditional method, this adaptive perception algorithm not only considers the average energy of the signal, but also pays more attention to the energy distribution of each frequency band. Therefore, the adaptive communication spectrum sensing algorithm based on energy detection can improve the detection performance and reduce the misjudgment rate, as well as improve the anti-noise ability.

5 Simulation Experiment and Result Analysis

In order to verify the effectiveness of the theory and algorithm in this paper, the related simulation experiments and result analysis are given in this section. The channel is assumed to be AWGN channel, the noise is additive Gaussian white noise, and all signals are independent of each other.

Figure 2 shows the misclassification rate of the traditional energy detection method and the adaptive communication spectrum sensing algorithm based on energy detection at the same signal-to-noise ratio.

Figure 3 shows the effect of the number of sub-band divisions of the same signal on the detection performance of the adaptive spectrum sensing algorithm based on energy detection.

As can be seen in Fig. 2, the misjudgment rate of the adaptive spectrum sensing algorithm is reduced by about 3% to 5% compared to the traditional energy detection algorithm in both low and high SNR cases. This is because in the case of low SNR, the traditional energy detection algorithm may be interfered by noise due to the dominance of noise, while the adaptive spectrum sensing algorithm benefits from the adaptive perception coefficients for the correction of the judgment threshold value, which results in a lower misjudgment rate.

As can be seen from Fig. 3, the number of sub-band divisions has a certain effect on the detection performance. The number of sub-band divisions has a certain impact on the detection performance, when the number of sub-bands is between $f_i = 12 - 20$, the detection performance of the adaptive spectrum sensing algorithm is better, and as the number of sub-band divisions increases, the detection performance of the algorithm is decreasing. This is because when the number of sub-band divisions is high, the signal information contained in each sub-band is less, which leads to the signal sub-band can not accurately reflect the relationship between the local and the whole; the signal energy in each sub-band may only reflect a very small range of frequency characteristics, and the computed adaptive perception coefficients can not be well corrected for the

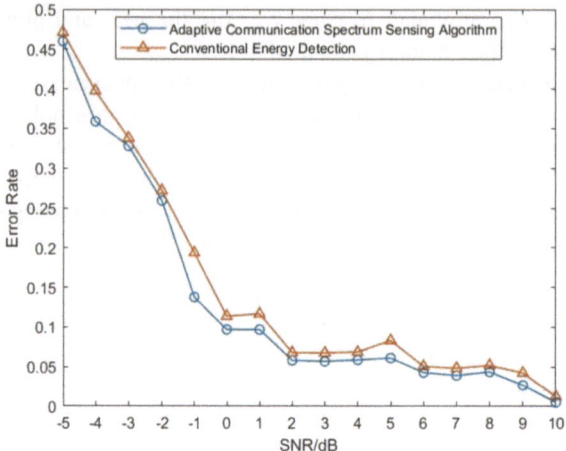

Fig. 2. Misclassification rate at different signal-to-noise ratios

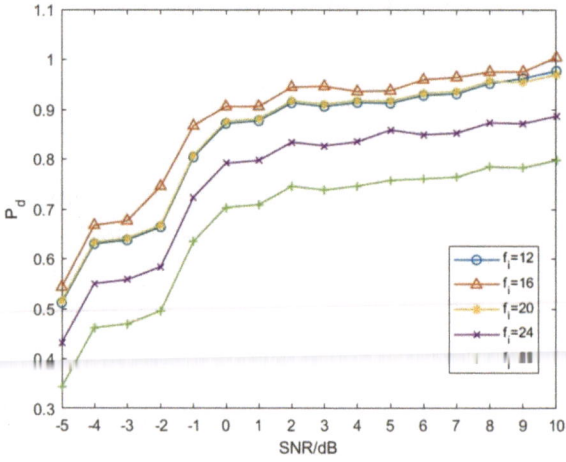

Fig. 3. Impact of the number of sub-band divisions on detection performance

judgment threshold value. Therefore, a reasonable selection of the number of sub-band divisions helps to improve the detection performance of the adaptive spectrum perception algorithm.

6 Conclusions

In this paper, an adaptive communication spectrum sensing algorithm based on energy detection is proposed. For the signals that cannot accurately estimate the noise energy, adaptive communication spectrum sensing coefficients are used to correct the judgment threshold value, so as to carry out adaptive communication spectrum sensing for signals with different noise energies. Theoretical analysis and simulation experimental results show that this algorithm can effectively improve the detection performance. In addition, the influence of the compensation coefficient on the detection performance of the algorithm is also analyzed, and when the compensation coefficient is within a reasonable range, the detection performance of the algorithm is significantly improved.

Acknowledgment. This work is supported by NSFC under Grant 62201132 and U23B2021.

References

1. Yucek, T., Arslan, H.: A survey of spectrum sensing algorithms for cognitive radio applications. In: IEEE Communications Surveys & Tutorials **11**(1), 116–130 (2009). https://doi.org/10.1109/SURV.2009.090109
2. Wang, B., Liu, K.J.R.: Advances in cognitive radio networks: a survey. IEEE Journal of Selected Topics in Signal Processing **5**(1), 5–23 (2011). https://doi.org/10.1109/JSTSP.2010.2093210
3. Ali, A., Hamouda, W.: Advances on spectrum sensing for cognitive radio networks: theory and applications. In: IEEE Communications Surveys & Tutorials **19**(2), 1277–1304 (2017). https://doi.org/10.1109/COMST.2016
4. Liang, Y.-C., Chen, K.-C., Li, G.Y., Mahonen, P.: Cognitive radio networking and communications: an overview. IEEE Trans. Veh. Technol. **60**(7), 3386–3407 (2011). https://doi.org/10.1109/TVT.2011.2158673
5. Bagwari, A., Singh, B.: Comparative performance evaluation of spectrum sensing techniques for cognitive radio networks. In: 2012 Fourth International Conference on Computational Intelligence and Communication Networks, Mathura, India, pp. 98–105 (2012). https://doi.org/10.1109/CICN.2012.66
6. Yang, L., Fang, J., Duan, H., Li, H.: Fast compressed power spectrum estimation: toward a practical solution for wideband spectrum sensing. IEEE Trans. Wireless Commun. **19**(1), 520–532 (2020). https://doi.org/10.1109/TWC.2019.2946805
7. Sarikhani, R., Keynia, F.: Cooperative spectrum sensing meets machine learning: deep reinforcement learning approach. IEEE Commun. Lett. **24**(7), 1459–1462 (2020). https://doi.org/10.1109/LCOMM.2020.2984430
8. Gao, A., Du, C., Ng, S.X., Liang, W.: A cooperative spectrum sensing with multi-agent reinforcement learning approach in cognitive radio networks. IEEE Commun. Lett. **25**(8), 2604–2608 (2021). https://doi.org/10.1109/LCOMM.2021.3078442
9. Sofotasios, P.C., Rebeiz, E., Zhang, L., Tsiftsis, T.A., Cabric, D., Freear, S.: Energy Detection based spectrum sensing over $\kappa-\mu$ and $\kappa-\mu$ extreme fading channels. IEEE Trans. Veh. Technol. **62**(3), 1031–1040 (2013). https://doi.org/10.1109/TVT.2012.2228680
10. Sobron, I., Diniz, P.S.R., Martins, W.A., Velez, M.: Energy detection technique for adaptive spectrum sensing. IEEE Trans. Commun. **63**(3), 617–627 (2015). https://doi.org/10.1109/TCOMM.2015.2394436
11. Gao, R., Li, Z., Qi, P., Li, H.: A robust cooperative spectrum sensing method in cognitive radio networks. IEEE Commun. Lett. **18**(11), 1987–1990 (2014). https://doi.org/10.1109/LCOMM.2014.2361851
12. Muchandi, N., Khanai, R.: Cognitive radio spectrum sensing: a survey. In: 2016 International Conference on Electrical, Electronics, and Optimization Techniques (ICEEOT), Chennai, India, pp. 3233–3237 (2016). https://doi.org/10.1109/ICEEOT.2016.7755301

13. Patil, V.M., Patil, S.R.: A survey on spectrum sensing algorithms for cognitive radio. In: 2016 International Conference on Advances in Human Machine Interaction (HMI), Kodigehalli, India, pp. 1–5 (2016). https://doi.org/10.1109/HMI.2016.7449196
14. Axell, E., Leus, G., Larsson, E.G., Poor, H.V.: Spectrum sensing for cognitive radio : state-of-the-art and recent advances. IEEE Signal Process. Mag. **29**(3), 101–116 (2012). https://doi.org/10.1109/MSP.2012.2183771
15. Atapattu, S., Tellambura, C., Jiang, H.: Energy detection based cooperative spectrum sensing in cognitive radio networks. IEEE Trans. Wireless Commun. **10**(4), 1232–1241 (2011). https://doi.org/10.1109/TWC.2011.012411.100611

A Cache Scheduling Method Based on Adaptive Expiration for Data Process System

Bin Fang[1(✉)], Yibo Zhong[1], Yizhen Sun[1], Jinjin Tu[2], Guang Jiang[1], and Yao Xiao[3]

[1] Information and Communication Branch of State Grid, Hunan Electric Power Company Limited, Changsha 410007, China
403183314@qq.com

[2] Nanjing NARI Information and Communication Technology, Nanjing City 210000, China
tujinjin@sgepri.sgcc.com.cn

[3] School of Software, Xinjiang University, Urumqi 830091, China

Abstract. Cache data plays a crucial role in the operation of the power system. In practical applications, the demand for retrieval and queries has surged due to the vast amount of terminal data collected throughout the province. Placing data in the cache effectively improves query speed [1]. However, cache space is limited, and the amount of storable data is small [2]. Without proper control over the data placed in the cache, the hit rate may decrease [3]. Thus this paper proposes a new control method through adaptively setting the expiration time of cached data, which considers the historical traffic distribution and real-time query distribution. The simulation results show that the proposed method can significantly reduce the cache hit rate, which leads to the improvement of cache utilization. In addition, it avoids frequent cache replacements and reduces I/O overhead.

Keywords: Cache data · Memory Scheduling Algorithm · Citation Heat · Cache Replacements

1 Introduction

With the technology development of the Internet of Things (IoT), the increasing number of underlying sensing terminals continuously contributes the stable operation of the power grid. For example, billions of intelligent sensing terminals keep collecting measurement data like instantaneous current, voltage, power and etc. from electric power customer. And these terminals would transmit the collected electricity measurement data with a fixed interval to the central processing system, which provides query and statistical analysis service [4] for the power grid. Specifically, the measurement center associates the collected data with user profiles based on device IDs, resulting in correlated data. Subsequently, this correlated data is stored in a database for further querying and statistical analysis by other applications. For example, it may be used to analyze changes in the electricity load for residential users or changes in the electricity load for specific power distribution areas [5].

However, due to the vast number of collection terminals across the entire province, the generated data and the demand for retrieval and queries are extremely large. Although

© The Author(s) 2025
P. Siarry et al. (Eds.): WCNA 2023, LNEE 1361, pp. 45–55, 2025.
https://doi.org/10.1007/978-981-96-2409-6_5

placing data in a cache can effectively improve query speed, current issues exist in the caching scheduling system:

- Due to limited cache space, it is not feasible to store all data in the cache space.
- On the other hand, if content is placed in the cache space indiscriminately, the hit rate may be low, leading to inefficient cache performance.

Therefore, it is necessary to adopt a certain strategy to determine the data stored in the cache, those classical methods often overlook the distribution of data query demands and cannot guarantee a high cache hit rate. Additionally, frequent cache content replacement can result in significant I/O overhead. In this regard, we propose a method to address the technical issue of low cache hit rates in existing cache data replacement methods. Our main contributions are as follows:

- We have designed a method based on historical data statistics, collecting historical data and calculating the historical citation impact of each piece of data to obtain the historical query distribution of the data.
- Our adaptive cache data expiration method has the capability to dynamically adjust the expiration period of current cache data based on the historical query distribution and real-time query situations.
- We have implemented a system that can perform data replacement. When the runtime of the current cache data reaches its expiration period, the system replaces the current cache data. Results indicate that the adaptive setting of cache data expiration periods can more conveniently and accurately achieve cache data scheduling and querying.

The structure of this paper is as follows: Sect. 1 serves as the introduction, Sect. 2 covers related work, Sect. 3 outlines the overall architecture, Sect. 4 presents the solution methods, Section 5 shows our experimental results, and Sect. 6 concludes the paper.

2 Related Work

2.1 Cache Scheduling

In computer systems, cache scheduling refers to the effective utilization of cache to enhance the performance of program execution. Cache, being a rapid yet expensive storage device, is typically employed to store recently accessed data, allowing for quicker access to this data in the near future. The primary objective of cache scheduling is to maximize the cache hit rate, ensuring the retrieval of the required data from the cache [6].

2.2 Common Cache Scheduling Algorithm

The core idea of Least Recently Used (LRU) is to replace the cache block that has not been used for the longest time. First-In-First-Out (FIFO) strategy replaces data that entered the cache earliest [7]. Optimal Replacement (OR) represents a theoretically optimal cache replacement strategy, selecting data for replacement that is predicted to be unused for the longest time in the future [8]. In practical applications, due to the difficulty of accurately predicting future access patterns, the Optimal Replacement strategy is rarely used in practice. Another method called Least Frequently Used (LFU) assumes that the least access frequency data would be used less in the future, and selects this type data for replacement in the following period.

2.3 Hotspot Caching

"Hotspot caching" typically refers to the data blocks or objects in the cache [9] which is accessed frequently within a time period. In computer systems, the "Hotspot caching" is usually used to optimize the scheduling technique, leads to enhancing the frequent cache data access performance and reducing the access latency.

Thus this paper proposed a cache data scheduling method, which considers the historical traffic distribution and real-time query distribution. The proposed method and analysis would be illustrated in the following chapters.

3 Overall Architecture

This paper considers a cache data scheduling framework for realtime measurement data process system. The framework comprises a data statistics module, an adaptive setting module, and a data replacement module, as shown in Fig. 1.

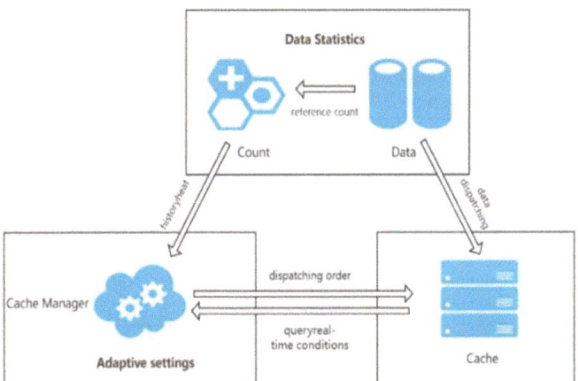

Fig. 1. Overall Architecture

3.1 The Data Statistics Module

The data statistics module is designed to compute the historical data query distribution, through collecting historical data and calculating the historical citation impact for each piece of data.

3.2 The Adaptive Setting Module

The adaptive setting module is designed to dynamically adjust the expiration period of the current cache data, which based on the historical data query distribution and real-time data query situation.

3.3 The Data Replacement Module

Designed to replace the current cache data once its runtime reaches the expiration period.

$$v_{\tilde{h}_n, \tilde{h}_k} = \sum_{\tilde{i}=0}^{win} \gamma_{\tilde{h}_n, \tilde{h}_k}^{\tilde{i}} \tag{1}$$

4 Methods

In this subsection, we will articulate the specific details of the proposed solution design.

The overall process of our adaptive setting of cache data expiration method is illustrated in Fig. 2 overall Process. Compared to traditional cache scheduling algorithms, the introduction of the adaptive method incorporating historical citation impact and real-time query situations effectively enhances the cache hit rate and significantly reduces I/O overhead. The detailed description of each part of the adaptive setting of cache data expiration method is provided in the subsequent sections.

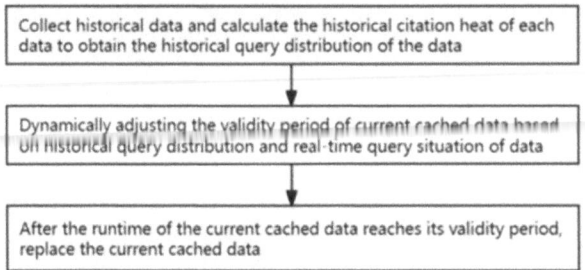

Fig. 2. Overall Process

4.1 Dynamic Adjustment of Cache Expiration Periods

In this research project, it is assumed that there are N smart sensing devices in the system totally. Each device collects K data points at a fixed frequency every day and sends them to the measurement center. This implies that the measurement center generates N * K associated data points daily, which are then stored in the database. Due to the temporal nature of measurement data, the value of data diminishes over time, meaning that the probability of querying data that is further in the past is lower. Assuming that the shared cache capacity of the measurement center can cache up to H associated data points at any given moment, to maximize the utilization of the cache space, it is assumed that the cache is always populated with H associated data points.

When statistically counting historical reference counts, assume that the current cache space's data set is represented as $D = \{d1, d2, ..., d_h\}$, where d_h represents the $\tilde{h}_k - th \in \{1, 2, ..., K\}$ data item from device $\tilde{h}_n \in \{1, 2, ..., N\}$. Additionally, define a cache hit as when a queried data item is stored in the cache, allowing it to be directly extracted from the cache without querying the database. The reference intensity of a data item at time t is defined as the number of times the data item is queried at time t. Therefore, after collecting a sufficient amount of historical query data logs, it is possible to calculate the historical reference intensity $\gamma_{n,k}^{t}$ for device n and data item k at time t. This represents the mean number of times the data item is queried at time t, providing insight into the expected query count for that data item at time t.

Here, as shown in Fig. 3, the figure illustrates the specific process of dynamically adjusting the expiration period of the current cache data based on the historical query distribution and real-time query situations.

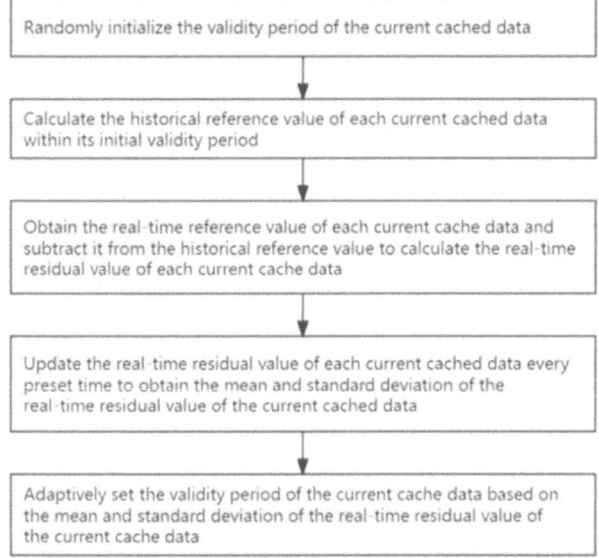

Fig. 3. Dynamic adjustment of cache expiration periods

Firstly, randomly initialize the expiration period of the current cache data as win, where win $\in$ [win$_{min}$, win$_{max}$]. Here, win$_{min}$ and win$_{max}$ represent the minimum and maximum values of the expiration period, respectively. Let $\tilde{t} = 0$ denote the current runtime of the cache data. If $\tilde{t} <$ win, it implies that the cache data is within its expiration period. Otherwise, it indicates that the cache data has expired, and cache data replacement is required to ensure the efficiency of data queries. Then, calculate the historical reference value within the initial expiration period for each current cache data based on the following formula:

$$v_{\tilde{h}_n,\tilde{h}_k} = \sum_{\tilde{t}=0}^{win} \gamma_{\tilde{h}_n,\tilde{h}_k}^{\tilde{t}} \tag{2}$$

where $v_{\tilde{h}_n,\tilde{h}_k}$ represents the historical reference value for the $\tilde{h}_k - th$ data item from device $\tilde{h}_n$, win denotes the initial expiration period, $\tilde{t}$ represents the current runtime of the cache data, and $\gamma_{\tilde{h}_n,\tilde{h}_k}^{\tilde{t}}$ represents the historical reference intensity of the $\tilde{h}_k - th$ data item from device $\tilde{h}_n$ at time $\tilde{t}$.

Next, calculate the real-time reference value of cache data. The real-time reference value of d_h is expressed as $\hat{v}_{\tilde{h}_n,\tilde{h}_k}$, which is the total number of times the data has been queried in real-time since it was placed in the cache, that is, the number of hits of the data within the current cache runtime. Afterwards, the historical reference value $v_{\tilde{h}_n,\tilde{h}_k}$ of each cache data is subtracted from the real-time reference value, and the real-time residual value of each current cache data is calculated. Subsequently, the real-time residual value of each current cache data is calculated based on the following formula:

$$RT_{\tilde{h}_n,\tilde{h}_k} = v_{\tilde{h}_n,\tilde{h}_k} - \hat{v}_{\tilde{h}_n,\tilde{h}_k} \tag{3}$$

Here, $RT_{\tilde{h}_n,\tilde{h}_k}$ represents the real-time residual value of the $\tilde{h}_k$-th data of device $\tilde{h}_n$, and $\hat{v}_{\tilde{h}_n,\tilde{h}_k}$ represents the real-time reference value of the $\tilde{h}_k$-th data of device $\tilde{h}_n$. Obviously, historical citation value represents the number of times data may be queried during its validity period, which can be regarded as a predicted value of the number of times a piece of data has been queried during its validity period based on massive historical data. Real time citation value represents the real-time cumulative number of times the data has been queried. Therefore, the real-time residual value of a piece of data obtained after the difference between the two is the predicted value of the number of times the data is queried in the subsequent validity period. The higher the value, the higher the probability of the data being queried in the subsequent time. Conversely, the lower the value, the lower the probability of being queried in the subsequent time.

Subsequently, to reduce computational complexity, characteristics of the data in the cache space are statistically analyzed at predefined intervals. This involves updating the real-time remaining values for each cache data item periodically. The mean $\overline{RT}$ and standard deviation $\overline{\sigma}$ of the real-time remaining values are calculated. After updating these features, the cache expiration periods are adaptively adjusted. Clearly, a larger mean $\overline{RT}$ indicates a higher overall remaining value for the cache data, suggesting that data query demands may be concentrated within the subsequent expiration period. In this case, extending the cache expiration period is appropriate. However, a larger standard deviation $\overline{\sigma}$ reveals greater volatility in the data, meaning that the distribution of remaining values is uneven. A small subset of data items may account for most of the remaining value, while the majority of data items have lower remaining values. In such situations, cache space utilization is low, and it is necessary to shorten the cache expiration period to promptly replace cache data, thereby improving cache space utilization and query efficiency.

Therefore, based on the following formula, this paper proposes an algorithm for adaptive setting of cache expiration periods:

$$win' = \min\{win_{\max}, \max(win_{\min}, win \times (1 + \theta))\} \tag{4}$$

Here, win' means the adaptive expiration period, win_{max} and win_{min} respectively represent the maximum and minimum values of the cache data expiration period, win represents the initial expiration period of the current cache data, and θ represents the scale of expiration period variation, $\theta = \lambda \times f_1(\frac{\overline{RT}}{RT}) - (1 - \lambda) \times f_2(\overline{\sigma}, \sigma)$, λ represents the weight, $\lambda = \frac{\overline{RT}}{\overline{RT}+\overline{\sigma}}$, $f_1(\frac{\overline{RT}}{RT})$ represents the mean function, $f_2(\overline{\sigma}, \sigma)$ represents the standard deviation function, $\overline{RT}$ and $\overline{\sigma}$ represent the mean and standard deviation of the real-time remaining value of the current cache data, RT and σ represent the mean and standard deviation of the historical reference value of the current cache data within the expiration period. The weight λ represents the strength of the influence of the mean and standard deviation on the scale of expiration period variation. If the mean is greater than the standard deviation, the distinctiveness of the mean is stronger. Therefore, its impact on the scale of expiration period variation is greater. Conversely, if the standard deviation is greater, its influence on the scale of expiration period variation is greater.

In addition, the definitions of the mean function F1 and the standard deviation function F2 are as follows:

$$f_1(x) = \max\upsilon - \frac{2 \times \max\upsilon}{1 + e^{(w_1 \times (x \times (1-x) \times w_2 - thre1))}}$$

$$f_2(x, \sigma) = \begin{cases} \max\upsilon \times e^{(\frac{x \times \ln \frac{\varepsilon}{\max\upsilon}}{\sigma})}, & \textit{if } x \leq \sigma \\ \max\upsilon \times \left(\frac{1}{1+e^{\left(\frac{2}{\ln \sigma} \times (x-x_0)\right)}} - 1 \right), & \textit{else.} \end{cases} \tag{5}$$

Here, $\max\upsilon$ represents the maximum value of the expiration period variation scale, $\theta \in [- \max\upsilon, \max\upsilon]$, w1 represents the slope, w2 represents the weighting coefficient, both of which are constants, and w1 > 0, w2 > 1, thre1 represents the offset constant, ε represents a very small positive constant close to 0, $x_0 = \sigma - \ln \sigma \times \ln \frac{\varepsilon}{\max \upsilon - \varepsilon}$.

If only the mean of the real-time remaining value is considered, it may not accurately capture the overall trend of data being queried. This is because if the real-time hit rate of the data is consistently low, the remaining value of the data will remain relatively high. In this case, even if the real-time remaining value of the data is high, the low hit rate implies that the data is rarely used in actual queries. Thus, despite having a high remaining value within the expiration period, the data may not fully realize its potential. Therefore, in this invention, to more accurately assess the value of the data, the mean function $f_1(x)$ takes into account both the real-time remaining value and the real-time cache hit rate. This provides an overall perspective for the adaptive adjustment of the cache expiration period. If both the cache value and the cache hit rate are high, the function value is larger, indicating an extension of the cache expiration period. Conversely, a decrease in either value has a negative impact, reducing the function value. As for the standard deviation function $f_2(x, \sigma)$, it ensures that the change scale of the standard deviation σ, which is close to the historical reference value, approaches 0. Simultaneously, on the overall trend, a smaller standard deviation of the remaining value implies a larger value for the expiration period variation scale. Conversely, if the standard deviation of the remaining value is larger, meaning that the remaining value is concentrated in a small number of data items, the function value of the standard deviation decreases. This achieves adaptive cache expiration period adjustment from a local perspective. Finally, the adaptive setting of the cache expiration period is achieved through the weighted results of the functions $f_1(x)$ and $f_2(x, \sigma)$. This result balances the overall perspective and the local perspective, providing a more reasonable and accurate adaptive expiration period.

During the above process, the cache validity period is continuously iterated and updated. When the running time $\tilde{t}$ of the cache data exceeds the adaptive cache validity period win', it will proceed to the next step. Otherwise, the current process will continue. In addition, due to the continuous increase in runtime $\tilde{t}$, although the validity period is adjusted adaptively at intervals, whether before or after the adaptive adjustment, as long as the runtime $\tilde{t}$ exceeds the adaptive validity period win', it will proceed to the next process.

4.2 Cache Replacement

After the above adaptive adjustments, the basic process of cache replacement is shown in Fig. 4:

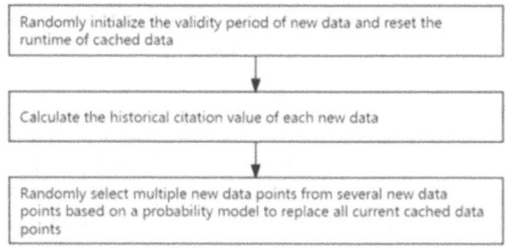

Fig. 4. Cache Replacement Flowchart

Since new data is constantly being generated, within the runtime $\tilde{t}$ of the current cache data, several new data items that have not been cache in the cache space may be generated. These new data items can be represented as a set $M = \{b_1, b_2, ..., b_M\}$, where b_m represents the $\tilde{m}_k$-th data item of device $\tilde{m}_n$. The expiration period win for new data is first randomly initialized in the range $[win_{min}, win_{max}]$, and the runtime of cache data $\tilde{t}$ is reset to 0.

Next, calculate the historical reference value of each new data item in the set M within the initialized expiration period win. Specifically, the historical reference value υ_m for data item b_m is calculated based on its historical reference heat, and can be represented as:

$$\upsilon_m = \upsilon_{\tilde{m}_n, \tilde{m}_k} = \sum_{\tilde{t}=0}^{win} \gamma_{\tilde{m}_n, \tilde{m}_k}^{\tilde{t}} \tag{6}$$

Due to cache data has real-time relevance, in general, the value of new data is much higher than that of old data stored in the cache. Therefore, it is necessary to consider replacing cache data when the cache validity period expires. In order to accelerate the replacement speed and under the above assumptions, all old data in the cache will be discarded, which means that the cache space will empty H data bits.

The historical citation popularity to some extent characterizes the degree to which data is used in subsequent actual queries, but a high historical citation popularity does not necessarily mean that real-time citation popularity will be high. Therefore, this study based on probability models M Randomly [10] select H data to replace the current cache data in the cache space. The process of randomly selecting multiple new data pieces from several new data pieces based on a probability model to replace all the current cache data is shown in Fig. 5:

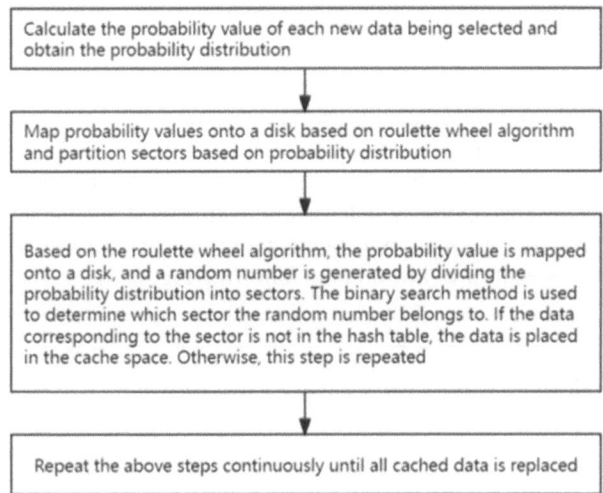

Fig. 5. Probability Model Flowchart

Specifically, calculate the probability of selecting each of the M new data items, with a time complexity of O(M). The calculation formula is:

$$P_{select}(v_m) = \frac{v_m}{\sum_{i=1}^{M} v_i} \tag{7}$$

Based on the roulette wheel algorithm, map the probability values onto a disk, divide the disk into sectors based on the probability distribution, and choose the corresponding data based on which sector a random number falls into to place it in the cache space. For example, by simplifying the probability accumulation, obtain a set of boundaries for the sectors on the disk, with a length of M + 1, $P_S = \left\{0, \sum_{i=1}^{1} P_{select}(v_i), \sum_{i=2}^{2} P_{select}(v_i), \sum_{i=3}^{3} P_{select}(v_i), ..., 1\right\}$.Obviously, the size of the i-th sector is $P_S[i] - P_S[i-1] = P_{select}(v_i)$, Clearly, the time complexity of this step is also O(M).

Subsequently, generate a random number and determine the sector where the random number falls using binary search. If the data corresponding to the sector is not in the hash table, place the data into the cache space and add it to the hash table; otherwise, repeat this step. It is understood that, to avoid caching the same data, a hash table of length H is maintained in this study. If the randomly obtained data already exists in the hash table, it will not be placed in the cache space. Therefore, the time complexity of this step is O(log2M).

Finally, repeat the last step until the number of data placed in the cache space reaches H. At this point, the length of the hash table also reaches H. Therefore, the overall time complexity is O(Hlog2M). Since M is generally much larger than H, i.e., M > Hlog2M, the cache replacement algorithm proposed in this paper has an overall time complexity of O(M) and a space complexity of O(M).

5 Experiment

According to the above chapters, it can be inferred that the proposed method in this study adapts and dynamically adjusts the cache expiration period by combining historical reference heat and real-time query situations. The probability model is then employed to randomly select data for caching, achieving optimal cache hit rates with low time complexity. Because this approach balances time efficiency and cache efficiency. We compared our proposed method with traditional cache replacement methods through simulation, and the results are shown in the following Fig. 6:

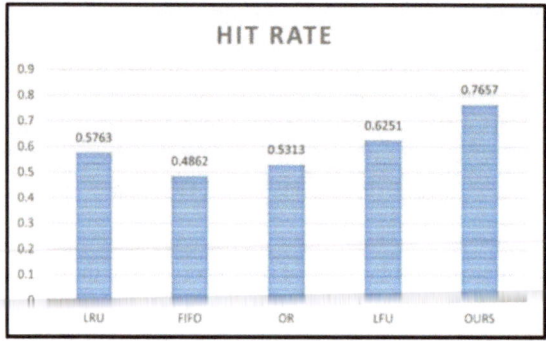

Fig. 6. Comparison Results

6 Conclusion

In summary, the adaptive expiration period method proposed in this study dynamically adjusts the expiration period of current cache data by combining the historical query distribution and real-time query distribution of the current cache data. It comprehensively considers the distribution features of real-time hit rate and real-time residual value of cache data for adaptive adjustment, making the adaptive settings more accurate and significantly improving cache hit rate. As a result, the efficiency of cache utilization is greatly enhanced, and frequent cache replacements are avoided, achieving the goal of reducing I/O overhead.

Acknowledgment. This work is supported by State Grid Hunan Electric Power Co., Ltd research project No. 5216A8220004 and Hunan Key Laboratory for Internet of Things in Electricity, P.R.China No. 2019TP1016.

References

1. Sonia, A., Alsharef, P., Jain, M., Arora, S., Zahra, R., Gupta, G.: Cache memory: an analysis on performance issues. In: 2021 8th International Conference on Computing for Sustainable Global Development (INDIACom), New Delhi, India, pp. 184–188 (2021)
2. Abrams, M., Standridge, C.R., Abdulla, G., Williams, S., Fox, E.: Caching proxies: limitations and potentials. In Proceedings of the 4th International World Wide Web Conference (1995)
3. Stone, H., Wolf, J., Turek, J.: Optimal partitioning of cache memory. IEEE Trans. Comput. **41**(09), 1054–1068 (1992). https://doi.org/10.1109/12.165388
4. Hui, P., Hongzhu, T., Yaqin, Y., et al.: Database management technology for smart grid dispatch control system. Power System Automation **39**(01), 19–25 (2015)
5. Zhai, M., Wang, J., Wu, Q., et al.: Architecture and key technologies of wide area distributed real time database system for power grid dispatching. Power System Automation **37**(02), 67–71 (2013)
6. Rixner, S., Dally, W.J., Kapasi, U.J., Mattson, P., Owens, J.D.: Memory access scheduling. In: Proceedings of 27th International Symposium on Computer Architecture (IEEE Cat. No.RS00201), Vancouver, BC, Canada, pp. 128–138 (2000). https://doi.org/10.1145/339647.339668
7. Tanwir, G.H., Affandi, A.: Early result from adaptive combination of LRU, LFU and FIFO to improve cache server performance in telecommunication network. In: 2015 International Seminar on Intelligent Technology and Its Applications (ISITIA), Surabaya, Indonesia, pp. 429–432 (2015). https://doi.org/10.1109/ISITIA.2015.7220019
8. Finkelstein, M., Shafiee, M., Kotchap, A.N.: Classical optimal replacement strategies revisited. IEEE Trans. Reliab. **65**(2), 540–546 (2016). https://doi.org/10.1109/TR.2016.2515591
9. Yang, C..L, Lee, C.H.: Hotspot cache: joint temporal and spatial locality exploitation for i-cache energy reduction. In: Proceedings of the 2004 International Symposium on Low Power Electronics and Design, pp. 114–119 (2004)
10. Ross, S.M.: Introduction to Probability Models. Academic Press (2014)

Research on Airport Communication and Navigation Equipment Inspection Assistance System Based on AR Technology

Weijia Ye[1], Rubiao Han[2], Peng Gao[2], and Ning He[1(✉)]

[1] R&D Center, The Second Research Institute of CAAC, Chengdu, China
{yeweijia,hening}@caacsri.com
[2] China West Airport Group, China West Airport Group Qinghai Airport Co., Ltd., Xining, China
{hanrb,gaopeng}@westaport.com

Abstract. Aviation transportation is an important component of our country's transportation system. With the rapid improvement of the national economy, the civil aviation industry in our country has experienced rapid development, and the number and scale of airports are also constantly increasing. Communication and navigation systems play an important role in the transportation of aircraft and the scale operation of airports. Efficient and accurate communication and navigation system can not only effectively ensure air transportation safety but also improve overall operational efficiency, making airport operations more rational and secure. This paper combines augmented reality technology with VSLAM (Visual Simultaneous Localization and Mapping) technology to assist personnel in daily communication and navigation equipment inspection tasks. By providing some functions to improve the efficiency of airport personnel, the function such as positioning, some auxiliary information, and remote assistances.

Keywords: Augmented Reality · VSLAM · Civil Aviation · Equipment inspect

1 Introduction

With the development of the aviation industry, the airport's communication and navigation equipment have become one of the important for aviation safety. Which system have some radar, navigational instruments, and communication devices, all of those are very import for the aircraft's takeoff. Therefore, the airport personnel need checkout that equipment working well every day. Consequently, the regular inspections are necessary. However, those daily inspect works is time-consuming.

The airport communication and navigation equipment's daily works are not easily, the works such as check out the device's status whether or not is normally, if the device has some problem, the airport personnel need to deal with or reported. These procedures are very import for the safety of the communication and navigation systems at airports [1].

P. Siarry et al. (Eds.): WCNA 2023, LNEE 1361, pp. 56–64, 2025.
https://doi.org/10.1007/978-981-96-2409-6_6

Nowadays, many airports are use the mobile phones for regular inspections, which is used for record problem and no tools for deal with it, and which is also low-technology and low-efficiency. Therefore, to improve the efficiency of inspection task. We use the augmented reality (AR) technology can offer new possibilities to overcome these challenges. By superimposing the virtual information onto the real world, enhancing both the efficiency and accuracy of inspections. Which can visually present information and fault to inspectors, helping them quickly address the device's issues, thus keep the daily operation of equipment and aviation safety. In addition to overlaying auxiliary information, our system also provides remote expert assistance functionality. By utilizing anchor technology, it enables spatial annotations to be displayed simultaneously on both the glasses end and the expert end screens, effectively improving the efficiency of airport personnel in inspecting and troubleshooting.

Therefore, the application of augmented reality technology in the daily inspection of airport communication and navigation equipment not only improves inspection efficiency and accuracy but also reduces operational errors and enhances aviation safety. This technological application is expected to bring new breakthroughs and advancements to the field of aviation safety, providing a more solid guarantee for the sustainable development of the aviation industry.

2 Introduction of Inspection and Troublesh Shooting Auxiliary System

The inspection and troubleshooting auxiliary system consist of four parts, the Management platform, AR wearable device, Remote Assistant Service and WeChat applet. The inspection and troubleshooting auxiliary system flowchart show in Fig. 1.

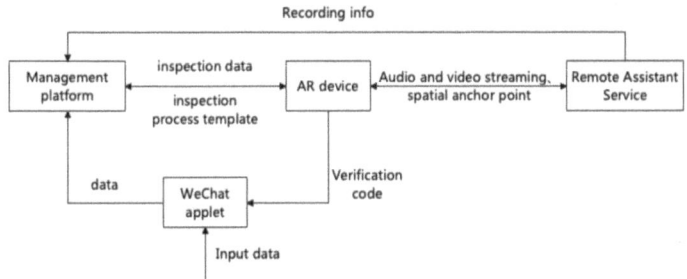

Fig. 1. Inspection and troubleshooting auxiliary system

The management platform within the system primarily function contains user login management, AR device management, inspection data management, remote video data management, etc. it is also responsible for transferring the daily inspection template content of communication and navigation equipment to AR glasses and saving the inspection data returned by the AR device, which includes text, images, and other information.

Subsequently, when airport personnel log into the AR glasses, the corresponding inspection process template is automatically updated and presented. Personnel only need to fellow the displayed process content to conduct their work, thus avoiding omissions in inspection tasks. Upon arrival at the site of the communication and navigation equipment, personnel use the AR device's camera to capture environmental information. We employ VSLAM (Visual Simultaneous Localization and Mapping) algorithm to construct real-world scene maps, and based on augmented reality technology, we overlay virtual information in the real environment [2]. This better assistants' personnel in their inspection tasks of communication and navigation equipment.

After completing the inspection template content process on the AR end, personnel upload location information and handwritten signature images through WeChat program, ultimately completing the entire process.

In addition to following the routine inspection process, personnel may encounter situations where communication and navigation equipment malfunctions. We provide two methods to assist personnel in troubleshooting and handing these issues. The first method allows personnel to search for relevant information through voice or text input on AR glasses and complete troubleshooting with the help of images or documents. The second method involves calling remote experts and establishing an audio-video connection with them. Using spatial anchor technology, annotations from the remote expert can be displayed in real-time on the AR glasses, thereby enhancing the efficiency of fault resolution.

Obviously, this process is not simple linear, from the Fig. 2 the sequence diagram between AR and Management platform, we can know the data transferred direction in those two parts.

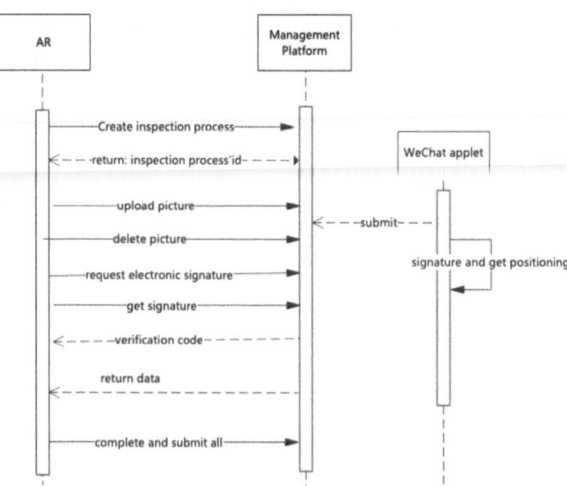

Fig. 2. Sequence diagram of system

3   Related Work

We first use three-dimensional scene reconstruction technology to complete the map construction on real environment point cloud data. We obtain the three-dimensional model of the target through surface reconstruction, while also storing the point cloud dataset. Then, we capture images through the camera module, convert the captured image data into suitable OpenGL ES rendering, and use the transformed data to call the database through the tracking module to identify and track the target content inside. We design virtual scenes in Unity, and finally render the images captured by camera overlaid with the virtual images. The final effect is presented on AR glasses. Airport personnel can then see virtual augmented information overlaid on real devices in the real environment and interact with it (Fig. 3).

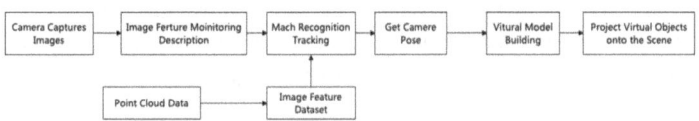

Fig. 3. System technology flow chart

3.1   Visual Simultaneous Localization and Mapping

To better assist the airport personnel, we utilize vSLAM (visual simultaneous localization and mapping) technology, which allows AR device to locate themselves and construct environmental maps in unknown environments [3]. This is achieved by gathering environmental through sensors such as cameras and inertial measurement units, and then utilizing algorithms for information fusion to determine the device's position. SLAM algorithms are divided into frontend and backend components, with the frontend responsible for calculating the robot's position and pose changes based on the matching results of adjacent images or point clouds. as shown in Fig. 4.

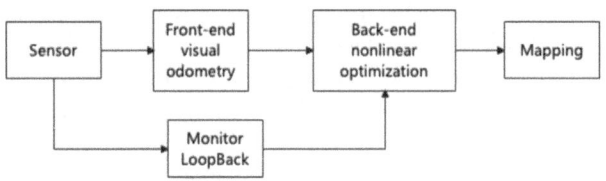

Fig. 4. The workflow of the vSLAM

Our AR device consist of four environmental perception cameras to perceive the offset of device relative position, one depth camera and an inertial measurement unit responsible for sensing device orientation.

We utilize slam technology, combining monocular vision and RGBD methods. Initially, monocular vision is employed to extract features from images, match features between adjacent frames, and calculate the relative position of the camera to the scene based on matched key features points. We use the RGBD method, employing the Time-of-Flight principle, directly measures the distance of each pixel in the image from the camera. We utilize the Kinect Fusion framework for mapping and tracking, while employing environmental perception cameras to replace the point-to-plane ICP algorithm in the Kinect Fusion structure for pose calculation through feature matching, facilitating point cloud fusion optimization [4, 5].

The point cloud reconstruction is as follows: first, the input depth image is converted into a three-dimensional point cloud, and then the normal vector of each point is calculated. To calculate the position of the AR device, based on the camera's position, the point cloud of the current frame is fused into the network model. Finally, using ray projection algorithms based on the current frame camera position, the point cloud from the model is projected onto the current frame's perspective, and its normal vectors are computed. These vectors are used to register the device move, the camera captures points clouds from different perspectives of the scene, reconstructing the complete surface of the scene. The workflow of the algorithm shown as Fig. 5.

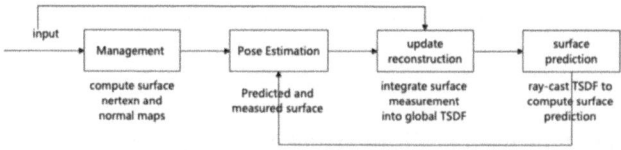

Fig. 5. The workflow of the KF algorithm

First, the original depth data undergoes bilateral filtering processing, aiming to preserve clear boundaries. Traditional filtering involves weighted averaging in the spatial domain, where pixels closer to the center have higher weights. Bilateral filtering, however, adds weighted averaging in the value domain on top of spatial domain weighted averaging. In other words, pixels with grayscale values closer to the grayscale value of the center pixel have higher weights. Near boundaries, there are significant differences in grayscale values. Therefore, although pixels on both sides of the boundary are close in the spatial domain, their weights for each other are low due to the large differences in grayscale values. This preserves clear boundaries.

In traditional 3D reconstruction, camera pose estimation is based on Structure from Motion (SfM), while Kinect Fusion uses ICP (Iterative Closest Point) to align point clouds. The ICP algorithm aligns point cloud data from different coordinate systems into a single coordinate system. Initially, it finds a suitable transformation; the registration operation essentially involves finding a rigid transformation from one coordinate system to another.

The map is updated using TSDF (Truncated Signed Distance Function). SDF considers only the SDF values within the vicinity of the surface. If the maximum value within the vicinity is the max truncation, the actual distance is divided by this value. This normalization ensures that the TSDF values range between -1 and $+1$. After updating the TSDF values, we can use them to estimate the voxel/normal map.

Then by scanning the real environment, we first selected an office scene for testing the three-dimensional reconstruction, and we can get the point cloud data firstly, then obtaining the three-dimensional model of the target through surface reconstruction, the picture shown in Fig. 6.

Fig. 6. The 3D reconstruction model diagram

The import the model data into Unity and, with the use of augmented reality SDK, overlay the virtual model and information in the Unity scene. The display effect is as shown in the Fig. 7. We can see that in three-dimensional reconstruction image, some identification arrows and informational panels have been added. Once the airport personnel put on the AR glasses, all this information can be overlaid and displayed in the real environment [6].

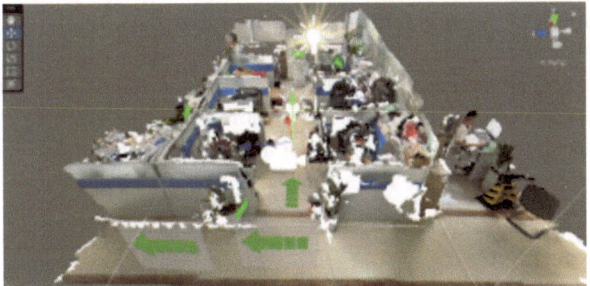

Fig. 7. The 3D model adding the virtual things in Unity

3.2 Tracking Registration Technology

Computer vision-based tracking registration algorithms are divided into two types: marker-based tracking registration and natural feature-based tracking registration. The marker-based method involves identifying a marker, recording its features, and then projecting objects onto the corresponding marker by recognizing scenes captured through cameras. The focus of this paper is to achieve mutual positioning between three-dimensional objects and the natural world using natural marker-based technology [7].

When the airport personnel arrive at the inspection site and puts on the AR glasses, the glasses will prompt. 'Please scan the positioning marker', as shown as Fig. 8, to complete the positioning of the virtual and real scenes by scanning the marker.

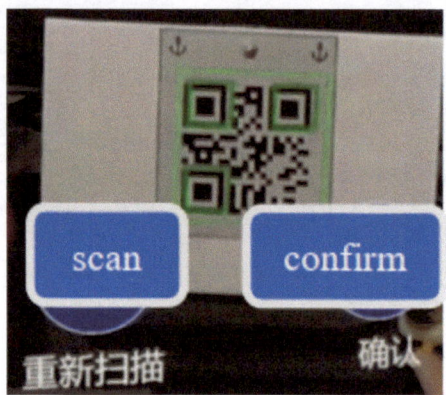

Fig. 8. The marker-based

After the positioning is completed, airport personnel can use AR glasses to see the equipment maintenance information overlaid on the real scene, as shown as Fig. 9. Due to cinfidentiality concerns regarding airport equipment,we attempted to use normal robotic equipment as an example. The image depicts wearing AR glasses to view an information panel overlaid in the real environment. The panel is divided into two sections: the left side displays the normal equipment status, while the right side shows the inspection process information. By equating airport equipment with robots, this panel can provide similar information for daily inspections by airport staff. The auxiliary inofrmation also can include images, textual guidance, model animations, and more, which depends on what designning in Unity [8].

3.3 Remote Expert Sytem

Our system not only provides information assistance for daily inspections by airport employees but also has a system to assist in troubleshooting. The airport personnel can connect with remote experts through AR glasses, where the experts can synchronize and share the staff's first-person perspective. Additionally, experts can provide troubleshooting suggestions remotely through spatial annotations [9].

The AR Remote Expert system enables users to connect and converse with remote experts in mixed reality. Through remote assistance features, users can resolve issues more quickly, even when they are not in the same location as the experts.

AR smart glasses capture audio and video, which are then forwarded to the terminal device via servers. Experts input instructional content on the terminal device, which is then forwarded to the AR smart glasses and displayed in the user's field of view. The

Fig. 9. The information panel

remotely annotated information is accurately positioned in spatial locations, making expert guidance more vivid and improving communication efficiency and accuracy. Users can interact with the AR glasses through voice commands, gestures, etc., freeing up their hands.

Compared to existing technologies, the AR-based Remote Expert system transforms "human-machine-human remote virtual interaction" into "human-human remote on-site interaction," offering the following advantages: a) Real-time presentation of device information. b) Remote experts can seamlessly access text and graphical information in real time.

c) Intuitive conveyance of detailed and complex instructions: Instead of conveying instructions through audio calls, on-site staff and remote collaborators can use graphics and arrows to annotate specific parts. These annotations remain fixed in the remote user's space and do not change position as the operator moves around.

4 Conclusion

In this paper, we primarily introduced the display of augmented reality (AR) virtual information overlaid on the real environment through AR devices, as well as the remote expert system providing auxiliary technological means for airport staff to conduct daily equipment inspections and troubleshoot issues. However, there are still some shortcomings at present, such as unstable overlay display of information and positioning inaccuracies. We will continue to improve these issues in our future research.

Acknowledgment. This work is supported by the Innovative Task in Artificial Intelligence Industry for 2021: Intelligent Navigation Products for the visually impaired and Optimization Research on Shift Scheduling for Remote Tower Control Center Controllers for Multi-Airport Collaborative Command under Grant No. MZGC20230074.

References

1. Ye, W., Yu, Q., He, N., Pang, L.: Augmented Reallity for Assist Apron Operation Safety Supervisor in Aircraft Inspection Process
2. Gu, Z., Liu, H.: A survey of monocular simultaneous localization and mapping. CAAI Trans. Intelligent Syst. **10**(4), 499–507 (2015)
3. Salas-Moreno, R.F., Glocken, B., Kelly, P.H., et al.: Dense planner slam, in Mixed and Augement Reality (ISMAR). In: 2014 IEEE International Symposium on, pp. 157–164, IEEE (2014)
4. Chow, J.C., Lichti, D.D.: Photogrammetric bundle adjustment with self-calibration of the primesense 3D camera technology: microsoft kinect. Access, IEEE **1**, 465–474 (2013)
5. Newcombe, R.A., et al.: KinectFusion: real-time dense surface mapping and tracking. Proc. IEEE ISMAR (2011)
6. Richards-Rissetto, H., Remondino, F., Agugiaro, G., Robertsson, J., von-Schwerin, J., Girardi, G.: Kinect and 3D GIS in archaeology. In: Proceedings 18th IEEE International Conference on Virtual Systems and MultiMedia (VSMM), Guidi, G., Addison, L., (eds.), pp. 331–338. Milan, Italy (2012)
7. Wagner, D., Langlotz, T., Schmalstieg, D.: Robust and unobtrusive marker tracking on mobile phones. In: IEEE International Symposium on Mixed and Augmented Reality (ISMAR), pp 121–124 (2008)
8. Ventura, J., Reitmayr, G.: Global localization from monocular SLAM on a mobile phone. IEEE Trans. Visual Comput. Graphics **20**(4), 531–539 (2014)
9. Azpiazu, J., Siltanen, S., Multanen, P.: Remote support for maintenance tasks by the use of augmented reality: the ManuVAR project. In: 9th Congress on Virtual Reality Applications (CARVI), Spain, pp. 10–11 (2011)

Application of Quality Function Deployment in Assessing Unmanned Assault Capabilities for Mountain Counter-Terrorism Operations

Jianrong Yang[1]([✉]), Qixiang Li[2], and Shubing Zhang[1]

[1] Armed Police Engineering University, Xi'an, China
1085798666@qq.com
[2] College of Equipment Management and Support, Armed Police Engineering University, Xi'an, China

Abstract. Employing the Quality Function Deployment (QFD) method, this study systematically analyzes the assault capabilities required of unmanned equipment in mountain counter-terrorism operations.The complex requirements of counter-terrorism missions were transformed into specific equipment capability requirements, and the main tactical indicators that should be possessed in unmanned equipment assault in mountain counter-terrorism were clarified. This method helps researchers to more accurately grasp the essence of anti-terrorism tasks and ensure that the equipment developed can meet the needs of actual combat.

Keywords: QFD · Counter-Terrorism · Unmanned Systems · Assault Capabilities · Mountainous Operations

This study draws parallels between unmanned equipment development and product design, underscoring the importance of customer-centric design in both domains.First, the design characteristics of the "product" are determined by customer needs; Secondly, customer needs are hierarchical, and with the continuous progress of design, customer needs are constantly concretized and clarified. Therefore, QFD method can be used to analyze the demand of unmanned equipment capability just like designing products, and the functional characteristics of unmanned equipment for mountain anti-terrorism are regarded as "customers", and different functions are scored by questionnaire method. As the "customer demand weight", that is, the content of the left wall of the quality house; According to the technical requirements corresponding to different functions, the demand index of assault capability of mountain anti-terrorism unmanned equipment is comprehensively analyzed, that is, the content on the wall of the quality house.

1 Transformation and Expression of Assault Function Characteristics of Unmanned Equipment for Mountain Anti-terrorism

In the face of complex and arduous tasks, the mountain anti-terrorism unmanned equipment construction needs are particularly urgent. Unmanned equipment with its unique advantages, such as high mobility, flexibility and precise strike capability, has become

P. Siarry et al. (Eds.): WCNA 2023, LNEE 1361, pp. 65–73, 2025.
https://doi.org/10.1007/978-981-96-2409-6_7

the key to improve the effectiveness of anti-terrorism operations. Surveillance drones can monitor terrorist activities in real time, intelligent analysis systems can quickly process a large amount of intelligence data, and unmanned combat platforms can perform search and strike tasks in high-risk environments, greatly reducing the risk of casualties and improving the success rate of operations. It can be seen that with the continuous evolution of the mountain anti-terrorism battlefield, the in-depth application of unmanned intelligent technology will provide the mission Department (branch) with more powerful combat capabilities, help it maintain strategic advantages in the vast and complex mountain environment, and effectively respond to various terrorist threats. However, the key to the demand analysis of mountain anti-terrorism unmanned equipment assault capability is to transform the anti-terrorism task requirements into equipment functional characteristics.

Characteristic analysis process of mountain anti-terrorism unmanned equipment assault capability: Whether the unmanned mountain anti-terrorism equipment can meet the operational needs must be based on the test criteria of "clear, enclosed, sealed, accurate, accessible, and eliminated". On this basis, the method of expert interview is adopted to obtain the functional requirements of the unmanned mountain anti-terrorism equipment under different types of anti-terrorism personnel and different mountain anti-terrorism combat forces to the maximum extent. Based on the decomposition results of the mountain anti-terrorism assault mission, the interview outline is carefully designed to effectively fulfill the anti-terrorism assault combat mission.

Analysis of expert interview results: In order to more scientifically and reasonably convert mission requirements into functional characteristics of equipment, through interviews with experts and scholars in related fields, functions of mountain anti-terrorism unmanned equipment should be determined, problems such as quality and operation that may be encountered in the use of equipment, and expectations of equipment, etc., pain points and itching points of existing mountain anti-terrorism unmanned equipment assault are introduced. Then, according to the task characteristics and pain points, the questionnaire is designed as the basis for further launching the functional characteristics of the army's unmanned anti-terrorism equipment in the mountain. Through the interview, it is preliminarily determined that the mountain anti-terrorism unmanned equipment should have the following functions: First, rapid isolation and evacuation. In anti-terrorist raids, rapid isolation and evacuation missions are essential. Unmanned intelligent equipment can accurately judge the scope of the dangerous area in a short time, quickly deploy isolation measures, effectively prevent terrorists from escaping or reinforcements, and ensure the safe evacuation of surrounding people. This requires a high degree of flexibility and the ability to process complex environmental information in real time. Second, the analysis and judgment of the situation should be targeted. Give full play to the advantages of unmanned intelligent equipment, monitor the target area all weather and obtain intelligence from it in real time, and provide useful intelligence for combat commanders through deep learning algorithms. At the same time, it is necessary to take into account the content of terrorist communications, interception and disclosure of their evil intentions. Third, under the conditions of information-based war, whoever has the initiative has the priority. Through the scientific combination of intelligent equipment and drones, the enemy communication system can be effectively shielded and interfered with, and once the communication between terrorists is cut off, they are like a scattered sand, a blow. In addition, using such intelligent electronic equipment to fight, while hiding itself, it can also have a greater interference effect on enemy radar systems. Therefore, in the battlefield, we must firmly control the information advantage. Fourth, the criterion for weighing the combat capability of UAV intelligent equipment is its ability to complete tasks. At this point, how to quickly and stealthily approach the target is crucial. With its excellent stealth technology and flexible mobility, the unmanned mountain anti-terrorism equipment can quickly meet the enemy in the complex and changeable battlefield environment. These equipment can use the terrain, weather and other natural conditions, skillfully avoid enemy detection, quiet like a virgin, moving like a rabbit, to create favorable conditions for the execution of assault missions. Fifth, carry out precision strikes. Precision strike capability is one of the important indexes to measure the unmanned equipment of mountain anti-terrorism. With high-precision positioning systems and advanced guidance technology, these devices can precisely target terrorist hideouts, minimizing damage to surrounding people and infrastructure. Whether it is long-range missiles or close-range drone attacks, unmanned intelligent equipment can ensure efficient and accurate strike effects, highlighting the power of modern technology.

2   Building the House of Quality Model

Construct the House of Quality of functional characteristics and capability demand indicators for mountain anti-terrorism unmanned equipment assault, and transform the importance evaluation of functional characteristics into the importance evaluation of capability demand indicators through the relationship transformation matrix. (See Table 1).

Table 1. Mountain anti-terrorism unmanned equipment assault function characteristics and capability requirements index House of Quality

functional characteristics	Capability index Importance degree	B1	B2	B3	B4	..	Bn
A1	W1						
A2	W2						
A3	W3						
...	...						
Am	Wm						
	Score						
	Percent						

3 Mountain Anti-terrorism Unmanned Equipment Assault Function Characteristic Indicators and Weights

The functional characteristics of mountain anti-terrorism unmanned equipment can be summarized as rapid isolation and evacuation, analysis and acquisition of intelligence, communication shielding interference, covert and rapid enemy contact, and precision strike. Through necessary means, the functional importance of the above equipment is scored, and after summarizing the data, the functional characteristics and importance of mountain anti-terrorism unmanned equipment are sorted out Authors and Affiliations. Detailed presentation is shown in Table 2.

Table 2. Attack function characteristics and importance of unmanned mountain anti-terrorism equipment

Designation	Functional characteristics of unmanned mountain anti-terrorism equipment	Importance degree
A1	Rapid isolation evacuation	0.08
A2	Analysis for intelligence	0.12
A3	Communication shielding interference	0.14
A4	Covert quick contact with enemy	0.21
A5	Carry out precision strikes	0.45

Fill the functional characteristic indicators and weights of the above mountain anti-terrorism unmanned equipment into the left wall of the quality house.

4 The Design of Mountain Anti-terrorism Unmanned Equipment Capability Demand Indicators

Through precise requirements analysis, we ensure that every technology development is closely aligned with the ultimate operational objective. The HOQ transformation matrix between functional characteristics and capability indicators of mountain anti-terrorism unmanned equipment is established. The expert evaluation method is adopted to sort out the capability items required by each function and determine the main capability demand indicators. Detailed presentation is shown in Table 3.

The demand index of mountain anti-terrorism unmanned equipment capability is filled into the wall of the quality house.

5 Application of Analytic Hierarchy Process (AHP) in QFD Requirement Analysis

Because the mapping relationship between different capability indicators and functional requirements is not one-to-one, the same function often needs multiple capabilities to realize. In the application of QFD method, it is often difficult to obtain accurate and

Table 3. Demand index of mountain anti-terrorism unmanned equipment assault capability

Designation	Mountain anti-terrorism unmanned equipment capability demand index
B1	Target recognition and tracking capability
B2	Environmental awareness and obstacle avoidance
B3	Autonomous navigation and positioning capability
B4	Intelligent decision and control ability
B5	Communication and data processing capabilities
B6	Intelligent fire control capability

quantified evaluation data. Considering this complex mapping relationship and many uncertain factors faced by unmanned mountain anti-terrorism equipment in combat, this paper combines the analytic Hierarchy process (AHP) to determine the QFD correlation matrix and output the capability index weight.

Analytic hierarchy Process (AHP) is a simple and convenient method for quantitative analysis of non-quantitative events in system engineering, and it is also an effective method for describing people's subjective judgment. It provides people with easy to understand thinking mode to solve various complex problems. The principle of analytic hierarchy process (AHP) is to dissect a complex problem layer by layer and establish an analytical model of "goal-criterion-scheme", which is similar to the analytical idea of "customer needs-performance index" in QFD. Using analytic hierarchy Process (AHP) to output QFD importance ranking avoids the limitation and fuzziness of expert judgment and takes into account the consistency problem. Make the result more objective and accurate.

Due to the different focus and research fields of different experts, there will be different opinions on the factors affecting the completion of combat missions and their importance, so it is necessary to comprehensively deal with the opinions of the expert group. The commonly used theoretical methods are time group decision method and opinion aggregation method with ordered weighted average operator as the core. When constructing the original judgment matrix, the 1–9 scale method is adopted, and the average scores of several experts are taken as the final judgment matrix for each index. The following takes the judgment matrix C of the priority of each capability indicator Bn relative to the function feature A1 in the index layer as an example to analyze the weight value of each indicator relative to A1(See Table 4).

The judgment matrix is normalized by column, and the formula is $h_{ij} = \frac{b_{ij}}{\sum_{i=1}^{6} b_{ij}}$, i $= 1,2,3,4,5,6$ j $= 1,2,3,4,5,6$

Where hij is the normalized column element and bij is the relevant parameter in the original judgment matrix. The normalized standard matrix is then summed by column. (See Table 5).

The data after summing rows are converted into percentage weights, and the output is the weight of each capability index relative to the functional feature A1 (See Table 6).

Table 4. Judgment matrix C

	B1	B2	B3	B4	B5	B6
B1	1	2	3	5	6	6
B2	1/2	1	3	2	4	2
B3	1/3	1/3	1	1/2	1	1/2
B4	1/5	1/2	2	1	3	1
B5	1/6	1/4	1	1/3	1	2
B6	1/6	1/2	2	1	1/2	1

Table 5. Normalized judgment matrix C

	B_1	B_2	B_3	B_4	B_5	B_6	$F_i = \sum h_i$
B_1	0.4225	0.4364	0.2500	0.5085	0.3871	0.4800	2.4845
B_2	0.2113	0.2182	0.2500	0.2034	0.2581	0.1600	1.3009
B_3	0.1408	0.0727	0.0833	0.0508	0.0645	0.0400	0.4523
B_4	0.0845	0.1091	0.1667	0.1017	0.1935	0.0800	0.7355
B_5	0.0704	0.0545	0.0833	0.0339	0.0645	0.1600	0.4667
B_6	0.0704	0.1091	0.1667	0.1017	0.0323	0.0800	0.5601

Table 6. Weight of indicator layer relative to functional feature A1

Index	B_1	B_2	B_3	B_4	B_5	B_6
Weight W	41.41%	21.60%	7.54 %	12.20%	7.18%	9.34%

Since the comparison matrix is evaluated by human beings, it is difficult to make accurate judgments on all factors, and there are certain errors. However, when the comparison matrix is actually constructed, it is difficult to ensure such logical consistency, so consistency test should be conducted. Consistency test is judged by the consistency ratio index CR. If CR < 0.1, the degree of inconsistency is within the allowable range among $CR = \frac{CI}{RI}$.

CI is the consistency judgment index, RI is the average random consistency index, and RI can be judged by the order n of the matrix and the correspondence table between RI (See Table 7).

In this case, the order of the judgment matrix is 6, corresponding to RI = 1.24.

The calculation method of CI is $CI = \frac{\lambda_{max} - n}{n - 1}$.

Where λ_{max} is the largest eigenroot of the judgment matrix C, $\lambda_{max} = \frac{1}{n} \times \sum \frac{C \times W}{W}$

By bringing the data into the formula, it is easy to obtain the consistency test result of the index layer judgment matrix C (See Table 8):

Table 7. Mapping between n and RI

n	1	2	3	4	5	6	7	8	9
RI	0	0	0.58	0.90	1.12	1.24	1.32	1.41	1.45

Table 8. Consistency test results

	λ_{max}	CI	RI	CR	Test result (CR<0.1?)
Judgement matrix C of index layer relative to function characteristic A1	6.4019	0.0804	1.24	0.0648	Pass

The weight of the index layer relative to the judgment matrix of other functional features (A2,A3,A4,A5) can be obtained by the same steps. Each weight obtained from the analysis is taken as the QFD relationship matrix, combined with the weight of each functional feature of the demand layer, and the weighted average algorithm is used.

$R_n = \sum_{m=1}^{5} w_m f_{mn}$

QFD importance output can be obtained. Finally, the importance score is summarized as a percentage, and the output result is B1 > B2 > B3 > B6 > B4 > B5, that is, target recognition and tracking ability > environmental perception and obstacle avoidance ability > autonomous navigation and positioning ability > intelligent fire control ability > intelligent decision and control ability > communication and data processing ability. See chart below.

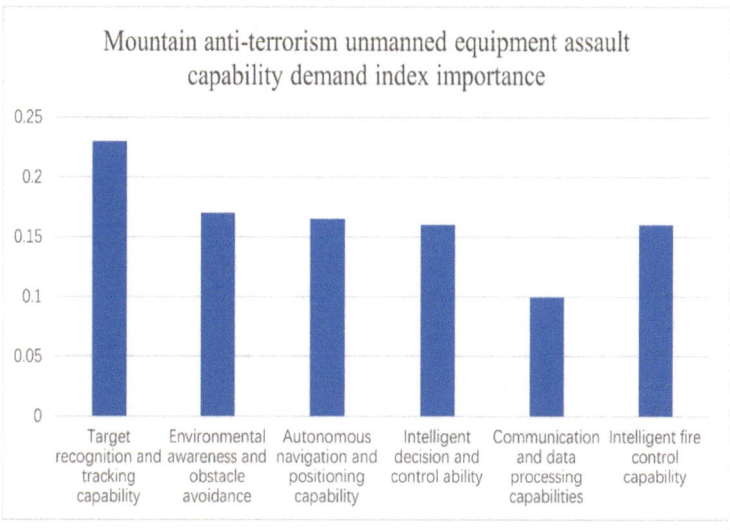

Therefore, the importance evaluation of assault functional characteristics of mountain anti-terrorism unmanned equipment is transformed into the importance evaluation of capability demand index, indicating that the ability index that has the greatest impact on the realization of assault functional characteristics of mountain anti-terrorism unmanned equipment is target identification and tracking ability, which provides certain ideas for the future research and development of unmanned intelligent anti-terrorism assault equipment.

6 Challenge and Countermeasure

Although the QFD method has shown remarkable effect in the capability requirement analysis of unmanned intelligent anti-terrorism assault equipment, it also faces some challenges in the practical application process. The most important challenges include the dynamic change of the mission landscape and the limitations of the technology implementation.

In view of the dynamic changes in the mission situation, the equipment research and development team needs to establish a flexible demand update mechanism, regularly collect troop feedback and anti-terrorism situation information, and timely adjust the design scheme. At the same time, strengthen pre-research and technical reserves, improve the ability to predict future trends, and ensure the forward-looking design of equipment. The limitations of technology implementation require the equipment research and development team to fully consider the limitations of existing technology at the beginning of the design, and pay attention to cost control and practical operation while seeking technological innovation. Through the cooperation with scientific research institutions and high-tech enterprises, the introduction of the latest scientific and technological achievements can effectively overcome technical obstacles and improve the performance and practicality of equipment.

Through the application of QFD method, the research and development of unmanned intelligent anti-terrorism assault equipment has become more systematic and efficient. Although there may be various challenges in the development process, as long as we adopt effective strategies, we can ensure the smooth progress of research work.

References

1. Xu, Z.: Creating a new form of human civilization with the mind of the world – accurately grasping the new responsibility of the report of the 20th National Congress of the Party. Learning and Exploration **11**, 1–15 (2022)
2. Sun, C.: Artificial Intelligence-based Humanoid Target Equivalent Damage Perception and Evaluation. Xi'an Technological University (2023)
3. Xie, S.: Technology development and influence analysis of unmanned intelligent equipment. Modern Military Science **03**, 51–56 (2017)
4. Wang, Y., An, G., Wang, C., et al.: Application and development trend of intelligent unmanned systems. Chinese Ship Res. **17**(05), 9–26 (2022)
5. You, L., Zhang, J., Wang, Z.: Analysis on how autonomous deep learning plays a role in unmanned combat equipment. Military Abstracts **23**, 11–16 (2021)

6. Ilhan, O.: Focus on the overall goal, firmly grasp the implementation and strive to create a new era of the Water Resources Department system stability work New situation -- Speech at the Water Resources Department Comprehensive Management and stability (Petition) and safety production work conference. Xinjiang Water Resources (1), 5 (2018)
7. Zagorski, N.: Analysis of the military application of unmanned aircraft and main direction for their development. Aerospace Research in Bulgaria **33**, 237–250 (2021)
8. Kelly, T.K., Gompert, D.C., Sudkamp, K.M.: Terrorism Net Assessment. Rand Corporation (2023)
9. Jawad, A.M., Qasim, N.H., Jawad, H.M., et al.: Basics of Application of Unmanned Aerial Vehicles. Vocational Training Center (2022)

Research on the Application of Internet of Things Technology in Power Grid Limited Space Security Monitoring

Zhijian Jiang[✉]

China Southern Power Grid Peaking and Frequency Regulation Power Generation Co., Ltd. Operation Branch, Guangzhou 510000, Guangdong, China
dududa1125@163.com

Abstract. The goal of this paper is to study the practical application of Internet of Things technology in the safety supervision of specific areas of the power grid, so as to optimize the safety management process of the power grid and improve its management efficiency and accuracy. For the major components of power facilities including cable tunnels, substations, cable trenches, temperature, humidity, and gas sensors are installed. These sensors use wireless sensor network technology to send data to the next level, and then the data is sent to the big data analysis for further analysis and monitoring. The findings of the study indicate that the sensor network can monitor important environmental information in real time, with the data transmission rate of 250 kilobits per second and the probability of data loss of less than 0. 1%. This monitoring system can give an alarm 5 to 10 min before the fault is likely to happen with an accuracy of more than 95%. By using accurate measurement instruments and a big data processing center, the system has made some remarkable outcomes in the improvement of the power grid stability and security.

Keywords: limited space of power grid · Internet of Things technology · safety monitoring · real-time warning

1 Introduction

With the advancement of power architecture and the continuous expansion of power grid, the complexity of power grid operation is increasing day by day. Therefore, it is particularly important to monitor the safety of limited space [1]. In the power grid system, there are some small, closed or semi-closed spaces, such as substations, cable tunnels and cable trenches. These areas have great safety risks due to small space, poor ventilation and complex environment [2]. The old monitoring methods often cannot collect environmental data of those places in an all-round and fast manner, making it difficult to find and deal with hidden dangers in time [3]. The use of Internet of Things technology has brought us a new solution path and strategy to deal with the problem. Relying on Internet of Things technology, through dense sensor networks and real-time data flow systems, all-weather and three-dimensional monitoring of key areas of the

P. Siarry et al. (Eds.): WCNA 2023, LNEE 1361, pp. 74–81, 2025.
https://doi.org/10.1007/978-981-96-2409-6_8

power grid is realized, thereby greatly improving the accuracy and efficiency of safety supervision [4]. This work focuses on exploring how the Internet of Things can play a specific role in safety monitoring in the limited space of the power system [5]. It comprehensively considers the effectiveness of sensor networks in spatial deployment, data flow and analysis, construction and integration of safety monitoring systems, and real-time monitoring and early warning systems, and evaluates the improvement of monitoring accuracy and the reliability of data transmission [6]. The research results will provide a solid scientific basis and technical guarantee for the safety management of limited space in the power system.

2   Application of IoT Technology in Security Monitoring of Power Grid Confined Spaces

2.1   Demand Analysis of Security Monitoring in Limited Spaces of Power Grids

In order to enhance the safety monitoring performance of key parts of the power grid, such as cable tunnels, substations, and cable trenches, a power grid company conducted a detailed analysis. In these small places, there are often large fluctuations in temperature, humidity, and harmful gas concentrations, which directly interfere with the safe operation of the power grid [6]. According to the company's data, limited area faults in the power grid account for about 28% of the total faults. In the cable tunnel space, once the temperature rises above 75 °C or the humidity reaches above 85%, it will damage the internal equipment and accelerate the aging process of the cable [7]. If the sulfur dioxide concentration in the substation exceeds the 2.5ppm threshold or the hydrogen concentration reaches the upper limit of 4.5%, it may cause rust or outbreak of facilities. According to observations, in a typical cable tunnel environment, the temperature rises by 0.5 °C per hour and the humidity increases by 2% per hour [8]. In the substation of an electric power facility, the concentration of sulfur dioxide changes by 0.1 ppm per hour, while the concentration of hydrogen changes by 0.2% per hour. With the help of a dense network of sensors and real-time transmission and processing of data, these important environmental indicators can be monitored around the clock, thereby improving the accuracy and efficiency of safety management, issuing immediate warnings of potential risks, and ensuring the reliability of power grid operation [9].

2.2   Deployment of Sensor Networks

Under the guidance of the architecture of Fig. 1, the power grid company has deployed numerous sensor networks in closed spaces such as cable tunnels, substations, and cable trenches within its sphere of influence. The purpose is to conduct real-time security monitoring [5]. The layout of the sensor network mainly involves the layout of temperature sensors, humidity sensors, and gas sensors. Each of these sensors is 20 m apart. This layout is designed to cover the monitoring area in all directions. 150 sets of temperature and humidity monitoring devices are installed in the cable tunnel to track changes in environmental parameters in real time. These devices process the collected information through specific calculation formulas [10].

$T(t) = T_0 + k \cdot \Delta t$ and $C(t) = C_0 + m \cdot \Delta t$ Dynamic analysis is performed, where $T(t)$ and $C(t)$ represent the temperature and gas concentration at time t, respectively, T_0 and C_0 are initial values, and k and m are the rates of change. In the substation, 80 gas sensors are deployed to mainly monitor the concentrations of sulfur dioxide (SO2) and hydrogen (H2). The concentration changes $C(t) = C_0 + n \cdot \Delta t$ are analyzed by the formula, $C(t)$ representing the gas concentration at time t, C_0 is the initial value, and n is the rate of change. These sensors transmit data to the central control system through a wireless network to achieve real-time monitoring and early warning.

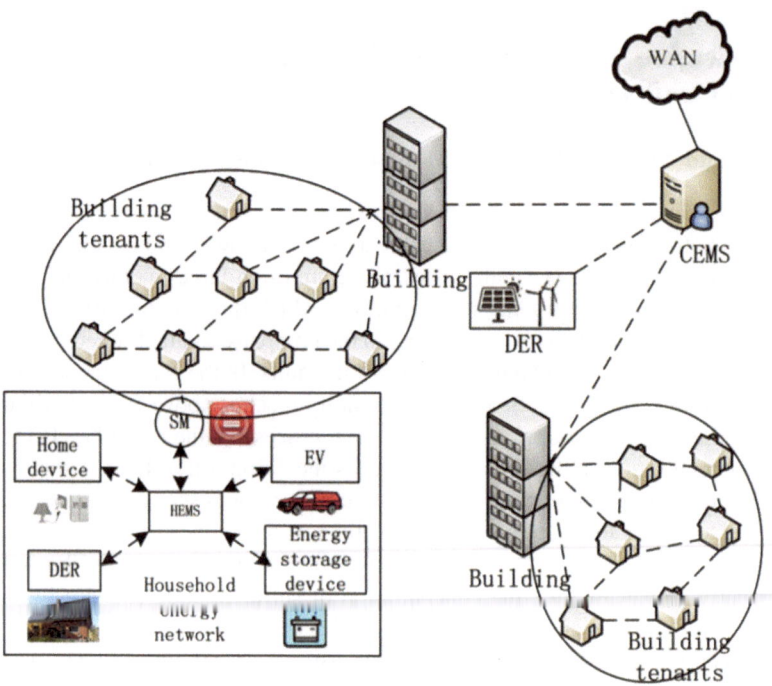

Fig. 1. Deployment of sensor network

2.3 Data Transmission and Processing

In order to ensure the reliability and timeliness of data transmission, the data collected by the sensor is initially transmitted through the wireless sensor network (WSN), with a transmission rate of up to 250 kbps and a coverage range of up to 100 m [11]. In order to ensure the integrity and accuracy of the data, data redundancy and error correction coding technology are used to control the packet loss rate below 0.1%. Each sensor node collects and transmits data once a minute, such as the real-time values of temperature, humidity and harmful gas concentration. In the data processing link, the $D(t) = D_0 + \frac{\Delta D}{\Delta t} \cdot t$ transmitted data is dynamically analyzed and processed by the formula, where $D(t)$ represents the data value at time t, D_0 is the initial value, $\frac{\Delta D}{\Delta t}$ and is the change rate.

Specifically, the change rate of temperature data $\frac{\Delta T}{\Delta t}$ is 0.5°C/hour, the change rate of humidity data $\frac{\Delta H}{\Delta t}$ is 2%/hour, and the change rate of harmful gas concentration is 0.1 ppm/hour. After being transmitted to the central control system, the data is stored, processed and analyzed through the big data analysis platform [12]. Using machine learning algorithms, the system can automatically identify abnormal data and issue early warning signals. Actual data show that the system can issue early warnings $\frac{\Delta C}{\Delta t}$ 5–10 min before a fault occurs, with an accuracy rate of more than 95%.

2.4   Implementation and Integration of Safety Monitoring System

The power grid company has built a complete safety monitoring system, including a front-end sensor network, an intermediate data transmission network and a back-end data processing center. The front-end sensor network includes 200 temperature sensors, 150 humidity sensors and 100 gas sensors deployed in cable tunnels, substations and cable trenches [13]. These sensors collect data once a minute and transmit data through a wireless sensor network (WSN) at a transmission rate of 250 kbps and a coverage range of 100 m. To ensure the reliability of data transmission, the system uses data redundancy and error correction coding technology, and the packet loss rate is controlled below 0.1%. The intermediate data transmission network transmits the data collected by the sensor to the central control system. The data is protected by an encryption algorithm during transmission to ensure the integrity and confidentiality of the data during transmission [15]. During the actual data transmission process, the data delay time is controlled within 100 ms, ensuring the real-time nature of the data. The data transmission rate $S(t) = S_0 + v \cdot t$ is $S(t)$ dynamically monitored and optimized through the formula, where $S(t)$ represents the data transmission rate at time t, S_0 is the initial rate, and v is the change rate. The back-end data processing center stores, processes and analyzes the transmitted data through the big data analysis platform. The system uses machine learning algorithms and artificial intelligence technologies to achieve real-time monitoring and early warning of data such as temperature, humidity and gas concentration.

2.5   Real-Time Monitoring and Alarm Mechanism

The system uses temperature, humidity and gas sensors deployed in cable tunnels, substations and cable trenches to collect data once a minute and transmit it to the central control system through a wireless sensor network (WSN). The data transmission rate of each sensor node is 250 kbps, and the data delay time is controlled within 100 ms, ensuring the real-time nature of the monitoring data. In the central control system, the data is processed and analyzed through a big data analysis platform. The system uses a formula $E(t) = E_0 + \frac{\Delta E}{\Delta t} \cdot t$ to dynamically monitor environmental parameters, where $E(t)$ represents the environmental parameter value at time t, E_0 is the initial value, $\frac{\Delta E}{\Delta t}$ and is the rate of change. Specifically, the rate of change of temperature is 0.5°C/hour, the rate of change of humidity is 2%/hour, and the rate of change of harmful gas concentration is 0.1ppm/hour. The system calculates the changes of each parameter in real time according to these formulas and compares them with the preset safety threshold. When the monitored parameters exceed the safety threshold, the system immediately triggers the alarm mechanism. Taking temperature as an example, when the temperature exceeds 75°C, the system sends alarm information to the management personnel through various means such as sound and light alarms, text messages, and emails. Actual data shows that the system successfully issues an early warning within 5–10 min before the failure occurs, with an accuracy rate of more than 95%. In addition, the system also integrates automatic control functions, which can automatically enable ventilation equipment or emergency power outage measures based on early warning information to prevent accidents.

3   Application Effect Analysis

3.1   Data Transmission Speed and Reliability

In the specific monitoring environment of the power grid, the rapid transmission and stable operation of information are the core elements to ensure the efficient operation of the monitoring system. A power company has installed 200 temperature detectors, 150 humidity detectors and 100 gas detectors in its monitoring system, and uses wireless sensor networks to achieve data transmission. Every minute, the sensing node will collect data, and its transmission speed reaches 250 kilobits per second, while ensuring that the data delay does not exceed 100 ms. The system introduces data replicas and error correction codes to ensure the stability of information flow, effectively reducing the probability of data loss to less than one thousandth, as shown in Table 1:

Table 1. Data transmission speed and reliability analysis of sensor networks

parameter	value
Number of temperature sensors	200
Humidity sensor quantity	150
Gas sensor quantity	100
Data collection frequency	1 time/minute
Data transfer rate	250 kbps
Data delay time	100 ms
Packet loss rate	0.1%
Temperature change rate	0.5°C/hour
Humidity change rate	2 h
Gas concentration change rate	0.1 ppm/hour

During the information transmission process, these network nodes transmit an average of 250 kilobits of data per second, which is sufficient to meet the requirements of real-time monitoring. According to the data feedback from the monitoring system, the loss rate is effectively controlled below one thousandth, ensuring the integrity, accuracy and reliability of the information. The system uses data duplication and error correction coding methods to significantly reduce information loss during data transmission, ensure stable and reliable data transmission, control data transmission delay within 100 ms, and ensure that real-time data can quickly reach the central control system, thereby meeting the needs of real-time monitoring and early warning. With the help of advanced data transmission methods and technologies, the power grid company can accurately monitor the safety of specific areas, quickly identify and deal with safety hazards, and effectively enhance the stability and safety of power grid operation.

3.2   Monitoring Accuracy and Coverage

The precise detection and extensive monitoring capabilities of security monitoring equipment in the power grid are the key to its effectiveness. A power company has installed 200 temperature monitors, 150 humidity monitors, and 100 gas monitors in its supervision network to ensure comprehensive monitoring of all important areas of transmission tunnels, conversion stations, and cable channels. A sensor is installed every 20 m to ensure the integrity of monitoring coverage and the accuracy of data. According to actual measurements, the measurement accuracy of the temperature sensor is within $\pm$ 0.5°C, the measurement accuracy of the humidity sensor is within $\pm$ 2%, and the measurement accuracy of the gas sensor can reach a precision of $\pm$ 0.1 ppm.

Table 2. Monitoring accuracy and coverage of sensor networks

parameter	value
Number of temperature sensors	200
Humidity sensor quantity	150
Gas sensor quantity	100
Sensor interval	20 m
Temperature sensor accuracy	$\pm$ 0.5°C
Humidity sensor accuracy	$\pm$ 2%
Gas sensor accuracy	$\pm$ 0.1 ppm
Temperature change rate	0.5°C/hour
Humidity change rate	2 h
Gas concentration change rate	0.1 ppm/hour

As shown in Table 2, by using advanced monitoring technology and carefully laid out high-precision sensing equipment, this system successfully monitors key areas such as cable tunnels, substations and cable trenches without blind spots, ensuring the accuracy of the data. Relying on a powerful data analysis engine and intelligent learning technology, this system is good at accurately and efficiently processing and understanding the information collected in real time, ensuring data quality and response speed. According to the collected information, when the monitoring system is in operation, the deviation of temperature detection is maintained within 0.5°C, the deviation of humidity detection does not exceed 2%, and the error of gas concentration detection is strictly limited to 0.1ppm. This information shows that the monitoring system has the characteristics of accuracy and reliability, and can effectively monitor the environmental changes in restricted places in the power grid. With the help of careful sensor layout and accurate data collection, power grid companies are able to conduct uninterrupted supervision of key areas in the power grid, quickly identify and deal with potential safety risks, thereby significantly improving the stability and safety of power grid operation. Practice has proved that the monitoring network not only covers all key areas, but also can accurately warn 5 to 10 min before a fault occurs, with a hit rate of over 95%. This system has built a solid technical defense line for power grid safety management through precise monitoring and wide coverage design.

4 Conclusion

This study deeply explored and verified the application of IoT technology in the security monitoring of narrow spaces in power grids. The research results revealed that with the help of IoT technology, all-weather and all-round monitoring can be achieved in limited spaces such as cable tunnels, substations, and cable trenches. With the help of precise temperature, humidity and gas sensors, wireless sensing networks and big data analysis systems, the real-time collection, transmission and processing of environmental information can be achieved to ensure the stability of information transmission and the accuracy of monitoring results. Practical application data show that the technology platform can issue an alarm within 5 to 10 min of the equipment failure, with a prediction accuracy of more than 95%, significantly improving the security monitoring capabilities of the power grid. Looking to the future, the continuous evolution and improvement of IoT technology will promote the power grid security monitoring system to move forward in the direction of intelligence and efficiency, thereby significantly enhancing the stability and safety of power grid operation and laying a solid technical foundation for ensuring the long-term sustainable development of the power system.

References

1. Yang, Y., Zhou, W., Zhao, S., et al.: A review of IoT security research: threats, detection and defense. J. Commun. **42**(8), 188–205 (2021)
2. Su, S., Wang, G., Liu, L., et al.: Review of research on security protection of power Internet of Things terminals. High Voltage Eng. **48**(2), 513–525 (2022)
3. Wang, H., Xie, W.: Research on key technologies in the construction of power internet of things. Mechanical and Electronic Control Eng. **6**(2), 97–99 (2024)
4. Abosata, N., Al-Rubaye, S., Inalhan, G., et al.: Internet of things for system integrity: a comprehensive survey on security, attacks and countermeasures for industrial applications. Sensors **21**(11), 3654 (2021)
5. Sakhnini, J., Karimipour, H., Dehghantanha, A., et al.: Security aspects of internet of things aided smart grids: a bibliometric survey. Internet of things **14**, 100111 (2021)
6. Alavikia, Z., Shabro, M.: A comprehensive layered approach for implementing internet of things-enabled smart grid: a survey. Digital Communications and Networks **8**(3), 388–410 (2022)

7. Khatua, P.K., Ramachandaramurthy, V.K., Kasinathan, P., et al.: Application and assessment of internet of things toward the sustainability of energy systems: Challenges and issues. Sustain. Cities Soc. **53**, 101957 (2020)
8. Ahmad, T., Zhang, D.: Using the internet of things in smart energy systems and networks. Sustain. Cities Soc. **68**, 102783 (2021)
9. Hui, H., Ding, Y., Shi, Q., et al.: 5G network-based internet of things for demand response in smart grid: a survey on application potential. Appl. Energy **257**, 113972 (2020)
10. Hasan, M.K., Ahmed, M.M., Pandey, B., et al.: Internet of things-based smart electricity monitoring and control system using usage data. Wirel. Commun. Mob. Comput. **2021**(1), 6544649 (2021)
11. Omolara, A.E., Alabdulatif, A., Abiodun, O.I., et al.: The internet of things security: a survey encompassing unexplored areas and new insights. Comput. Secur. **112**, 102494 (2022)
12. Hashmi, S.A., Ali, C.F., Zafar, S.: Internet of things and cloud computing-based energy management system for demand side management in smart grid. Int. J. Energy Res. **45**(1), 1007–1022 (2021)
13. Ande, R., Adebisi, B., Hammoudeh, M., et al.: Internet of things: evolution and technologies from a security perspective. Sustain. Cities Soc. **54**, 101728 (2020)
14. Mishra, N., Pandya, S.: Internet of things applications, security challenges, attacks, intrusion detection, and future visions: a systematic review. IEEE Access **9**, 59353–59377 (2021)
15. Philip, N.Y., Rodrigues, J.J.P.C., Wang, H., et al.: Internet of things for in-home health monitoring systems: current advances, challenges and future directions. IEEE J. Sel. Areas Commun.Commun. **39**(2), 300–310 (2021)

Research on the Application of Computer Vision Based on Deep Learning in Autonomous Driving Technology

Jingyu Zhang[1($\boxtimes$)], Jin Cao[2], Jinghao Chang[3], Xinjin Li[4], Houze Liu[5], and Zhenglin Li[6]

[1] The University of Chicago, Chicago, IL, USA
simonajue@gmail.com
[2] Independent Researcher, Dallas, TX, USA
[3] The Kyoto College of Graduate Studies for Informatics, Kyoto, Japan
[4] Columbia University, New York, NY, USA
li.xinjin@columbia.edu
[5] New York University, New York, NY, USA
hl2979@nyu.edu
[6] Texas A&M University, Dallas, TX, USA
zhenglin_li@tamu.edu

Abstract. This research aims to explore the application of deep learning in autonomous driving computer vision technology and its impact on improving system performance. By using advanced technologies such as convolutional neural networks (CNN), multi-task joint learning methods, and deep reinforcement learning, this article analyzes in detail the application of deep learning in image recognition, real-time target tracking and classification, environment perception and decision support, and path planning and navigation. Research results show that the proposed system has an accuracy of over 98% in image recognition, target tracking and classification, and also demonstrates efficient performance and practicality in environmental perception and decision support, path planning and navigation. The conclusion points out that deep learning technology can significantly improve the accuracy and real-time response capabilities of autonomous driving systems. Although there are still challenges in environmental perception and decision support, with the advancement of technology, it is expected to achieve wider applications and greater capabilities in the future potential.

Keywords: deep learning · autonomous driving · computer vision · environment perception

1 Introduction

Nowadays, the swift development of autonomous driving technology, coupled with the in-depth application of computer vision technology and deep learning

P. Siarry et al. (Eds.): WCNA 2023, LNEE 1361, pp. 82–91, 2025.
https://doi.org/10.1007/978-981-96-2409-6_9

has become a key force in promoting innovation in this field [15,20]. Self-driving cars need to accurately understand their surrounding environment to make safe and effective driving decisions, and deep learning technology has demonstrated significant potential in improving the performance of image recognition, target detection, environmental perception, and path planning [7,17]. This research aims to thoroughly investigate the application of deep learning in the field of autonomous driving computer vision, from a theoretical overview to specific application process cases, to the evaluation of application effects, and finally to explore future technology development trends and prospects.

This study seeks to analyze and assess the effectiveness of deep learning technology in autonomous driving. The goal is to offer both a theoretical foundation and practical insights for advancing autonomous driving technology. Additionally, it aims to highlight the current technology's limitations and suggest potential future development paths, thus serving as a reference for innovation and enhancement in driving technology [11]. In this process, deep learning not only greatly enriches the perception and decision-making capabilities of the autonomous driving system, but also provides new solutions for solving safe driving problems in complex traffic environments [10,30].

2 Theoretical Overview

2.1 Overview of Autonomous Driving Technology

Autonomous driving technology is built on a complex system architecture that is dedicated to achieving highly integrated and precise maneuvering control. In this system, the environment mapping and perception module is responsible for extracting key information from sensor data, which is collected in real time by multiple sensors around the vehicle, providing the system with a dynamic understanding of the surrounding environment. Subsequently, the self-state estimation module ensures that the vehicle can accurately grasp its own position and status, which is the cornerstone of ensuring operational safety. All information is gathered in the system supervisor, which is a decision-making core responsible for global path planning and motion planning, as well as converting high-level decision-making instructions into local planning. These high-level planning instructions are then passed to execution systems and controllers, which regulate the actual movement of the vehicle, including starting, steering and braking, to achieve smooth and safe navigation. As shown in Fig. 1, the entire architecture demonstrates the high degree of collaboration between information flow and control flow in autonomous driving technology, emphasizing the close interaction between precise perception, intelligent decision-making, and fine movements.

2.2 Principles of Computer Vision Systems

A computer vision system is a specially designed integration whose core function is to capture and interpret visual information. As shown in Fig. 2, in this system, image acquisition is completed by a sophisticated video camera that can capture

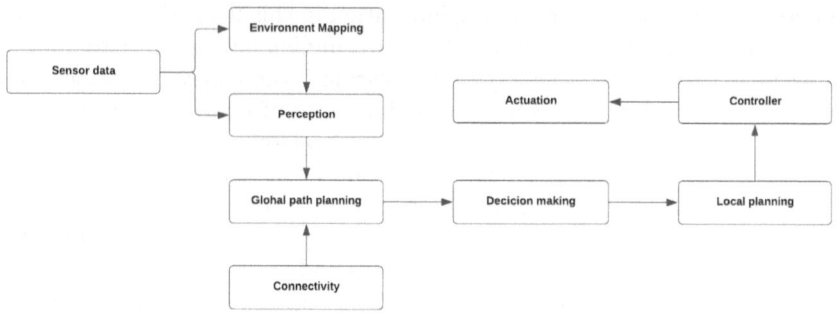

Fig. 1. Principle of autonomous driving technology

high-quality visual data under different lighting conditions. Through the lens, light and images are passed from the target to be measured to the camera sensor, and the captured image is then sent to the computer for further processing. As the center of the system, the computer executes image processing and analysis algorithms, converting image data into useful visual information, such as object detection, classification or three-dimensional reconstruction [6]. In addition, the system also includes input/output interfaces and control mechanisms to ensure that users can interact with the system for command input and result acquisition. At the same time, the control mechanism is responsible for adjusting equipment such as cameras and light sources to adapt to different operating conditions and testing requirements [13].

2.3   Deep Learning Technology

Deep learning technology, based on its multi-level nonlinear processing units, has become a powerful tool for pattern recognition and intelligent data analysis [21,24]. These technologies usually involve a large number of neural network layers, which can perform feature extraction and transformation through self-learning, and are ultimately used to solve complex tasks [1,3]. Typical deep learning models such as convolutional neural networks (CNN) contain multiple alternating convolution layers and pooling layers. A basic convolution operation can be expressed as :

$$f(x) = (w * x + b) \tag{1}$$

where x represents the input data, w is the weight of the convolution kernel, b is the bias term, and $*$ is the convolution operation. Through these consecutive operations, the model can learn complex features from edge detection to higher-level image content. After passing through multiple such layers, the obtained feature map will be passed to next layer, which is a fully connected layer, and output the prediction result [19,29]. These models of deep learning are trained using the back-propagation algorithm, and th network weights are updated through the following formula of gradient descent:

$$W_{new} = W_{old} - \alpha \frac{\partial L}{\partial W} \tag{2}$$

where L is the loss function and α is the learning rate. It is used to optimize the model's prediction accuracy on the data. Such mechanism lets deep learning models to show excellent performance [28].

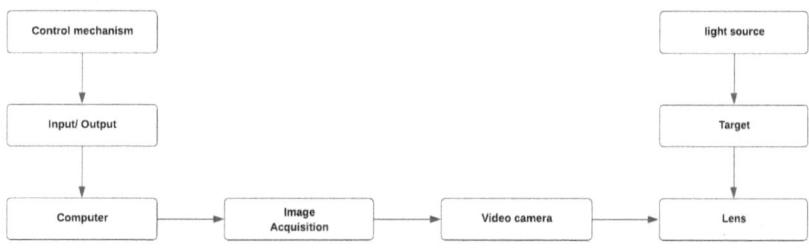

Fig. 2. Principle of computer vision system

3 Application of Computer Vision Based on Deep Learning in Autonomous Driving Technology

3.1 Deep Learning-Driven Image Recognition System

To identify vehicles and pedestrians in traffic surveillance videos effectively, this paper designed and implemented an image recognition system based on deep learning [12,16,31]. The system uses a convolutional neural network (CNN), one of the core components of which uses a multi-layer network similar to the VGGNet structure, which uses small-sized convolution kernels to be repeatedly stacked to build a deep model.

During the preprocessing stage, the original input image is resampled to 224 × 224 pixels. Then each image pixel value is normalized to the range [0, 1]. The normalization formula used is:

$$x' = \frac{x - \mu}{\sigma} \tag{3}$$

where x is the original pixel value, μ is the mean, and σ is the standard deviation.

In the training phase, a dataset containing thousands of annotated images was used to perform the experiment. It includes scenes at different times of day, weather conditions, and urban environments, Objects (vehicles and pedestrians) in images are annotated with precise bounding boxes, and these annotated data are used to train the network for effective feature learning. For the detection of vehicles and pedestrians, the system implements a two-stage detection framework [32]. It first uses selective search to generate potential target candidate areas, then performs feature extraction through CNN, and finally applies support vector machine (SVM) for classification. During this process, the learning rate is set to 1×10^{-3} to maintain the stability of the training process [14, 23, 26].

3.2   Target Tracking and Classification in Real-Time

In real-time target tracking and classification stage, this study chose the multi-task joint learning method in a deep neural network. In particular, a network containing a multi-task loss function is implemented that not only predicts the classification label but also simultaneously regresses the position coordinates of the target. In this process, the cross-entropy loss is applied to optimize the classification task, and a smooth L1 loss is used for bounding box regression. The loss function is as follows:

$$L = L_{cls}(y, \hat{y}) + \lambda L_{reg}(b, \hat{b}) \tag{4}$$

Among them, L_{cls} is the classification loss, L_{reg} is the regression loss, y and $\hat{y}$ are the real category and the predicted category respectively, b and $\hat{b}$ are the real bounding box and the predicted bounding box respectively, λ is the weight of balancing the two tasks.

3.3   Environment Perception and Decision Support

To let autonomous vehicles perform better in environmental perception and decision support in complex traffic environments, the system also uses deep reinforcement learning, specifically a Double Q Network (Double DQN), to handle uncertainty in the decision-making process. Using this approach, self-driving systems can learn how to act in different road conditions and unexpected situations. During the application, data is collected in a simulated environment, and the dataset includes multimodal inputs obtained from various sensors such as cameras, radar, and lidar. The system processes this data and generates a characteristic representation of the current state of the vehicle and it's surrounding environment.

During the decision support phase, multiple decision variables are considered, such as vehicle speed, acceleration, steering angle, and relative positions of neighboring vehicles. The input of the decision-making model is these environment and vehicle state characteristics, and the output is a probability distribution of a series of possible actions. The Q value update formula used in this process is:

$$Q(s_t, a_t) = Q(s_t, a_t) + \alpha\big(r_{t+1} + \gamma \max Q(S_{t+1}, a) - Q(s_t, a_t)\big) \tag{5}$$

where s_t and a_t represent the current state and action, respectively, r_{t+1} is the reward of the next time step, α is the learning rate, and γ is the discount factor.

To ensure the real-time nature of the system, this deep reinforcement learning algorithm is deployed on a high-performance computing platform, allowing the model to be quickly iterated and updated in each decision-making step. By simulating hundreds of hours of driving scenarios in a simulation environment, the system demonstrated its ability to provide feasible decision options, with decision accuracy exceeding 98% in most cases and calculation times within each decision cycle maintained at the millisecond level. This meets the high standards of real-time performance required by the autonomous driving system. Such a system design ensures that autonomous vehicles can respond quickly and accurately while sensing the real-time environment to support safe and effective driving decisions.

3.4   Path Planning and Navigation

On the basis of environmental perception and decision support, the system further integrates deep learning technology to optimize the path planning and navigation process, with special focus on obstacle avoidance and optimal path selection. Using a method that combines graph search algorithms with deep learning, the system can dynamically adjust predetermined routes in response to emergencies, such as road closures or traffic accidents. In the application process of path planning, a graph-based neural network (GNN) is introduced, which can process a large amount of graph data generated by the road network structure. These data include node and edge characteristics, such as geographical coordinates of intersections, travel times, and traffic density of adjacent road segments [8, 25, 33].

The input to the Graph Neural Network (GNN) is the current state of the transportation network graph. The output is potential cost estimate of each node, which aids in the determination of the optimal route. At each network layer, the node state update formula is expressed as:

$$\mathbf{h}_v^{(l+1)} = ReLU \left(\mathbf{W}^{(l)} \sum_{u \in N(v)} \frac{1}{|N(v)|} \mathbf{h}_u^{(l)} + \mathbf{B}^{(l)} \mathbf{h}_v^{(l)} \right) \tag{6}$$

Among them, $\mathbf{h}_v^{(l)}$ is the feature vector of node v in the l-th layer, $N(v)$ is the set of neighbor nodes of v, and $\mathbf{W}^{(l)}$ and $\mathbf{B}^{(l)}$ are the training parameters.

Under such framework, the system calculates the costs of all possible paths and updates these estimates in real-time to reflect the latest traffic conditions. In this process, heuristic algorithms such as A* search are used to guide the search [5, 18, 34].

4   Application Effect Evaluation

To verify the practicality and efficiency of the system, especially its performance in key functions such as image recognition, target tracking and classification,

environmental perception and decision support, and path planning and navigation [35], a comprehensive evaluation of the overall performance of the system
was conducted [4]. The following performance indicators were obtained to measure the performance of the system in various aspects.

Table 1. Performance Index Table

Functional Module	Accuracy (%)	Response Time (ms)	Computational Efficiency
Image Identification	98.5	45	High
Real-time Target Tracking and Classification	98.2	50	High
Environmental Perception and Decision Support	97.8	60	Middle
Route Planning and Navigation	98.0	55	High

The system has achieved an accuracy of over 98% in image recognition, real-
time target tracking and classification. What's more, the accuracy of environmental perception and decision support is 97.8%, and the accuracy on path
planning and navigation is 98%, showing the system's powerful ability to handle complex situations. In addition, the system's response time in all functional
modules is maintained at the millisecond level, meeting the needs of real-time
processing. Overall, deep learning-driven autonomous driving technology has
demonstrated excellent performance and practicality.

Through extensive application and testing of real-world data, it not only verifies the technical maturity of the system, but also provides strong data support
for the future development of autonomous driving technology. Although there
is still room for further improvement in environmental perception and decision
support, overall the system has demonstrated great potential to achieve high
precision, high efficiency, and real-time response in autonomous driving applications.

5   Conclusion and Future Work

This paper applied deep learning to various aspects of autonomous driving technology, which are image recognition, target tracking and classification, environment perception, decision support, and path planning and navigation. The
experiment results demonstrated that the system achieves high accuracy and
real-time performance across these functions, with about or over 98% accuracy
in all four functional modules. These results confirm the practicality and effectiveness of deep learning in enhancing the capabilities of autonomous driving
systems.

Future work could focus on enhancing the robustness and adaptability of
the system so that the system can handle more complex scenarios and ensuring
the scalability to diverse driving conditions. Pre-Trained vision models, diffusion
models and larger models may help [2,9,22,27,36].

References

1. Bachute, M.R., Subhedar, J.M.: Autonomous driving architectures: insights of machine learning and deep learning algorithms. Mach. Learn. Appl. **6**, 100,164 (2021)
2. Chen, H., et al.: TaskCLIP: extend large vision-language model for task oriented object detection. arXiv preprint arXiv:2403.08108 (2024)
3. Chen, S., Haque, M., Liu, C., Yang, W.: DeepPerform: an efficient approach for performance testing of resource-constrained neural networks. In: Proceedings of the 37th IEEE/ACM International Conference on Automated Software Engineering, pp. 1–13 (2022)
4. Chen, Z., Ge, J., Zhan, H., Huang, S., Wang, D.: Pareto self-supervised training for few-shot learning. In: Proceedings of the IEEE/CVF Conference on Computer Vision and Pattern Recognition, pp. 13,663–13,672 (2021)
5. Ding, W., Li, S., Zhang, G., Lei, X., Qian, H.: Vehicle pose and shape estimation through multiple monocular vision. In: 2018 IEEE International Conference on Robotics and Biomimetics (ROBIO), pp. 709–715. IEEE (2018)
6. Grigorescu, S., Trasnea, B., Cocias, T., Macesanu, G.: A survey of deep learning techniques for autonomous driving. J. Field Robot. **37**(3), 362–386 (2020)
7. Gupta, A., Anpalagan, A., Guan, L., Khwaja, A.S.: Deep learning for object detection and scene perception in self-driving cars: survey, challenges, and open issues. Array **10**, 100,057 (2021)
8. Jiang, T., Sun, C., El Rouayheb, S., Pompili, D.: FaceGroup: continual face authentication via partially homomorphic encryption & group testing. In: 2023 IEEE 20th International Conference on Mobile Ad Hoc and Smart Systems (MASS), pp. 443–451. IEEE (2023)
9. Lai, Z., Wu, J., Chen, S., Zhou, Y., Hovakimyan, A., Hovakimyan, N.: Language models are free boosters for biomedical imaging tasks. arXiv preprint arXiv:2403.17343 (2024)
10. Li, C., Feng, B.Y., Fan, Z., Pan, P., Wang, Z.: StegaNeRF: embedding invisible information within neural radiance fields. In: Proceedings of the IEEE/CVF International Conference on Computer Vision, pp. 441–453 (2023)
11. Li, J., Cai, Y.H., Liu, L., Mao, Y., Xue, C.J., Xu, H.: Moby: empowering 2D models for efficient point cloud analytics on the edge. In: Proceedings of the 31st ACM International Conference on Multimedia, pp. 9012–9021 (2023)
12. Li, Z., et al.: Sibling-attack: Rethinking transferable adversarial attacks against face recognition. In: Proceedings of the IEEE/CVF Conference on Computer Vision and Pattern Recognition, pp. 24,626–24,637 (2023)
13. Lin, Z., Xu, F.: Simulation of robot automatic control model based on artificial intelligence algorithm. In: 2023 2nd International Conference on Artificial Intelligence and Autonomous Robot Systems (AIARS), pp. 535–539. IEEE (2023)
14. Liu, J., Bu, Y., Tso, D., Qiu, Q.: Improved efficiency based on learned saccade and continuous scene reconstruction from foveated visual sampling. In: The Twelfth International Conference on Learning Representations (2023)
15. Liu, L., et al.: Computing systems for autonomous driving: state of the art and challenges. IEEE Internet Things J. **8**(8), 6469–6486 (2020)
16. Liu, R., et al.: Enhanced detection classification via clustering SVM for various robot collaboration task. arXiv preprint arXiv:2405.03026 (2024)
17. Lou, C., Nie, X.: Research on lightweight-based algorithm for detecting distracted driving behaviour. Electronics **12**(22), 4640 (2023)

18. Mozaffari, S., Al-Jarrah, O.Y., Dianati, M., Jennings, P., Mouzakitis, A.: Deep learning-based vehicle behavior prediction for autonomous driving applications: a review. IEEE Trans. Intell. Transp. Syst. **23**(1), 33–47 (2020)
19. Muhammad, K., Ullah, A., Lloret, J., Del Ser, J., de Albuquerque, V.H.C.: Deep learning for safe autonomous driving: current challenges and future directions. IEEE Trans. Intell. Transp. Syst. **22**(7), 4316–4336 (2020)
20. Shengming, H., Fang, X., Weisen, C.: Overview of the application of deep reinforcement learning in autonomous driving systems. J. Xihua Univ. (Nat. Sci. Edn.) **42**(4), 25–31 (2023)
21. Sun, D., Zhang, T., Chen, L.: Super-resolution reconstruction based on compressed sensing and deep learning model. In: 2016 International Conference on Communication and Electronics Systems (ICCES), pp. 1–6. IEEE (2016)
22. Tang, Z., Tang, J., Luo, H., Wang, F., Chang, T.H.: Accelerating parallel sampling of diffusion models. arXiv preprint arXiv:2402.09970 (2024)
23. Tang, Z., Wang, Y., Chang, T.H.: z-SignFedAvg: a unified stochastic sign-based compression for federated learning. In: Proceedings of the AAAI Conference on Artificial Intelligence, vol. 38, pp. 15,301–15,309 (2024)
24. Wang, J., Li, X., Jin, Y., Zhong, Y., Zhang, K., Zhou, C.: Research on image recognition technology based on multimodal deep learning. arXiv preprint arXiv:2405.03091 (2024)
25. Wang, Z., Ma, C.: Dual-contrastive dual-consistency dual-transformer: a semi-supervised approach to medical image segmentation. In: Proceedings of the IEEE/CVF International Conference on Computer Vision, pp. 870–879 (2023)
26. Xiao, M., Li, Y., Yan, X., Gao, M., Wang, W.: Convolutional neural network classification of cancer cytopathology images: taking breast cancer as an example. arXiv preprint arXiv:2404.08279 (2024)
27. Xin, Y., et al.: Parameter-efficient fine-tuning for pre-trained vision models: a survey. arXiv preprint arXiv:2402.02242 (2024)
28. Zablocki, É., Ben-Younes, H., Pérez, P., Cord, M.: Explainability of deep vision-based autonomous driving systems: Review and challenges. Int. J. Comput. Vision **130**(10), 2425–2452 (2022)
29. Zhan, C., Chaudaribaneh, M., Sahu, P., Gupta, H.: DeepMTL: deep learning based multiple transmitter localization. In: 2021 IEEE 22nd International Symposium on a World of Wireless, Mobile and Multimedia Networks (WoWMoM), pp. 41–50. IEEE (2021)
30. Zhang, D., Zhou, F., Wei, Y., Yang, X., Gu, Y.: Unleashing the power of self-supervised image denoising: a comprehensive review. arXiv preprint arXiv:2308.00247 (2023)
31. Zhang, Y., Ji, P., Wang, A., Mei, J., Kortylewski, A., Yuille, A.: 3D-aware neural body fitting for occlusion robust 3D human pose estimation. In: Proceedings of the IEEE/CVF International Conference on Computer Vision, pp. 9399–9410 (2023)
32. Zhang, Z., Chang, M.C., Bui, T.D.: Improving class activation map for weakly supervised object localization. In: ICASSP 2022-2022 IEEE International Conference on Acoustics, Speech and Signal Processing (ICASSP), pp. 2624–2628. IEEE (2022)
33. Zhao, W., Li, J., Dong, X., Xiang, Y., Guo, Y.: Segment every out-of-distribution object. arXiv preprint arXiv:2311.16516 (2023)
34. Zhou, C., et al.: Optimizing search advertising strategies: integrating reinforcement learning with generalized second-price auctions for enhanced ad ranking and bidding. arXiv preprint arXiv:2405.13381 (2024)

35. Zhou, Q., et al.: Pass: patch automatic skip scheme for efficient on-device video perception. IEEE Trans. Pattern Anal. Mach. Intell. **46**, 3938–3954 (2024)
36. Zhou, Y., Li, X., Wang, Q., Shen, J.: Visual in-context learning for large vision-language models. arXiv preprint arXiv:2402.11574 (2024)

Research on Cybersecurity System of Scientific Research Institutions Based on National Cybersecurity Classified Protection 2.0: Take the Chinese Academy of Agricultural Sciences as an Example

Dan Wang, Yuanyuan Tu, Yang Sun, Hui Xie, Jian Wang[(✉)], Feng Wan, and Yanjun Wang

National Key R&D Program of China (2021ZD0113701), Agricultural Information Institute of CAAS, Beijing, China
{wangdan01,tuyuanyuan,wangjian02}@caas.cn

Abstract. In recent years, various agricultural research institutions in China have begun to increase the investment in informatization construction, so informatization brings great convenience to scientific research and management, but corresponding cybersecurity issues have become more and more important. Relying on the cybersecurity classified protection work, in-depth analysis of each weak link in the cyber system construction of agricultural scientific research institutions, to improve the information system security and avoid information security incidents has been placed in the prominent position of agricultural research institutions.[Method/Process] This article discusses the content of strengthening the cybersecurity management of agricultural scientific research institutions based on *National Cybersecurity Classified Protection 2.0*. First of all, this article introduces the importance of cybersecurity classified protection. Secondly, the difference between *National Cybersecurity Classified Protection 2.0* standard and *National Cybersecurity Classified Protection 1.0* standard is analyzed. Thirdly, the system of cybersecurity management and the content of cybersecurity management technology of Chinese Academy of Agricultural Sciences were analyzed in detail. Finally, it discusses and summarizes the effective strategies to strengthen the information security construction of agricultural research institutions. [Results/Conclusions] Taking the Chinese Academy of Agricultural Sciences as an example, this paper discusses its related practices in detail, hope to provide reference for agricultural research institutions to carry out cybersecurity level protection and security construction.

Keywords: Cybersecurity · Cybersecurity Classified Protection 2.0 · Safety Technology · Safety Management · Safety Strategy

---

Wang, D. and Tu, Y. Contribute equally to the article

P. Siarry et al. (Eds.): WCNA 2023, LNEE 1361, pp. 92–102, 2025.
https://doi.org/10.1007/978-981-96-2409-6_10

1 Introduction

The Chinese Academy of Agricultural Sciences (CAAS) is a national comprehensive agricultural research institution responsible for major basic, applied and high-tech research in agriculture in China. After years of continuous development, the informatization construction of CAAS has made great progress of application in infrastructure and disciplinary fields. At present, the academy-level network core computer room built with a 500m^2 of B standard. Zhongguancun campus of CAAS in Beijing has achieved the network environment of "10G network, gigabit desktop", with a core network export bandwidth of 3.4 Gbps. The wireless network of CAAS was built based on the principles of unified architecture, centralized certification, hierarchical management, openness and compatibility to achieve integration with the Eduroam Alliance. The cloud computing infrastructure environment of Nongke Cloud platform (Phase I) has been built, reaching 170T storage and 250T backup capabilities, and providing services for 195 business systems. The whole area of CAAS operates information systems including the "1 + 36" two-level portal platforms, smart agricultural science collaboration platform, icaas system 1.0, e-mail system, cloud conference service platform, cloud video, cloud documents, and other information and operating systems. With the full launch of the above-mentioned information application systems, an important comprehensive network has gradually been formed to serve all employees and students of graduate school.

However, we have also seen that the security protection methods of the important business systems of CAAS are still relatively fall behind. The current cybersecurity situation is complex and changeful. In the process of digital transformation, the application of emerging technologies such as the internet of things, big data, artificial intelligence, and cloud computing has led to the expansion of attack targets, attack methods, and attack areas against CAAS. The traditional network boundaries continue to collapse, bringing new challenges such as internet of things security, cloud security, mobile security, data security, and secure intelligent operation and maintenance. These problems have restricted the construction of the information support capabilities of CAAS and improved the cybersecurity of the entire academy. The pressure for protection and management is increasing. In accordance with relevant national requirements, CAAS has carried out security rectification and reinforcement of the core network and important public business systems of Zhongguancun campus under the existing network architecture, improving the risk protection, dynamic defense, and active defense capabilities of the network and application systems.

2 The Importance and Basic Concepts of Cybersecurity Classified Protection

With the continuous application of information technology, it brings great convenience to scientific research and management of scientific research units, but it also brings hidden dangers in cybersecurity. For example, in the first case after the official implementation of *The Cybersecurity Law of the People's Republic of China* on June 1, 2017, *The Teacher Development Platform* in Cuiping District, Yibin City was attacked by hackers on July 22, 2017. It made teachers' personal information was leaked, causing extremely bad social impact. This shows how important cybersecurity is [1]. The Chinese Academy of Agricultural Sciences is a national agricultural scientific research institution. Its stable development plays a vital role in people's lives and social stability. Therefore, strengthening cybersecurity management in scientific research field is of great significance to promoting informatization construction.

In May 2019, the Chinese government issued the *Cybersecurity technology: baseline for classified protection of cybersecurity, Cybersecurity Technology Classified Protection Evaluation Requirements, Cybersecurity Technology: technical requirements of security design for classified protection of cybersecurity* in three areas of network security. The national standard marks the official arrival of *Cybersecurity Classified Protection 2.0*, and *Cybersecurity Classified Protection 2.0 standard* was officially implemented on December 1, 2019.

The *Cybersecurity Classified Protection 2.0* is an era that complies with the current national requirements for strengthening cybersecurity and combines new technologies and applications such as cloud computing, mobile internet, internet of things, industrial control, and big data to carry out comprehensive management, system supervision, and proactive prevention and control. The following Table 1. Shows the comparison between *Cybersecurity Classified Protection* 1.0 *Standard* and *Cybersecurity Classified Protection 2.0 standard* [2].

As can be seen from the table above, the protection objects of the Cybersecurity Classified Protection 2.0 standard have been expanded from a single information system to include information systems, basic information networks, big data platforms, cloud computing platforms, internet of things, industrial control systems, etc., including all emerging fields in recent years. It constitutes the security requirements of "general security requirements + new application security extension requirements". The classification framework of the technical part is more unified, forming a four-layer protection system architecture supported by security communication network, security area boundary, security computing environment and security management center. Each link strengthens the requirements for the use of trusted computing technology, and gradually proposes the main trust verification requirements for each link. It can be seen that the goal of *Cybersecurity Classified Protection 2.0* is to establish a comprehensive cybersecurity defense system integrating "prevention, protection, management and control" to enhance the overall national cybersecurity defense capabilities, to change passive protection into active protection, and static protection into dynamic protection, to focus on protecting key information infrastructure, important information systems and big data security.

3   Basic Contents of Cybersecurity Classified Protection of Chinese Academy of Agricultural Sciences

With the rapid development of information technology, the Chinese Academy of Agricultural Sciences has accelerated the process of information construction. However, in the application process of information technology, cybersecurity problems are likely to result in theft, destruction, and tampering of important information.

In accordance with the requirements for cybersecurity classified protection, the Chinese Academy of Agricultural Sciences takes important academic business systems such

Table 1. Comparison between Cybersecurity Classified Protection 1.0 standard and Cybersecurity Classified Protection 2.0 standard

	Cybersecurity Classified Protection 1.0 Standard	Cybersecurity Classified Protection 2.0 Standard
Cybersecurity Classified Protection Objects	· Information System	· Basic Information Network · Information System · Big Data Platform · Cloud Computing Platform · Internet of Things · Industrial Control System
Management	· Security Management System · Security Management Organization · Personnel Security Management · System Construction Management · System operation management	· Security Strategy and System · Security Management Organization and Personnel · Security Construction Management · Security Operation Management
Technology	· Physical Security · Cybersecurity · Host Security · Application Security · Data Security and Backup Recovery	· Physical and environmental security · Cyber and Communication Security · Equipment and computing security · Application and data security

as the website of CAAS and the institution-level site group system as the entry point, and carries out website approve, construction and cybersecurity classified evaluation around the academy-level business systems. In the rectification and grade guarantee evaluation work, we fully analyze the characteristics of the business system and solve them one by one starting from the difficulties of security protection [3]. The establishment and improvement of the safe physical environment, safe communication network, safe area boundaries, safe computing environment, safety management center, safety management system, safety management personnel, security organization management, etc. have effectively improved the network and business system security protection capabilities, hidden danger discovery capabilities, and emergency response capabilities, providing reliable guarantee for the healthy development of the informatization of CAAS.

Throughout the above construction and implementation, the Chinese Academy of Agricultural Sciences website system successfully passed the three-level system evaluation under the standard of *Cybersecurity Classified Protection 2.0*, and achieved good results.

The cybersecurity classified protection system of CAAS includes cybersecurity management system and cybersecurity technology.

3.1 Contents of the Cybersecurity Management System

The construction of the cybersecurity management system of CAAS went through three stages of current situation investigation and analysis, planning and design of cybersecurity management system, and implementation of cybersecurity management system, and finally formed the cybersecurity management system of CAAS [4] (Fig. 1).

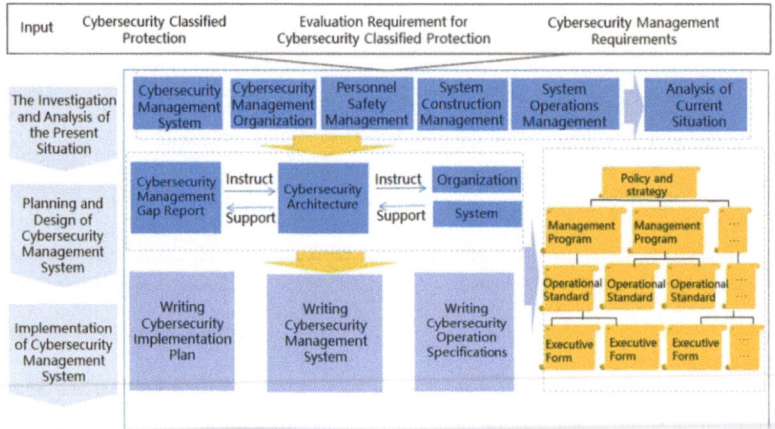

Fig. 1. The Technology Roadmap of Cybersecurity Management System of Chinese Academy of Agricultural Sciences

The cybersecurity management system of CAAS mainly includes the contents of four levels. The first level is a programmatic document, which is the overall strategy in various fields of information security and solves the "why" problem. The second level is specifications, procedures and management methods, which are specific requirements in various fields of information cybersecurity and solve the problem of "what to do". The third level is the rules, guidelines and manuals. It is a detailed rules in various fields of cybersecurity, which solves the problems of "how to do" and "how to achieve". The fourth level is records and forms. It is the trace of the actual implementation results of cybersecurity policies and standards, and solves the problem of "results". See Fig. 2. For details.

The cybersecurity management system of CAAS includes 23 management systems, 4 operating specifications and 34 execution forms. See Table 2.

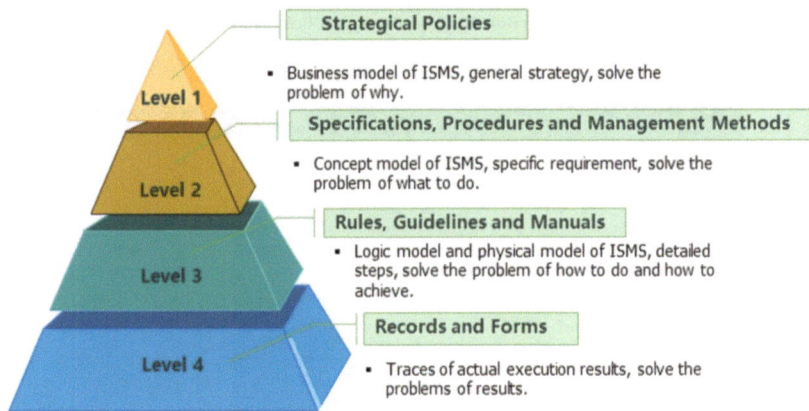

Fig. 2. Composition diagram of the cybersecurity management system of Chinese Academy of Agricultural Sciences

The Chinese Academy of Agricultural Sciences has further strengthened the following through the establishment and implementation of the cybersecurity management system.

1) *Strengthen cybersecurity work policies and strategies*

The Chinese Academy of Agricultural Sciences has further clarified the overall goals and strategies of cybersecurity across the institutions, formulated an overall cybersecurity plan and implementation route, and standardized and refined cybersecurity work guidelines and strategies. Use partition and classified protection for important business systems. Conduct standardized management of the recruitment, transfer and resignation of cybersecurity personnel. Conduct security management of the entire life cycle of system construction, including demand design, system development, testing, and online acceptance, and implement security control on each node during the operation and maintenance process. This effectively ensures the smooth implementation of the cybersecurity work of the Chinese Academy of Agricultural Sciences.

2) *Clarify the main responsibility system for cybersecurity*

The Chinese Academy of Agricultural Sciences has established the cybersecurity and informatization leading group. The main person in charge of the leading group is the leader group of CAAS, and there are management departments under it. Develop an annual cybersecurity work plan every year and ensure its implementation. In line with the principle of "who is in charge who is responsible for, who uses it who is responsible for, and who operates and maintains who is responsible for", clarify responsibilities, establish a cybersecurity responsibility system under the leading party group of CAAS and all units affiliated to the CAAS, and incorporate the cybersecurity responsibility system into the assessment mechanism of each unit will implement a one-vote veto when important cybersecurity incidents occur. The cybersecurity main responsibilities of each unit are consolidated layer by layer.

3) *Regularization of personnel training*

Table 2. Summary table of the cybersecurity management system of Chinese Academy of Agricultural Sciences

File Level	Manage Domains	Number of Files
Level 1	Work Policy and Strategy	1
Level 2	**Document Management System** • Document Management System **Safety Management Agency** • Information Security Organizational System and Responsibilities • Job Responsibility Management System **Personnel Safety Management** • Personnel Safety Management System • Cybersecurity Inspection Management System **System Construction Management** • Product Procurement and Usage Management System • Cybersecurity System Software Development Management System • Coding Safety Management System • Information Project Construction Management System • Cybersecurity Classified Protection Management System **System Operation Management** • Computer Room Safety Management Regulations • Office Environment Management System • Asset Safety Management System	22
Level 3	• Website Content Management System User Manual • Operating System Security Configuration Specifications • Database Security Configuration Specifications • Middleware Security Configuration Specifications	4
Level 4	• Safety Management Inspection Form • Safety Skills and Safety Awareness Assessment Records	34

Cybersecurity classic protection requires absolutely professional technical personnel to carry out it continuously for a long time. Currently, the personnel engaged in cybersecurity work of CAAS have lack of work experience and skills. Therefore, develop a training plan every year to provide cybersecurity-related personnel with knowledge and skills training. The cybersecurity-related personnel can master the corresponding skills as soon as possible and improve the unit's cybersecurity capabilities.

4) *Cybersecurity incident emergency response*

Facing the increasingly severe cybersecurity environment, it is very necessary to formulate contingency plans for cybersecurity emergencies in advance. The Chinese Academy of Agricultural Sciences adopts a combination of prevention and treatment, focusing on prevention. According to the length of system interruption, scope of impact, and the severity of data loss or tampering or theft that poses a threat to national and social stability or economic losses, CAAS adopts the event classification. Strengthening risk investigation, making various emergency response plans in advance, strictly implementing the monitoring and duty system, ensure detection of faults in time and 7*24-h emergency response.

4 Cybersecurity Technology Content

4.1 Cyber Data Security

The long-term accumulation of data by CAAS has become one of the most valuable assets of the organization, so important data needs to be protected [5]. The key business system data of CAAS shall be backed up at least once every working day, and the backup data shall be retained for no less than six months. Continuously improve the data backup and recovery mechanism, regularly verify the validity of backup data, ensure the integrity and availability of data, and avoid irreparable losses due to data loss. However, due to limited conditions, data is not backed up in the same city or off-site.

4.2 Business Application Security

From the perspective of business needs and system security, enable identity and authentication modules for network equipment, security equipment, servers and databases involved in important systems, and develop application access control policies and access permissions. Repair system vulnerability timely and conduct monitoring and security audits of important systems. Adhere to the principle of least authorization for business system administrator permissions. Conduct vulnerability scanning monthly and patch security vulnerabilities timely. To avoid direct installation in the formal environment and ensure the security of business applications, install security patches in the test environment and pass the test.

4.3 Communication Cybersecurity

The communication cybersecurity of CAAS has the characteristics of large coverage area and many nodes, which is the top priority of the informatization work. According to business conditions, different cyber segments have been divided, and important cyber segments have been protected [6]. Core switches, protecting walls, front servers of important business system, and database servers are adopted by redundant designs. Use VPN to log in remotely from the external network to access internal network resources to ensure the security of data during transmission. Implement 7*24-h network monitoring, keep log records of important network nodes for at least 6 months, regularly analyze logs and handle any abnormalities in time. Ultimately, improve cybersecurity and operational efficiency [7].

4.4 Computing Environment Security

The Chinese Academy of Agricultural Sciences sets passwords for all servers and changes the passwords every three months. The passwords must be complex. Regularly conduct network anti-virus and vulnerability scanning, and close the unnecessary service ports. Conduct security configurations in accordance with the minimum service principle, and enable corresponding security audit functions and monitor servers to ensure their safe operation. All devices and operating systems are remotely managed by using security protocols.

4.5 Regional Border Security

The Chinese Academy of Agricultural Sciences adopts different protection strategies for different cyber areas in accordance with the requirements of graded protection. For example, IPS devices and WAF devices are arranged in the system area of level-3 cybersecurity classified protection to monitor intrusion behaviors; WEB application firewalls are arranged to formulate reasonable cybersecurity access strategies [8]; necessary bastion machines, security audit equipment, security management platform, log audits and other cybersecurity equipment are set up, so that the high availability of cybersecurity in the system area of level-3 cybersecurity classified protection get better guaranteed.

4.6 Physical Environment Security

The physical environment of the core network computer room of CAAS has the requirements of dual-channel power supply, UPS power supply, away from water sources and electromagnetic interference, and meets the B-class computer room standards. The core network computer room is well monitored, alarmed, fire-fighting, and protected against lightning strikes, anti-static, power supply protection to ensure the safety of sites. At the same time, strict duty management of the computer room is set up, with dedicated personnel on duty. Strictly establish corresponding access control and access control measures to register and supervise the entry and exit of personnel to provide a reliable physical environment for the stable operation of the business system.

5 Effective Strategic Suggestions for Strengthening Cybersecurity Construction in Scientific Research Institutions

The research and construction of cybersecurity classified protection of CAAS has general reference significance for the cybersecurity construction of agricultural scientific research units. Therefore, the following strategic suggestions are put forward for the construction of cybersecurity classified protection.

5.1 Improve the Safety Management Organization and Management System

Establishing and improving security management institutions and strictly implementing management systems play a ballast role in ensuring the cybersecurity of scientific research institutions. The cybersecurity management organization of scientific research institutions consists of decision-making body, management body and executive body. The establishment of this 3-layer security management organization is conducive to clarify the corresponding responsibilities of each institution and each position step by step, and consolidating the main responsibilities. At the same time, scientific research institutions must establish and improve security management systems, and strictly organize and implement them. The above two aspects are of great significance to promote the cybersecurity construction of scientific research institutions [9].

5.2   Increase Investment in Cybersecurity

Over the years, agricultural scientific research institutions have invested very little money in cybersecurity protection, resulting in security risks. When planning and building information systems, security protection measures should be based on the three synchronizations principle, which is synchronous planning, synchronous construction and synchronous operation with the construction of information systems. Cybersecurity protection measures should be implemented throughout.

5.3   Strengthen the Introduction and Training of Professional Talents

Cybersecurity management requires quite professional technical talents. Therefore, scientific research institutions must do the following when introducing and training talents. First of all, strictly conduct qualification reviews for the introduction of professional and technical talents. Key positions should have qualifications such as CISP or PMP certificate that can prove their business capabilities [10]. After deciding on employment, a confidentiality agreement must be signed with the institutions. When resigning, all access rights, certificates, software and hardware equipment used must be recovered timely. Secondly, scientific research institutions should conduct training and assessment for business personnel at least once a year to improve the safety skills and safety awareness of business personnel gradually.

5.4   Strengthen Scientific Research Data Security Protection

The data accumulated over a long period of time by scientific research institutions has become one of the most valuable assets of the institution. Agricultural scientific research data are mostly composed of experimental data, monitoring data, image data, etc., which can easily be hundreds of gigabytes or even tens of terabytes. The research team does not have that much space to back up, so it is basically stored on the research team's server, which poses a great security risk. Formal data backup mostly uses database backup, network data backup, tape backup and mirror backup. Based on the importance of the data, the data backup strategy which is suitable for the institutions should be formulate to provide hardware redundancy for important network equipment, communication lines and servers. Prevent data loss due to system failure and irreparable losses.

6   Summary

In the era of big data, agricultural scientific research institutions in China are rapidly moving towards informatization, and at the same time they are also facing severe challenges in cybersecurity. Agricultural scientific research institutions are gradually paying more attention to cybersecurity classified protection, constantly summing up experience from practice, and further improving their own cybersecurity protection measures, thus significantly improving the overall standards of agricultural scientific research cybersecurity in China.

Acknowledgment. The authors gratefully acknowledge the financial supports by Research on the Long Term Monitoring, Analysis and Communication Models of National Agricultural Science and Technology Innovation Alliance based on All Media and Network's Big Data (grant no. Y2024JC12) and the National Key Research and Development Program of China (grant no. 2021ZD0113701).

References

1. Yang, C., Xu, W., Xia, P.: Security design of information system based on network security level protection. Modern Industrial Economy and Informationization (11), 68–69 (2019)
2. He, Z., Wang, Y., Liu, J.: Research on the status and 2.0 standard system of network security classified protection in China. Cyber Security And Data Governance (3), 9–19 (2019)
3. Luan, Q., Zhang, Z., Wu, J.: Evaluation and rectification of university information system based on classified protection. Office Informatization (4), 26–28 (2019)
4. Zhou, D.: The scheme of campus network security planning based on hierarchical protection. J. Fujian Computer **3**, 126–128 (2019)
5. Yuan, H.: Research and discussion on network security level protection 2.0 system. China Computer & Communication (1), 223–224 (2020)
6. Zhang, Z., Zhang, Z., Wang, R.: Model of cloud computing security and compliance capability for classified protection of cybersecurity 2.0. Netinfo Security (11), 1–7 (2019)
7. Yan, Z., et al.: Research on blockchain evaluation methods under the classified protection of cybersecurity. Chinese J. Eng. **10**, 1267–1285 (2020)
8. Feng, K., Zhang, D., Chen, X., Wang, H.: Railway network security level protection management system. Design and Dev. **8**, 66–70 (2020)
9. Zhang, Y., Ma, Y., Si, Q.: Research on key technology of security risk assessment based on classified cybersecurity protection idea. Classified Protection 2.0 (8), 28–32 (2020)
10. Zhang, X., Lu, L., Luo, H.: Research on the construction of network security level protection in universities. Computer Knowledge and Technol. (22), 71–73 (2020)

Research on DOA Estimation Method of Underwater Acoustic Signal Based on Machine Learning

Feiyu Zhao[✉]

Beijing Institute of Technology, Beijing, China
newemailforarticle@163.com

Abstract. This project is concerned with the development of a machine learning-based DOA estimation of underwater acoustic signals with the view of improving the efficiency and reliability of the signal processing in complex underwater environments. The feature analysis and classification of the underwater acoustic signals are done with the help of convolutional neural networks (CNNs). Through multi-step purification of the signal data and lowering the sampling frequency of the pre-processed signal data, a high-performance arrival angle assessment framework is constructed. The results indicate that the algorithm achieves high accuracy, recall and F1 in different noise intensity levels; particularly, low noise environment with an accuracy of 95%. 2%. The application verification of practical application also confirms that this technology has strong applicability and stability in complex seabed conditions.

Keywords: acoustic signal in underwater environment · direction of arrival estimation · convolutional neural network · machine learning

1 Introduction

As the machine learning technology continues to improve, especially in deep understanding, new methods have been offered to solve the nonlinear and complex data types problem for underwater acoustic signal DOA estimation. Machine learning can learn the arrival angle when environment is unknown by analyzing the characteristic and architecture of a large data set [1]. For instance, the application of deep machine learning algorithms like convolutional neural networks (CNN) and recurrent neural networks (RNN) in image processing and time series data analysis shows that it can be used in underwater acoustic signal processing. Studying the method of underwater acoustic signal arrival angle estimation through machine learning is not only conducive to enhancing the precision and anti-interference capability of the directional arrival angle judgment but also can open up new research perspectives and technological advancements for underwater acoustic signal processing [2]. The purpose of this research is to propose a method of DOA estimation of underwater acoustic signals using machine learning. Through the detailed analysis of the main links including data preprocessing, feature

P. Siarry et al. (Eds.): WCNA 2023, LNEE 1361, pp. 103–111, 2025.
https://doi.org/10.1007/978-981-96-2409-6_11

extraction, selection and optimization of the model, the application of the technology is systematically explained. In this case, the advantages of the proposed method are illustrated through the use of specific examples, and the applicability of this method in complex underwater scenes is shown, and the effectiveness of this method is proved by the use of effect confirmation.

2 Theoretical Overview

2.1 Basic Concepts of Underwater Acoustic Signals and DOA Estimation

Hydroacoustic signals refer to the acoustic wave patterns that diffuse in water. They are widely used in many fields such as ocean exploration, fishery resource surveys, underwater communications, and military reconnaissance. Due to the various forms of the underwater world, the propagation path of hydroacoustic signals will be interfered by many factors, such as water temperature, salinity, depth changes, and the influence of ocean noise. These factors make the propagation path of hydroacoustic signals complex and changeable [3]. Sound source localization refers to the use of signal processing methods to infer the position coordinates of the sound source relative to the microphone group. Positioning perspective estimation plays a key role in underwater positioning and navigation. Its accuracy directly affects the underwater instruments and their performance. Traditional arrival angle quantization methods mainly include beam tracking, characteristic evaluation, time interval estimation, etc. These methods are relatively mature in theory, but in practical contexts, especially when dealing with complex underwater conditions, they often encounter problems such as insufficient accuracy and heavy computational burden. In order to find solutions to these problems, in recent years, the direction angle estimation technology based on machine learning has gradually attracted the attention of the research community. Machine learning significantly improves the accuracy and robustness of azimuth prediction by understanding the characteristics and patterns of data in large data sets [4]. In particular, with the support of deep learning technology, it is able to effectively perform DOA estimation in complex underwater environments, significantly enhancing the capabilities and application effectiveness of underwater acoustic signal processing.

2.2 Basic Theory of Machine Learning

2.2.1 Supervised Learning and Unsupervised Learning

In machine learning, supervised learning and unsupervised learning play two crucial learning modes, dealing with various challenging problems respectively. Guided exploration (example a) belongs to a type of guided learning method, which inherently maps all incoming information to exact results. The algorithm process relies on learning the correlation between input and output to achieve the prediction value of new data. Supervision training mainly focuses on training with rich labeled data sets. General techniques include linear regression analysis, support vector machine (SVM), decision tree and artificial neural network. This technique is widely used in classification and prediction problems, such as image recognition, voiceprint recognition and medical judgment [5].

In contrast, unsupervised learning (see Figure b) belongs to a learning strategy that does not require guidance. It aims to explore the potential structure and rules of unlabeled data sets. The goal of the self-guided learning process is to discover the implicit features or stable structures of the data by parsing unclassified information [2]. These strategies are applied to classification, dimensionality simplification and anomaly recognition, showing significant advantages in these fields. The broad spontaneous learning computational framework covers clustering algorithms (such as K-means, hierarchical clustering), principal component analysis (MPA) and independent factor analysis (IFA), and unsupervised training is widely used in a wide range of fields such as market definition, image improvement and gene interpretation expression [6] (Fig. 1).

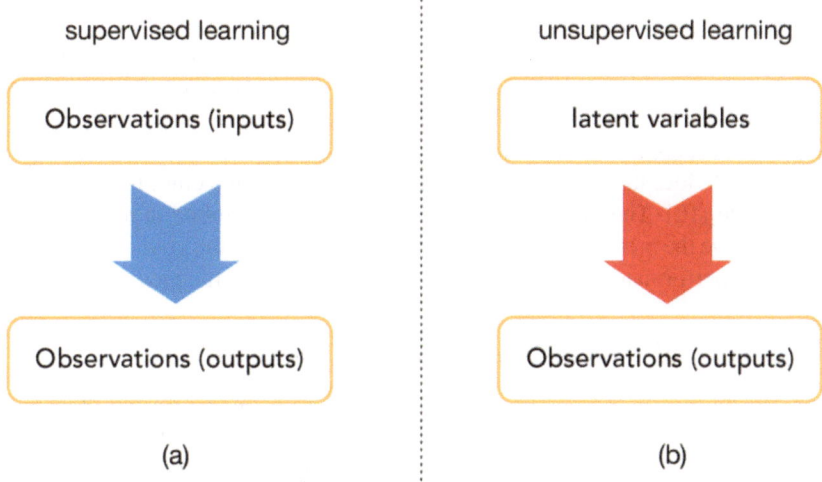

Fig. 1. The difference between supervised learning and unsupervised learning

2.2.2 Typical Algorithms

In the study of DOA estimation of underwater acoustic signals, machine learning algorithms provide solutions to the challenges of prediction and evaluation of direction in changing environments and achieve efficient strategy implementation [7]. As a highly machine learning architecture, convolutional neural network (CNN) shows excellent image and sequence data processing capabilities and is now widely used for feature extraction of underwater acoustic signals for analysis and processing. Convolutional neural network uses multi-scale convolution kernels to implement local recognition and feature abstraction of input information streams, and has the ability to accurately capture the spatial attributes of signals and enhance the accuracy and robustness of arrival angle estimation. Recurrent neural network (RNN) and long short-term memory network (LSTM) show unique advantages in processing sequence data. These two networks can remember and use information with long time spans, so as to better deal with the temporal correlation and multipath effects in underwater acoustic signals [8]. The optimal margin classifier (maximum margin machine) is a classification algorithm designed based on statistical theory. It uses the construction of the optimal classification hyperplane to complete the accurate classification of various signal samples. Support vector machine performs well in dealing with the challenges of complex data sets and scarce samples, and is suitable for the prediction of direction angles of underwater acoustic signals in high-latitude feature spaces [9]. The similarity-based evaluation method judges the proximity between the test data point and each data point in the training sample set, selects the nearest K matching sample points and applies the majority voting principle to determine the result. The KNN algorithm is easy to use and is suitable for the azimuth evaluation of small data sets, but the computational cost increases in the context of large-scale data sets. Classification and regression decision trees and their integrated methods (such as random decision forests and gradient boosting machines) perform hierarchical cutting of variables based on multiple evaluation criteria, and then complete the detailed interpretation of classification and regression analysis. These methods have significant explanatory power and robustness, and are applied to complex decision problems within the estimation of the arrival angle of underwater acoustic signals. Specifically, the decision model continuously divides the data set and establishes a model architecture for predicting the target variable [10]. The random decision forest relies on combining many tree structures to improve the prediction accuracy and robustness, while the gradient boosting tree enhances the performance of the model by gradually reducing the prediction error [11].

2.3 Application of Machine Learning in Signal Processing

Machine learning is widely and deeply applied in information processing. With the help of a series of algorithms, complex signal data is converted into useful information, and then efficient classification and prediction are completed. The figure shows a set of standard machine learning signal processing processes, covering four stages: data collection, pre-processing, feature analysis and classification evaluation. The initial stage is information collection, which covers the collection of basic signal data from many sensors or devices. The second stage is preliminary processing, which uses technologies such as screening information, ensuring uniformity and standardization to purify and standardize the initial data for further analysis. The key link is key data mining, which extracts core information from pre-processed data based on quantitative standards (covering maximum value, minimum value, mean, sum and integral, etc.). These features reflect the key attributes and patterns of the signal and are an important basis for the classifier to effectively learn and distinguish signals. The fourth stage is classification evaluation, which usually uses supervised learning methods such as support vector machine (SVM) for classification. The image reveals the working process of the SVM classifier. After the training stage, it is capable of performing verification stages to perform classification tasks on new types of data and determine whether the input data belongs to a specific category, such as distinguishing gas state and air pressure in the field of olfactory recognition [12].

By adopting this orderly signal processing method, machine learning can efficiently process complex signals and complete accurate classification and prediction [13]. Especially in the DOA estimation of underwater acoustic signals, these technical means can be used to extract key attributes, and then use advanced machine learning algorithms to accurately predict the target direction, thereby enhancing the accuracy and reliability of underwater target detection and positioning.

3 DOA Estimation Method of Underwater Acoustic Signal Based on Machine Learning

3.1 Application Case Background

During the marine resource exploration mission, several underwater sound sources must be accurately located in the designated area to assist in monitoring the state of the ocean and evaluating the resources. Therefore, the research team built an array of 64 underwater sound receivers with a diameter of two meters and a frequency range of 10 Hz to 20 kHz. It is located in the North Atlantic region, with a temperature range of 4 to 18 degrees Celsius, a salinity of 34.5 PSU, and a water depth of 200 to 800 m, accompanied by significant background sound and multipath effects. The data acquisition module ran continuously for 96 h and collected more than two TB of first-hand underwater acoustic data, which included complex multipath interference, noise, and signal attenuation. These basic premises indicate that in complex underwater environments, it is urgent to apply machine learning-based DOA estimation methods to enhance the accuracy and robustness of underwater acoustic signal processing, thereby ensuring accurate identification of underwater sound source locations in a changing environment.

3.2 Demand Analysis

During this period, the above-mentioned marine resource exploration mission was carried out. Given the many challenges such as the dramatic change in water depth in the target area, significant environmental noise, and the challenging sound wave reflection, the traditional wave direction inference method could not meet the needs of precise positioning. In detail, the 2 TB of raw audio data obtained covered frequency fluctuations in the frequency range of 10 Hz to 20 kHz. In an environment where the noise level exceeded the 50 dB threshold, it was necessary to accurately track various acoustic origins scattered in the waters at a depth of 200 m to 800 m. In addition, the information transmission encountered interference factors caused by temperature fluctuations (4 degrees Celsius to 18 degrees Celsius), salinity changes (34.5 PSU), and signal reflection and refraction during the transmission process, making the signal characteristics complex and changeable. In order to achieve high-precision vertex positioning accuracy assessment in such a changing background, it is necessary to use machine learning-based methods to abstractly extract signal features and structures to improve the accuracy and robustness of positioning. In particular, it is necessary to use advanced algorithms that can process large amounts of data, overcome the influence of multiple paths and noise environments, to ensure the accurate positioning of underwater sound bodies in actual applications, thereby promoting the effective development of seabed resources and environmental supervision.

3.3 Data Preprocessing and Feature Extraction

Before performing the DOA estimation of the underwater acoustic signal, it is necessary to perform comprehensive data cleaning and feature analysis on the 2 TB of initial underwater acoustic data. In the initial stage, a band-limited filter is used to limit the signal to the critical frequency band of 10 Hz to 20 kHz to eliminate unnecessary frequency interference. Then, wavelet transform technology is used to perform noise reduction processing to remove high-frequency noise interference. Data normalization is the next step. By performing normalization operations on the data, multi-dimensional consistency verification of the data is guaranteed. For feature capture, it mainly relies on short-time Fourier transform (STFT) to convert dynamic data into time-frequency diagrams and identify frequency components. The formula is as follows:

$$X(t,f) = \int_{-\infty}^{+\infty} x(\tau)w(\tau - t)e^{-j2\pi f\tau} d\tau$$

Among them, $x(\tau)$ is the original signal, $w(\tau - t)$ is the window function, and $X(t,f)$ is the time-frequency domain representation. In addition, the spectral features of the signal are further extracted using Mel-frequency cepstral coefficients (MFCCs). These features have good performance in underwater acoustic signal processing. The formula is as follows:

$$MFCC(n) = \sum_{k=1}^{K} \log(|X(k)|) \cos\left(n(k - 0.5)\frac{\pi}{K}\right)$$

Among them, $|X(k)|$ is the amplitude spectrum after Fourier transform, K is the number of Mel filter banks, and n is the serial number of the cepstrum coefficient.

3.4 Machine Learning Model Construction and Training

In the paper, it is decided to use convolutional neural network (CNN) as the core architecture for building and training the DOA estimation model of underwater acoustic signals [14]. In the initial stage, the preprocessed data set is fed into the convolutional neural network model. The information reception stage of the model carries the spatiotemporal data structure, followed by multiple convolution processes to achieve feature extraction. The filter element scale used in the filter layer is 3D with a step size of 1 and a padding method of "same" to maintain the stability of the feature map size. The convolution operation formula is as follows:

$$(X * W)(i,j) = \sum_{m=0}^{M-1}\sum_{n=0}^{N-1} X(i+m, j+n)W(m, n)$$

Among them, X is the input feature map, W is the convolution kernel, (i,j) is the output position, M and N are the height and width of the convolution kernel respectively. Next, the pooling layer (such as maximum pooling) is used for dimensionality reduction and feature selection. The pooling window size is 2×2 and the step size is 2. The formula is as follows:

$$Y(i,j) = \max_{0\leq m<2, 0\leq n<2} X(2i+m, 2j+n)$$

Among them, Y is the feature map after pooling, and X is the input feature map.

After feature extraction and scale adjustment, the feature sequence is reduced to a series of sequences and sent to the densely connected network to perform the recognition task. The output scale of the full-link layer is consistent with the type of output direction. The softmax function is used to calculate the probability distribution of each output direction. The formula is:

$$\sigma(z)_j = \frac{e^{z_j}}{\sum_{k=1}^{K} e^{z_k}}$$

Among them, z is the output vector of the fully connected layer, K is the number of categories, and $\sigma(z)_j$ is the probability of the jth category.

The model training uses the cross entropy loss function and the Adam optimization algorithm for iterative updates, and the training set and the validation set are divided in a ratio of 80:20. The formula of the cross entropy loss function is as follows:

$$L = -\sum_{i=1}^{N} y_i \log(\hat{y}_i)$$

Among them, y_i is the true label, $\hat{y}_i$ is the predicted probability, and N is the number of samples. Through multiple rounds of iterative training, the model gradually optimizes its parameters to improve the accuracy of DOA estimation on the validation set.

4   Effect Evaluation

4.1   Results and Analysis

After the model training phase, the performance of the model was evaluated using the validation dataset. The following table shows the accuracy, true positive rate, and F1 score during the direction angle prediction challenge under different noise levels.

Table 1. Analysis of model evaluation results

Noise level (dB)	Number of samples	Accuracy(%)	Recall rate (%)	F1 value (%)
0	500	95.2	94.8	95.0
10	500	93.6	93.2	93.4
20	500	90.4	89.9	90.2
30	500	87.1	86.5	86.8
40	500	83.3	82.7	83.0
50	500	78.5	77.9	78.2

As shown in Table 1, when the noise increases, the accuracy, retrieval rate and F1 score of the model decrease simultaneously. Even so, even in the noise background

(50dB), the calculation algorithm still maintains excellent performance parameters, with an accuracy of 78.5%, a true positive rate of 77.9%, and an F1 value of 78.2% for precision and recall, which shows that the directional angle prediction technology based on the convolutional neural network shows excellent adaptability and practical value in the noise-affected environment. Especially under quiet acoustic conditions (0 dB and 10 dB), the model shows excellent prediction accuracy, up to 95.2% and 93.6% respectively. In-depth analysis shows that the algorithm design has significant processing performance in the field of multipath attenuation of signals, and has the characteristics of accurate signal capture in complex underwater environments, and can perform accurate path prediction.

4.2   Performance of the Model in Practical Applications

The underwater acoustic signal directional positioning evaluation algorithm based on convolutional neural network shows excellent performance and strong adaptability in the actual application environment. The test results in the validation set show that the model maintains a high accuracy, recall rate and F1 value under various noise levels, especially in the environment with low noise level, the accuracy rate reaches 95.2%. In the implementation stage, this model is applied to the marine exploration mission in the North Atlantic Ocean to accurately locate multiple underwater sound sources. During the execution of the mission, the algorithm architecture is in the waters of 200 m to 800 m deep, and the accuracy level of the sound source positioning function maintains an accuracy of 87.1% in a noise environment of up to 30 decibels, confirming and demonstrating its stability in the actual environment. After three days of data monitoring and evaluation, artificial intelligence accurately identifies most of the preset sound sources, significantly improving work efficiency and accuracy. This data reveals that the machine learning-based directional angle estimation technology has the effect of showing in actual applications under complex sea conditions, providing a solid technical support for the environmental monitoring task of marine resource exploration operations.

5   Conclusion

This study has effectively verified the effectiveness and performance of high-precision horizontal positioning in complex underwater environments by creating a model structure for estimating the direction angle of underwater acoustic signals based on convolutional neural networks. The experimental results show that the model can maintain high accuracy, recall rate and F1 value under various noise levels, especially in low-noise environments. The test of practical application further proves the stability and flexibility of the technology in complex marine environments, greatly improving the efficiency and accuracy of marine exploration operations. This research has opened up a new technical path for the field of underwater acoustic signal processing, and at the same time revealed the huge application prospects of machine learning in solving complex signal processing challenges. Future academic exploration will continue to promote model effectiveness and develop more cutting-edge machine learning algorithms to improve the accuracy and application scope of prediction directions and promote underwater acoustic signal processing technology to improve its practical value.

References

1. Li, X., Chen, J., Bai, J., et al.: Deep learning-based DOA estimation using CRNN for underwater acoustic arrays. Front. Mar. Sci. **9**, 1027830 (2022)
2. Li, P., Tian, Y.: DOA estimation of underwater acoustic signals based on deep learning. In: 2021 2nd International Seminar on Artificial Intelligence, Networking and Information Technology (AINIT). IEEE, pp. 221–225 (2021)
3. Liu, Y., Chen, H., Wang, B.: DOA estimation based on CNN for underwater acoustic array. Appl. Acoust. **172**, 107594 (2021)
4. Xie, Y., Wang, B.: Data-driven DOA estimation methods based on deep learning for underwater acoustic vector sensor array. Mar. Technol. Soc. J. **57**(3), 16–29 (2023)
5. Liu, Y., Chen, H., Wang, B.: DOA estimation of underwater acoustic signals based on PCA-kNN algorithm. In: 2020 International Conference on Computer Information and Big Data Applications (CIBDA). IEEE, pp. 486–490 (2020)
6. Peng, J., Nie, W., Li, T., et al.: An end-to-end DOA estimation method based on deep learning for underwater acoustic array. In: OCEANS 2022, Hampton Roads. IEEE, pp. 1–6 (2022)
7. Xie, Y., Wang, B.: Direction-of-arrival estimation method based on neural network with temporal structure for underwater acoustic vector sensor array. Sensors **23**(10), 4919 (2023)
8. Yang, H., Lee, K., Choo, Y., et al.: Underwater acoustic research trends with machine learning: general background. J. Ocean Eng. Technol. **34**(2), 147–154 (2020)
9. Jiao, L., Yang, X., Quan, T., et al.: High-precision DOA estimation for underwater acoustic signals based on sparsity adaptation. Front. Mar. Sci. **9**, 1022494 (2022)
10. Lan, C., Chen, H., Zhang, L., et al.: Underwater Acoustic DOA Estimation of incoherent signal based on improved GA-MUSIC. IEEE Access (2023)
11. Cao, H., Wang, W., Su, L., et al.: Deep transfer learning for underwater direction of arrival using one vector sensor. J. Acoustical Society of America **149**(3), 1699–1711 (2021)
12. Nie, W., Zhang, X., Xu, J., et al.: Adaptive direction-of-arrival estimation using deep neural network in marine acoustic environment. IEEE Sensors Journal (2023)
13. Wu, Y., Li, X., Cao, Z.: A novel CCA-NMF whitening method for practical machine learning based underwater direction of arrival estimation. J. Beijing Institute of Technol. **33**(2), 163–174 (2024)
14. Ahmed, N., Wang, H., Raja, M.A.Z., et al.: Performance analysis of efficient computing techniques for direction of arrival estimation of underwater multi targets. IEEE Access **9**, 33284–33298 (2021)

Design and Implementation of Smart Home Control System Based on STM32

Lang Gao and Guo Li[✉]

College of Computer Science and Engineering, Guangxi Normal University, Guilin 541004, China
liguo@gxnu.edu.cn

Abstract. With the rapid development of Internet of Things (IoT) technology, smart home control has become an indispensable part of modern life. This study designs a smart home control system based on the STM32 microcontroller, combined with Tencent CloudBase and WiFi communication technology, to achieve remote control of home devices and real-time monitoring of home environments. The system uses the MQTT protocol to transmit data and send control commands. At the same time, a WeChat mini-program is developed as the client for controlling devices. The feasibility of the system has been verified through testing.

Keywords: STM32 microcontroller · Tencent CloudBase · WiFi · smart home · MQTT

1 Introduction

With the rapid development of technologies such as the Internet of Things and cloud computing, there is a growing demand for quality of life and convenience. Smart home devices such as robotic vacuum cleaners, automatic dishwashers, and smart speakers have become part of modern life due to their safety, convenience, energy efficiency, and automation features, providing people with a convenient, comfortable, and safe living environment [1]. According to reports [2], by 2025, 21.3% of households globally will use smart home devices, demonstrating the rapid growth trend of the smart home market. However, traditional smart home control systems currently face problems such as high cost, poor user experience, low penetration rate, and complexity. At the same time, many people still rely on manual mechanical control of smart homes, failing to fully utilize the remote-control capabilities of these devices.

The study describes the design of a smart home control WeChat mini-program based on the STM32 microcontroller. This program aims to remotely control multiple smart home devices via a mobile phone and monitor the home environment without the need to download additional applications, providing a convenient user experience. The mini-program achieves remote monitoring and control of home devices through sensor detection and user control requirements, thereby achieving functions such as [3] automation, reliability, and convenience. The main design features include:

© The Author(s) 2025
P. Siarry et al. (Eds.): WCNA 2023, LNEE 1361, pp. 112–125, 2025.
https://doi.org/10.1007/978-981-96-2409-6_12

- Utilize MQTT as the server to achieve bidirectional information transmission between the microcontroller and the mini-program.
- Implement automatic control of different devices based on user requirements and sensor monitoring information.
- Implement the functionality of manually controlling device status on the WeChat mini-program side.
- Implement the functionality of storing information and operation records of different users through CloudBase.

2   System Analysis and Designe

2.1   System Requirements Analysis

In order to design a better smart home control system, research and analyze the requirements of this realization system from different perspectives to determine the functions that the system should achieve [4]. Here, we focus on functionality, performance, and security considerations.

Functional Requirements Analysis
The system should consider the implementation of the following features:

1) Automation control and manual control [5, 6]: *Automation control involves the automatic operation of devices, such as alarms, based on environmental information gathered by sensors. Users need to control the on and off of devices manually through manual operations on a mobile terminal.*
2) Remote monitoring of home device and environmental status [7]: *Users expect to monitor home device status in real-time, so the system should enable viewing of home device status and environmental information such as light intensity and temperature on a mobile terminal.*
3) Data storage and management: *Users want data to be stored for easy access and recording at any time. At the same time, the system needs to be able to manage the user's operation records and home equipment status information, and differentiate the data information of different users.*
4) Visual interface: *The system should use a visual interface to display information about users and devices in a visual way. It is also necessary to keep the interface simple as well as to meet the demand for easy operation.*

Performance Requirements Analysis
The performance that the system should achieve includes:

1) Real-time performance: T*he system should achieve the performance of real-time updating of home equipment status, environmental status and operation records to ensure that users can access real-time information.*
2) Stability and reliability: *The system needs to operate stably and reliably and should avoid failures and interruptions due to network and other factors as much as possible.*
3) Security requirements analysis

4) The requirements of security are also part of the system that needs to be considered.
5) User authentication: *In order to ensure the legitimacy of users' identities, they must complete user authentication prior to accessing the system; otherwise, they will not be granted full usage permissions.*
6) Privacy protection: *Throughout the user's interaction with the system, it is crucial to ensure the proper safeguarding of their privacy. This involves segregating individual users' account information, device specifics, and activity logs to prevent any unauthorized disclosure of such data.*

2.2 System Framework Design

1) System architecture overview
2) The system architecture consists of three basic modules:

The smart home mini-program, the microcontroller, and the MQTT server [8]. For a comprehensive architectural lay out, see Fig.1. Assuming that the mini-program is successfully connected to the Internet and the microcontroller is networked through the WiFi module, these components are able to perform bi-directional data transfers with the server through the MQTT communication protocol [9].

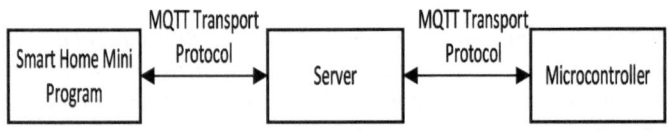

Fig. 1. System diagram

2.3 System Module Composition

Mini-Program Module
Both the front-end interface and back-end services play an integral role in the development of small programs. We use WeChat development tools to create the front-end interface to ensure that the design is user-friendly and enhances the overall user experience. For backend development, we utilize WeChat cloud development services to manage complex business logic, enable data transfer, and securely store key information such as user profiles and operation logs. Additionally, the backend leverages servers in conjunction with the MQTT protocol to transmit control instructions from the application to the microcontroller, effectively establishing communication between the two entities.

Server Module
The system adopts EMQX, an open-source message broker platform, as the core of its message proxy service, ensuring bidirectional transmission of information between

mini-programs and microcontroller endpoints[10]. Leveraging the built-in MQTT communication protocol interface of EMQX, the system efficiently facilitates stable and high-speed information exchange, thereby establishing a robust foundation for the system's reliable operation.

Microcontroller Module

At the microcontroller level, the STM32 microcontroller has been selected as the central component, with its primary function being the acquisition and display of environmental parameter data such as temperature and humidity. Moreover, by incorporating a WiFi module, wireless network connectivity is enabled for the microcontroller, facilitating the reception of device control commands transmitted from the app to the server. Leveraging rapid and reliable communication with the server, the microcontroller can efficiently achieve real-time monitoring of smart home devices' status and environmental parameters, alongside swift implementation of responsive actions.

Overall, the collaborative work of these three modules constitutes the core structure of the system, providing users with a feature-rich and stable smart home management platform. The system architecture diagram is shown in Fig. 2.

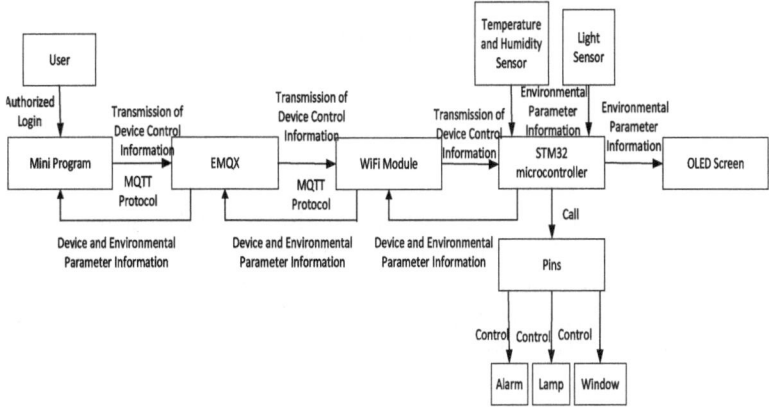

Fig. 2. System architecture diagram

3   The Hardware of the System

3.1   The Choice of Microcontroller

The choice of microcontroller is the most crucial part of the entire hardware design. In this system, we have opted for the STM32F103C8T6, which is the minimum system board of the STM32 series.

STM32F103C8T6 is indeed a powerful ARM Cortex-M3 microcontroller, with a maximum operating frequency of 72MHz, providing strong computing capability and responsiveness. It boasts a rich set of peripherals and powerful processing capabilities. Its peripherals include[11] multiple general-purpose timers, serial interfaces, SPI interfaces, I2C interfaces, etc., facilitating the connection of various sensors, actuators, and external devices. Below is a photo of the assembled STM32F103C8T6 for your reference (Fig. 3).

Fig. 3. STM32 microcontroller physical diagram

3.2 Display Screen

A 128 × 64 pixel display screen is used as the display for the microcontroller. The microcontroller can send data to the display screen, controlling the displayed content and refresh rate. In this system, the display screen is mainly used to show external environmental data collected by the microcontroller, such as temperature, humidity, WiFi connection status, etc. It provides clear and intuitive information display for the device, enhancing user experience and operational convenience. The effect of the display screen is shown in Fig. 4.

3.3 Sensor

During the entire system design process, temperature and humidity sensors, light sensors, and raindrop sensors were selected to monitor the indoor environmental conditions of the home.

1) Temperature and humidity sensor

The temperature and humidity sensor is a crucial component used to detect the ambient temperature and humidity. The system employs the DHT11 [12] as the temperature and humidity sensor due to its low power consumption, small size, and ease of use. The temperature and humidity parameters obtained from the sensor can influence the opening and closing of the alarm. If the temperature is too high or the humidity is too high, the alarm will respond immediately.

Fig. 4. Display screen display effect diagram

2) Light sensor

The BH1750 is a digital light intensity sensor capable of directly outputting digital signals, eliminating the need for complex calculation steps. It boasts high precision, a wide measurement range, and fast response times. The BH1750 communicates with microcontrollers via the I2C bus, enabling real-time monitoring of ambient light intensity. Ambient light intensity is utilized in the system to control the automatic on/off operation of lighting fixtures.

3) Raindrop sensor

The raindrop sensor is utilized to detect rain. In the system, it primarily serves the function of automatically closing windows when it rains. Upon detecting rainwater, the system can trigger corresponding automatic control actions, such as controlling a stepper motor to rotate in reverse, simulating window closure.

4) Stepper motor

For precise control of the window mechanism, a five-wire, four-phase 12V stepper motor was chosen to simulate the opening and closing of the window in the system design[13]. Since the microcontroller cannot provide a 12V power supply, a separate power supply was designed for the stepper motor during the system design phase. Its control method is simple: it rotates based on the given electrical pulse signals to control the angle and speed of rotation. On the user interface, numerical values representing the degree of window opening/closing are sent, and the stepper motor rotates the corresponding number of steps accordingly. In case of rain detection, if the window is open, the stepper motor automatically returns to the closed position.

3.4 WiFi Module

The ESP8266 [12, 14] is a commonly used WiFi module that can be controlled using an AT command set and is compatible with almost all types of microcontrollers. It provides a convenient way for microcontrollers to connect to wireless networks, thereby adding WiFi connectivity to the system. The ESP8266 transmits cloud data to the microcontroller via serial communication and also receives real-time data from sensors such as temperature and humidity sensors, light sensors, etc. After parsing, it transmits this data to the server. Additionally, communication between the ESP8266 and the server is typically done using the JSON data format.

4   The Implementation of the System

4.1   Selection of Communication Protocol and Data Transmission

MQTT Protocol

In this system, the data transmission and communication protocol chosen is MQTT (Message Queuing Telemetry Transport) protocol. MQTT is a lightweight, open-source publish/subscribe messaging transport protocol, initially developed by IBM, and has now become one of the communication standards in IoT field[15].

The main features of MQTT include[9, 16]:

1) Lightweight: *The MQTT protocol has minimal communication overhead, with small packet headers, allowing it to efficiently transmit data across different networks.*
2) Publish/Subscribe model: *The MQTT protocol employs a publish/subscribe communication model. Through this loosely coupled communication model between publishers and subscribers, messages are distributed and consumed. It enables one-to-many message publishing, where clients subscribing to the same topic can receive messages published to that topic.*
3) Reliability: *The MQTT protocol incorporates built-in reconnection mechanisms and Quality of Service (QoS) level control, ensuring the reliable transmission of messages. In the event of network disconnection, it can automatically reconnect, enhancing the system's stability.*
4) Asynchronous communication: *The MQTT protocol supports asynchronous communication, meaning that the sender and receiver do not need to be online simultaneously. Messages can be sent and received at any time, allowing for flexibility in communication timing.*

Data Transmission

1) In this system, the information we need to transmit includes:
   • From microcontroller to the mini-program: Transmit environmental parameter information and device information, such as temperature, humidity, illumination, and device status.
   • From the mini-program to the microcontroller:
     o When controlling the device, specifying the port number allows control of the device connected to that port on the microcontroller side.
     o Device switch status information: For lamps and alarms, there are two states: "on" and "off"; for windows, the device switch status is the degree of window opening from 0 to 100. When transmitting data, for lamps and alarms, transmit numerical

values of 0 or 1, where 0 indicates the device is off, and 1 indicates the device is on; for windows, transmit numerical values from 0 to 100 to represent the degree of window opening.

o Information about whether the device is set to automatic mode: If the user activates automatic mode, the device will automatically switch based on whether the environmental parameters are outside the set threshold. If the user deactivates automatic mode, the device's switch cannot be automatically controlled on the microcontroller side, and control can only be achieved through the user from the mini-program side or physical switch.

2) Data transmission formats

- From Microcontroller to Mini Program: Following the general communication transmission format of ESP8266, the system utilizes JSON data format for data transmission. "Hum" represents humidity, "Temp" represents temperature, and "Light" represents light intensity. "LED1", "LED2", "LED3", "BEEP1", and "CHUANG1" in this simulated data respectively indicate the status of devices connected to interfaces as "light", "alarm", and "window". The specific transmission format is illustrated in Fig. 5.

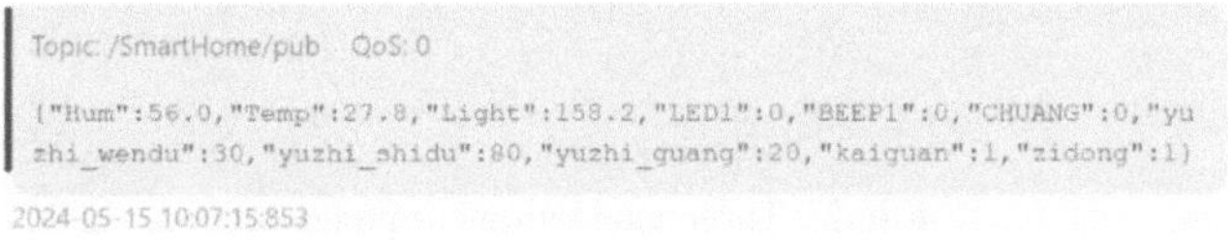

```
Topic: /SmartHome/pub   QoS: 0

{"Hum":56.0,"Temp":27.8,"Light":158.2,"LED1":0,"BEEP1":0,"CHUANG":0,"yu
zhi_wendu":30,"yuzhi_shidu":80,"yuzhi_quang":20,"kaiguan":1,"zidong":1}

2024-05-15 10:07:15:853
```

Fig. 5. Format for Data Transmission from the Microcontroller

- From Mini Program to Microcontroller: The transmission adopts the JSON data format. The keys consist of the interface numbers, while the values form an array. The first element indicates the device's on/off state, and the second element conveys the device's automatic information. The specific format is illustrated in Fig. 6.

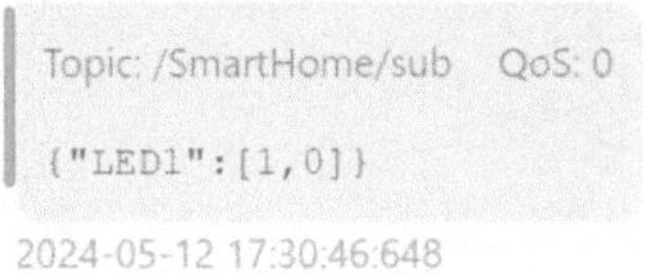

```
Topic: /SmartHome/sub   QoS: 0

{"LED1":[1,0]}

2024-05-12 17:30:46:648
```

Fig. 6. The format for data transmission in the Mini Program

4.2   User Interface Design

Page Layout and Design

The system adopts a clean and straightforward page layout and design to ensure users can easily navigate the mini program. It primarily consists of three main pages: the 'Home' page, 'Management' page, and 'My' page. In the 'My' page section, users can view their historical records and provide feedback. The 'Management' page includes pages for adding and deleting devices and scenes, as well as device details and scene modification pages (Fig. 7).

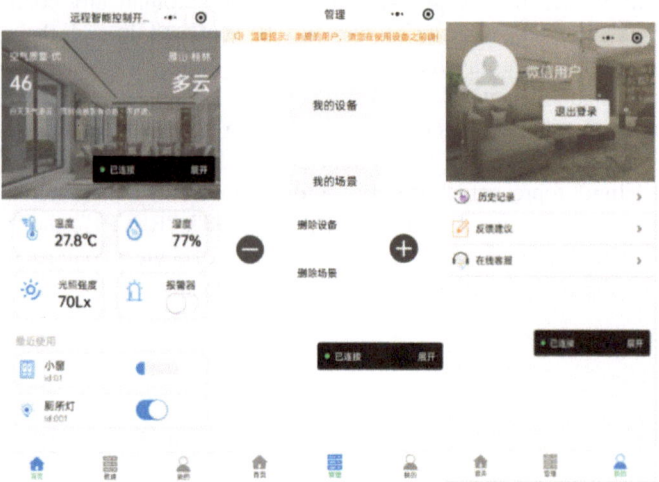

Fig. 7. Three navigation bar pages

In the 'Management' page, users can add or delete devices and scenes, as well as view and modify device details and scene settings (Fig. 8).

Fig. 8. Device Details Modification and Scene Modification Page

Real-Time Data Display

The user interface can display real-time data such as environmental parameters like temperature, humidity, recent usage records, and continuously updated weather information. These data points can help users better understand the current environment and system status.

4.3 WeChat Mini Program Development and Feature Implementation

Database Design

When designing the database, we considered factors such as user requirements, security, and simplicity, and established a reliable database structure. This primarily includes tables for device information, device types, scene information, historical records, microcontroller interfaces, and user information. The following table details the design of each table in the database and the fields included (Table 1).

Table 1. Details of the Data Table

table	usage	fields
smt-device	Storage for device information	Device name, Scene, number, Switch status, Image, Interface number, etc
smt-deviceType	Storage for device types	Device type, Picture, Creation time, etc
smt-History	Storage for operation history records	Device operation type, operation record, operation time, operation interface number, operation device number, etc
smt-Interface	Storage for microcontroller interface	Interface type, Interface number, Interface occupancy status, etc
smt-Scene	Storage for scene information	Scene Name, Number, etc
smt-UserAcc	Storage for user information	Username, ID, Avatar, Time

Cloud Function Design

To implement the various functions of the system, we designed and deployed multiple cloud functions to handle the logic part of the mini-program. These cloud functions include:

1) *Cloud functions for device and scene operations.*
2) *Cloud functions for user authorization login and storing user information.*
3) *Cloud functions for device status monitoring and operation recording.*
4) *Cloud functions for updating interface usage information.*
5) *Function implementation*

In this section, the system gradually develops and refines various functions based on requirements and design plans. The final designed system includes the following functions:

1) Device management: *Add and delete devices, Modify device information.*
2) Scene management: *Add and delete scenes, Modify scene information.*
3) Device control: *Provide functionality for automatic and manual device control.*
4) Data transmission: *Implement bidirectional transmission of information between microcontrollers and the mobile app.*
5) Data storage: *Store user information and operation records.*
6) Historical records display: *Design and display recent usage and historical records.*

5   System Testing

System testing aims to ensure that the system operates smoothly under various conditions, including both data transmission and device control aspects.

5.1   Data Transmission Test

The data transmission test aims to verify the reliability and accuracy of sensor data and control command transmission.

Test Content:.
By simulating user operations on the mini-program end, observe whether the MQTT protocol can correctly transmit control commands, and check if the environmental parameter information on the microcontroller end can be transmitted to the mini-program end in real-time via MQTT and displayed on the frontend interface.

Results Display
Figure 9.: User sends a command to turn on the light to the device connected to port 'LED1' on the mini-program end.

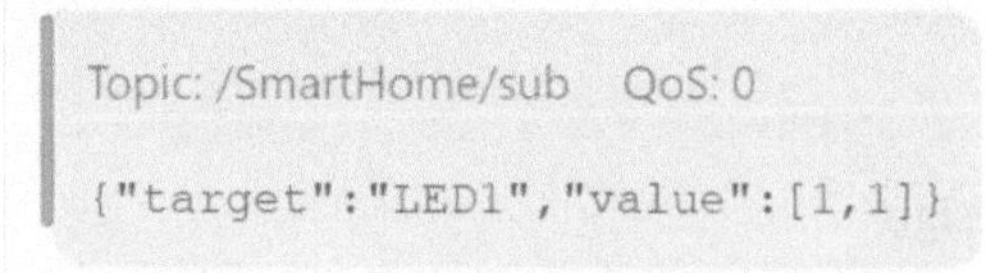

```
Topic: /SmartHome/sub    QoS: 0

{"target":"LED1","value":[1,1]}
```

Fig. 9. The mini-program end sends a command to turn on the light

Figure 10 Real-time transmission of control and environmental parameter information from the microcontroller end to the server.

The bidirectional transmission of information is both accurate and reliable, allowing for real-time updates and delivery based on user demand.

Topic: /SmartHome/pub   QoS: 0

{"Hum":56.0,"Temp":27.7,"Light":151.8,"LED1":1,"BEEP1":0,"CHUANG":0,"yu
zhi_wendu":30,"yuzhi_shidu":80,"yuzhi_guang":20,"kaiguan":1,"zidong":1}

Fig. 10. Transmit environmental parameters and display

5.2 Device Control Test

The device control test aims to verify the system's ability to control external devices, including both manual and automatic control scenarios.

Manually controlling device to turn on the light:. *The WeChat Mini Program sends an "turn on the light" command to the microcontroller end using the MQTT protocol, as shown in Fig. 11.*

Fig. 11. Send light on message, LED1 on

Automatically Controlling the Device to Turn on the Light:
Cover the light sensor to reduce the light intensity. When the device detects that the light intensity is below a certain threshold, it automatically turns on the light, as shown in Fig. 12.

Fig. 12. After covering, LED turns on

The testing conducted verified the system's reliable control capability over external devices, including both manual and automatic control modes, ensuring the system's functionality under various circumstances.

6   Conclusion

This paper primarily introduces the design of a smart home control mini-program system based on the STM32 microcontroller. The system covers two aspects: hardware selection and system implementation. In hardware selection, the STM32F103C8T6 microcontroller was chosen, along with the DHT11 temperature and humidity sensor, BH1750 light sensor, and a 12V stepper motor for temperature, humidity, light detection, and automatic window opening and closing functions. In terms of system implementation, the MQTT protocol was employed for bidirectional data transmission. The open-source platform EMQX was selected as the message broker server platform, and a mini-program was designed with developer tools, combined with CloudBase technology to achieve data visualization and remote control functionality on mobile terminals.

For future work, there are plans to further optimize system performance and expand its functionality and applicability, enhancing compatibility and scalability with external devices.

References

1. Li, M., Gu, W., Chen, W., et al.: Smart home: architecture, technologies and systems. Procedia Computer Sci. **131**, 393–400 (2018)

2. Chang, M.: Between convenience and security smart home and data protection. Computer & Network **46**(23), 56–57 (2020)

3. Wu, S., Wu, X.: Application and prospect of artificial intelligence technology in home security system. Network Security Technology & Application (4), 122–124 (2024)

4. Peng, Y.: Research on scenario-based smart home design. Furniture & Interior Design **31**(2), 52–57 (2024)

5. Wang, K., Wu, Y., Zhao, Q., et al.: Design of FPGA-based smart home control system. Laboratory Science **22**(3) (2019)

6. Chai, X., Shang, Y., Qin, X.: Design and realization of smart home system based on internet of things (IoT). Internet of Things Technologies **14**(2), 66–68 (2024)

7. Yan, M.: Hardware and software design of smart home monitoring system based on wi-fi control. Technology Innovation and Appl. **14**(5), 32–35 (2024)

8. Li, H.: Smart home control system based on MQTT. Telecom Power Technol. **41**(5), 4–6 (2024)

9. Chen, L., Lin, F., Guo, Q.: Intelligent home data transmission system based on MQTT. Digital Communication World (7), 52–54 (2023)

10. Yang, Y., Zhang, S., Liu, Z., et al.: Intelligent voice home control system based on ZigBee and WiFi. Computer Science and Appl. **14**, 41 (2024)

11. Gong, L.: Research on smart home system based on SC2440 and ZIGBEE protocol. MATEC Web of Conferences **228**, 02012 (2018)

12. Du, X., Wei, W., Zhao, Z.: Design and realization of smart home control system based on STM32.Practical Electronics **31**(23), 29–3215 (2023)

13. Wang, Y., Lv, H., Song, J.: Smart cloud Internet of things smart home system development based on STM32 microcontroller. Electronic Test **31**(18), 62–63126 (2020)

14. Cai, G.-P., Deng, T., Ni, J.: Design of smart home fire alarm system based on mobile client. Fire Science and Technol. **39**(3), 377–380 (2020)

15. Lakshminarayana, S., Praseed, A., Thilagam, P.S.: Securing the IoT application layer from an MQTT protocol perspective: challenges and research prospects. IEEE Communications Surveys & Tutorials (2024)

16. Buccafurri, F., De Angelis, V., Lazzaro, S.: MQTT-I: achieving end-to-end data flow integrity in MQTT. IEEE Trans. Dependable and Secure Computing (2024)

Communication Anti-jamming System Based on Deep Reinforcement Learning

Xuewei Feng, Hong Wen, Runhui Zhao$^{(\boxtimes)}$, Tao Tang, Weihong Shi, and Yulin Peng

Institute of Aeronautics and Astronautics, University of Electronic Science and Technology of China, Chengdu 611731, China

{fengxuewei,yulin}@std.uestc.edu.cn, sunlike@uestc.edu.cn, uestcrunhuiu@gmail.com

Abstract. Due to the open characteristics of the electromagnetic channel, there are malicious nodes that jam the normal data flow, prevent the legitimate receiver from obtaining information, and then intercept and tamper with the data, which makes the research of communication anti-jamming becomes more and more important. Traditional anti-jamming methods employ a single anti-jamming approach and cannot adaptively adjust their anti-jamming strategy based on the environment, often can't achieve a good anti-jamming effect in the complex communication environment. In order to solve these challenges, this paper studies the anti-jamming communication model based on deep reinforcement learning (DRL), and builds a simulation system to realize intelligent anti-jamming decision-making by using DRL algorithm. The simulation results show that the proposed intelligent anti-jamming decision can choose the best anti-jamming scheme according to the complex environment and effectively improve the communication quality.

Keywords: Communication anti-jamming · Anti-Jamming decision · Deep reinforcement learning

1 Introduction

Due to the electromagnetic openness of wireless communication technology, malicious nodes can not only determine the surrounding terminal signals and construct jamming signals to interfere with normal data flows but also intercept and tamper with communication data, posing a serious security threat to wireless communications. Therefore, to ensure the reliability of information transmission, communication systems must possess anti-jamming capabilities. Traditional communication systems achieve anti-jamming through methods such as frequency hopping [1], spread spectrum [2], time slot control, power control, rate control, anti-jamming case libraries, and particle swarm. Literature [3, 4] proposed a Multi-Sequence Frequency Hopping (MSFH) system, which reduced the bit error rate of the communication system. Some scholars have improved communication system performance using adaptive frequency hopping modes [5, 6]. Literature [7] proposed a new spread spectrum communication theory and anti-jamming technology, which has better feasibility and robustness in anti-jamming. Literature [8] applied differential frequency hopping communication systems to unmanned aerial vehicles (UAVs), significantly enhancing the UAVs' comprehensive anti-jamming performance.

© The Author(s) 2025
P. Siarry et al. (Eds.): WCNA 2023, LNEE 1361, pp. 126–133, 2025.
https://doi.org/10.1007/978-981-96-2409-6_13

However, these methods employ a single anti-jamming approach and cannot adaptively change the anti-jamming strategy according to the environment, making it difficult to protect communication links from attacks by variable dynamic jamming systems, failing to achieve optimal anti-jamming effects. Deep Reinforcement Learning (DRL), using deep learning networks, can discover and learn the best strategies in complex communication systems. Literature [9–12] applied Q-learning, SDRLA algorithm, and decentralized DRL algorithm to communication systems, significantly improving the communication quality of the systems. Literature [13] applied deep reinforcement learning algorithms to UAV swarms, ensuring effective communication among the UAVs. Literature [14] used DReL to implicitly learn residual noise to recover channel coefficients under noise interference. Literature [15] developed and implemented a novel distributed multi-agent RL algorithm. While these deep learning algorithms improve anti-jamming capabilities, their stability is greatly influenced by the samples. To enhance the stability of anti-jamming capabilities in communication systems, this paper proposes a design and system construction of an anti-jamming communication system model based on Deep Reinforcement Learning (DRL), utilizing algorithms such as DQN to achieve intelligent anti-jamming decisions, achieving stable anti-jamming performance.

2 Construction of Communication Anti-jamming System Model

The complete anti-jamming decision-making process for communication can be abstracted into three stages as shown in Fig. 1.

1) In the first stage, communication devices upload pilot data to the main station through the uplink. After receiving the information, the main station calculates the bit error rate or packet loss rate of the uploaded pilot data. If the uplink status is normal, the main station will notify the communication device to continue transmitting business data through a covert feedback link. If the uplink status is abnormal, i.e., subject to interference, the main station will instruct the communication device to reduce its transmission power and initiate the interference signal detection program of the second stage through a covert feedback link.
2) The second stage mainly describes the process of the main station extracting and detecting interference signals. As in the first stage, the main station notifies the communication device to stop uploading data through the feedback link. In the second stage, due to a certain delay in the spectrum monitoring by the interferer, the interferer continues to transmit interference signals to disrupt the uplink even after the communication device reduces its transmission power. Therefore, the main station can store the interference signal waveforms within this time window and extract features from the interference waveforms, then use machine learning methods to identify them. Then, the status of the interference signals is sent to the communication device through the covert feedback link. The communication device makes anti-jamming decisions based on its own status and the information of the interference signals.
3) In the third stage, the communication device updates its uplink data transmission strategy and retransmits the business data to the main station through the uplink.

Therefore, the data exchange and anti-jamming decision-making process between communication devices and the main station can be described as a Markov Decision

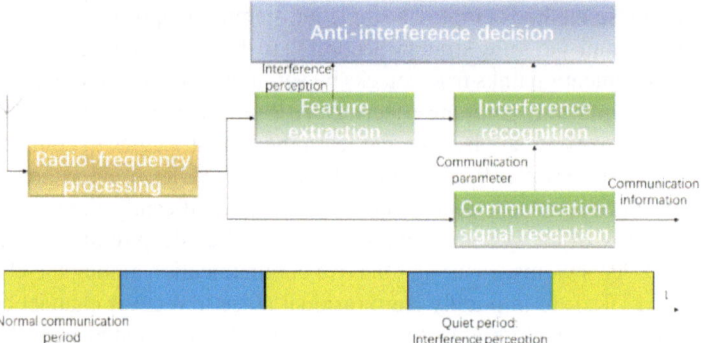

Fig. 1. Schematic Diagram of the Three Stages in the Anti-jamming Decision-Making Process for Communication Devices

Process (MDP), characterized by state space, action space, immediate rewards, and action functions.

2.1 State Space Modeling

In the environment we describe, communication devices can be abstracted as individual nodes. Therefore, the state space can be divided into two parts: states originating from the node itself, denoted as S_{self}, and states obtained from external sources, denoted as S_{ext}. The state of node i at the nth moment can be defined as $S_{i,n\tau} = \{s_{\text{self},i,n\tau}, s_{\text{ext},i,n\tau}\}$. Accordingly, the state space of node i at moment n can be defined as:

$$S_i = \{s_{i,\tau}, s_{i,2\tau}, \cdots, s_{i,N\tau}\}, s_{i,n\tau} \in S \tag{1}$$

Where S represents the set of available states for the node. We can further refine the definition of the node's own state $S_{\text{self},i,n\tau}$ as:

$$S_{\text{self},i,n\tau} = \{g_{i,n\tau}, f_{bi,n\tau}, p_{i,n\tau}, c_{i,n\tau}, a_{i,n\tau}\} \tag{2}$$

$g_{i,n\tau}$ represents the communication mode of node i at the nth moment. The states transmitted from the main station feedback link mainly include: interference method $j_{i,n\tau}$, interference power $p_{j,i,n\tau}$, signal power transmitted by node i as $p_{r,i,n\tau}$, transmission signal bandwidth $b_{i,n\tau}$, signal processing time $t_{i,n\tau}$, and communication quality $q_{i,n\tau}$. The external feedback state obtained by the node can be represented as:

$$s_{ext,i,n\tau} = \{j_{i,n\tau}, p_{j,i,n\tau}, p_{r,i,n\tau}, b_{i,n\tau}, t_{i,n\tau}, q_{i,n\tau}\} \tag{3}$$

2.2 Action Space Modeling

After receiving feedback from the main station at moment n, a node needs to decide whether to change its transmission state before the next transmission slot. This includes deciding whether to change the node's transmission signal mode $g_{i,(n+1)\tau}$, symbol rate $f_{b,I,(n+1)\tau}$, coding method $c_{i,(n+1)\tau}$, Signal-to-Noise and Interference Ratio (SINR) $sinr_{r,i,(n+1)\tau}$, and anti-jamming method $a_{i,(n+1)\tau}$. Here, it is assumed that the next state $S_{i,t+1}$ is determined only by the current state $S_{i,t}$ and action $act_{i,t}$, so the action space at moment n will have $M \times D \times J \times K \times L$ states. To ensure the stability of subsequent Deep Reinforcement Learning (DRL) algorithms, we design the action space as follows:

$$act = \{a_g, a_{f_b}, a_c, a_{\text{sinr}}, a_a\} \tag{4}$$

$a_g, a_{f_b}, a_c, a_{sinr}, a_a a_{sinr}, a_a$ correspond to signal mode, symbol rate, coding method, SINR, and anti-jamming method, respectively. '1' indicates adopting the next state in the state set, '0' means maintaining the current state unchanged, and '−1' indicates adopting the previous state in the state set.

2.3 Reward Space Modeling

The reward model influences the expected behavior of the agent. When the transmission link between the node and the main station is blocked by malicious interference signals, we want the node to complete anti-jamming decision-making based on complete state information and our set optimization objectives. We primarily divide the reward model into five parts: communication quality, communication rate, Signal-to-Noise and Interference Ratio (SINR), processing time, and signal bandwidth.

We have set thresholds for communication quality $q_{i,th}$, communication rate $f_{bi,th}$, and signal processing time $t_{i,th}$. Positive rewards are only generated when the communication quality and rate reach the preset thresholds. Meanwhile, for the SINR and signal bandwidth parts of the reward composition, no negative rewards are set while maintaining communication performance.

3 Anti-jamming Decision-Making Under Deep Reinforcement Learning

3.1 Anti-jamming Decision-Making Formulation Based on Deep Reinforcement Learning

DRL, which integrates Deep Learning (DL) and Reinforcement Learning (RL), can intelligently select the optimal strategy without the need to know all the state transition probabilities $P\left(s' \mid s, a\right)$ in advance. This significantly enhances resource utilization efficiency. In DRL, neural networks are used to approximate value functions or policy functions. It eliminates the need to store a specific value for every possible state or state-action pair. Instead, these values are predicted through the network, greatly reducing the required storage space and improving computational efficiency. There are also many algorithms within DRL suitable for different scenarios.

DQN, as a classic algorithm that combines reinforcement learning with deep learning, effectively solves decision-making problems in complex environments. DQN uses a target network Q with its parameters updated from the main network Q every T_a steps, making the learning values of $Q(s, a)$ closer to the optimal Q function. DQN trains by randomly sampling a batch of data from the experience replay pool (memory pool), standardizing, and normalizing the raw data. The model is optimized through continuous forward and backward propagation iterations until the main network converges. In the forward propagation (FP) process, this process computes the network output z_i layer by layer from input to output, with each layer undergoing a linear transformation through connection weights $W(c)$ and bias $\Theta(c)$, and passing through an activation function ϕ. The backward propagation (BP) process adjusts the network parameters based on the error of the output units. The error function $E[W(c), \Theta(c)]$ is

$$E[W(c), \Theta(c)] = \frac{1}{B} \sum_{i=1}^{B} (\hat{z}_i - z_i)^2 \tag{5}$$

where B represents the batch size.

Dueling DQN innovates in network architecture compared to traditional DQN, which estimates the $Q(s, a)$ values through a single process. In contrast, Dueling DQN splits this process into two independent paths: one estimates the state value function $V(s)$, which assesses the overall value of each state without considering specific actions, and the other estimates the advantage function $A(s, a)$ for each action, assessing the advantage of each action relative to others, i.e., the extra value of choosing that action over others in a given state. These two paths are then merged at the output layer to estimate the value of $Q(s, a)$:

$$\begin{aligned} Q(s, a) = V(s) + A(s, a) \\ - \frac{1}{|\mathcal{A}|} \sum_{a'} A(s, a') \end{aligned} \tag{6}$$

Here, $|\mathcal{A}|$ represents the number of available actions, used to calculate the average value of action advantages. $A(s, a)$ is the action advantage function, representing the extra value of taking action a in state s relative to the average action.

Double DQN addresses the issue of overestimation of Q-values by using two networks: one to select the best action and another to independently evaluate the value of that action. However, Double DQN introduces additional complexity, which may require more debugging and hyperparameter tuning to optimize performance. After transitioning to Double DQN, the formula for updating $Q(s, a)$ becomes:

$$\begin{aligned} Q(s, a) = r + \gamma Q_{target} \\ (s', argmax_{a'} Q_{online}(s', a')) \end{aligned} \tag{7}$$

Q_{online} represents the Q-value estimation for each action by the current policy network. Q_{target} represents the Q-value estimation for each action by the target policy network.

D3QN combines the characteristics of Double DQN (DDQN) and Dueling DQN. D3QN aims to improve learning efficiency and performance by leveraging the advantages of both techniques, especially when dealing with complex and high-dimensional tasks. D3QN addresses the overestimation issue present in standard DQN and the efficiency of state-action estimation to enhance overall performance. It adopts the network architecture of Dueling DQN, which splits the network into a value function part and an advantage function part. At the same time, it employs the approach from Double DQN, using two networks (an action network and a target network) to reduce the overestimation of Q-values. The update formula for $Q(s, a)$ becomes:

$$Q(s, a) = \int_{-\infty}^{\infty} z \cdot \mathbb{P}(Z = z \mid s, a)dz \tag{8}$$

$\mathbb{P}(Z = z \mid s, a)$denotes the distribution of Q-values obtained after taking action a in state s, where Z represents the random variable indicating the different possible values of the Q-value.

3.2 System Simulation and Results

Using Python 3.8, a custom reinforcement learning environment was developed through the gym package. A deep learning network is built and implemented using Pytorch1.12 framework. In this paper, the network parameters of DQN algorithm are (64, 512, 256, 128, 64), Double DQN algorithm are (384, 192, 96), Dueling DQN algorithm are (256, 128, 64), and D3QN algorithm are (384, 192, 96). Under fading channel conditions, the distribution of the reward function is determined by random testing of millions of points.

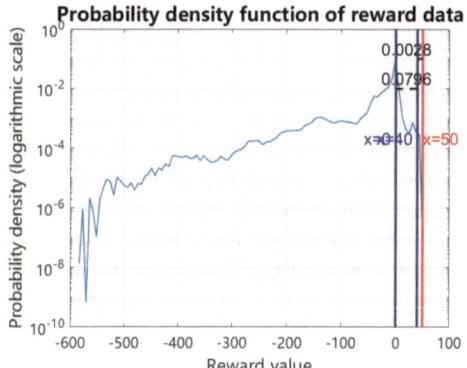

Fig. 2. Probability density function of reward data in fading channel

Figure 2 shows the reward distribution in our RL environment under fading channel conditions. It is observed that in the complex fading channel, the probability of getting a reward greater than 0 is 0.0796, while the probability of getting a reward greater than 40 is 0.0028. It is very challenging to make anti-jamming decisions in fading channels because many decisions are negative rewards. This results in less stability in the early stages of DRL training.

Figure 3 shows the training performance of various DRL algorithms in a fading channel environment. In the fading channel environment, we choose two representative training evaluation scenarios for each algorithm due to its complexity. DQN and DuelDQN showed relatively good training results. DQN was stable in different training processes, while DuelDQN showed consistent but unstable performance in the second training group, and the reward climbing time was longer. D3QN and DoubleDQN showed low training stability, large reward variance, and significant differences in training results between different groups.

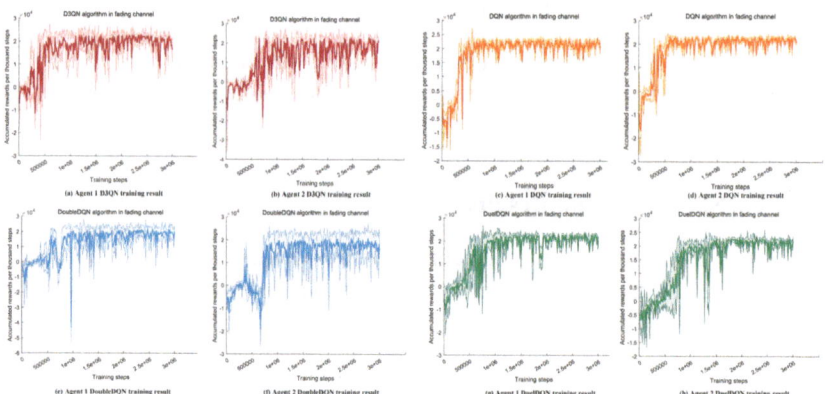

Fig. 3. The training and evaluation of different RDL algorithms are presented in the fading channel environment

4   Peroration

In order to solve the instability of anti-jamming algorithm under deep learning, this paper proposes an intelligent anti-jamming decision scheme based on deep reinforcement learning. Experiments show that the anti-jamming decision scheme proposed in this paper has a good effect, but different deep reinforcement learning algorithms still have different effects, among which the more classical DQN performs best. Its training efficiency and stability are more prominent in several algorithms, and it can adapt to various environments and sample differences to achieve good communication performance.

References

1. Qi, J., Zhang, H., Qi, X., Peng, M.: Deep reinforcement learning based hopping strategy for wideband anti-jamming wireless communications. IEEE Trans. Veh. Technol. (2023)
2. Li, X., Chen, J., Ling, X., Wu, T.: Deep reinforcement learning-based anti-jamming algorithm using dual action network. IEEE Trans. Wirel. Commun. **22**(7), 4625–4637 (2023)
3. Rui, W., Wei, W.: Research on anti-jamming method of frequency hopping communication based on blind source separation in complex electromagnetic environment. In: 2019 International Conference on Smart Grid and Electrical Automation (ICSGEA), Xiangtan, China, pp. 378–381 (2019)
4. Yang, S., Li, J., He, B.: A novel interference suppression method in spread spectrum communication based on blind source separation. In: 2017 3rd IEEE International Conference on Computer and Communications (ICCC), Chengdu, China, pp. 816–820 (2017)

5. Ma, X.H., Jing, W.F.: Performance analysis of a new spreading code. J. Time Freq. **42**(4), 345–356 (2019)
6. Zhao, L., Bao, L.Y.: S-T linear coupled cascade chaos spread spectrum code and its performance analysis. Telecommun. Eng. **61**(2), 218–223 (2021)
7. Huang, W., Zhang, S., Yan, R.H.: Novel spread spectrum communication theory and the anti-jamming applications. In: 2021 6th International Conference on Inventive Computation Technologies (ICICT), Coimbatore, India, pp. 56–61 (2021)
8. Xie, S., Qian, B.: Performance analysis of differential frequency hopping communication system over Rician channel. In: 2018 IEEE 4th Information Technology and Mechatronics Engineering Conference (ITOEC), Chongqing, China, pp. 1015–1019 (2018)
9. Zhang, Z., Wu, Q., Zhang, B., Peng, J.: Intelligent anti-jamming relay communication system based on reinforcement learning. In: 2019 2nd International Conference on Communication Engineering and Technology (ICCET), Nagoya, Japan, pp. 52–56 (2019)
10. Han, G., Xiao, L., Poor, H.V.: Two-dimensional anti-jamming communication based on deep reinforcement learning. In: 2017 IEEE International Conference on Acoustics, Speech and Signal Processing (ICASSP), New Orleans, LA, USA, pp. 2087–2091 (2017)
11. Liu, S., et al.: Pattern-aware intelligent anti-jamming communication: a sequential deep reinforcement learning approach. IEEE Access **7**, 169204–169216 (2019)
12. Wang, X., Chen, X., Wang, M., Dong, S.: Decentralized reinforcement learning based anti-jamming communication for self-organizing networks. In: 2021 IEEE Wireless Communications and Networking Conference (WCNC), Nanjing, China, pp. 1–6 (2021)
13. Zhang, J., Ding, W., Luo, Y., Wang, Y., Wang, C., Xiao, J.: Joint trajectory and power control design for UAV anti-jamming communication network. In: 2022 4th International Conference on Advances in Computer Technology, Information Science and Communications (CTISC), Suzhou, China, pp. 1–6 (2022)
14. Liu, C., Liu, X., Ng, D.W.K., Yuan, J.: Deep residual learning for channel estimation in intelligent reflecting surface-assisted multi-user communications. IEEE Trans. Wirel. Commun. **21**(2), 898–912 (2022)
15. Elleuch, I., Pourranjbar, A., Kaddoum, G.: A novel distributed multi-agent reinforcement learning algorithm against jamming attacks. IEEE Commun. Lett. **25**(10), 3204–3208 (2021)

Fault Diagnosis of Infrared Sensor Based on Convolutional Neural Network

Zhenghao Hu[1], Yingyi Liu[1(✉)], Lin Cheng[2], Yan Wang[2], Donglei Zhang[2], and Xun Tian[2]

[1] School of Automation Science and Electric Engineering, Beihang University, Beijing, China
liuyingyi@buaa.edu.cn
[2] State Grid Corporation of China, Beijing, China

Abstract. Infrared sensors are critical equipment for monitoring substations, but they are subjected to complex multi-physical field stresses in substation applications, leading to a significantly higher failure rate of infrared sensors in substations compared to other applications. Effective fault diagnosis of infrared sensors is of great importance for improving the safety of substations. This article proposes a method for fault diagnosis of infrared sensors. First, the network weights are trained based on the ImageNet dataset and kept fixed. Then, a transfer learning approach is used to fine-tune the model based on a dataset of infrared sensor failures, resulting in the classification results. To verify the effectiveness of the proposed method, the improved transfer learning model is compared to the model before improvement. Experimental results show that the proposed method greatly reduces training time and improves classification accuracy compared to the previous model. In conclusion, the method presented in this article offers a new approach for fault diagnosis of infrared sensors, which can contribute to enhancing the reliability of infrared sensors in substations.

Keywords: infrared sensor · image classification · transfer learning

1 Introduction

Infrared sensors are important equipment for monitoring substations and play a crucial role in the measurement and control devices, as well as the automation control and information systems of modern substations [1–3]. However, the working environment of a substation is a complex scenario where electromagnetic fields, humidity, heat, equipment vibration, and other physical factors are coupled, leading to a significantly higher failure rate of infrared sensors compared to other application scenarios [4]. Additionally, in substation applications, sensors also experience complex multi-physical field stress. If the performance of the sensors deteriorates or malfunction, it is impossible to prevent failures from occurring without quickly identifying the cause of the problem. In severe cases, this can result in the entire system being paralyzed. Therefore, a major concern both domestically and internationally is how to diagnose faults in infrared sensors [5].

Many scholars from various countries have proposed different targeted fault diagnosis method of sensors, based on the characteristics of different sensors [6–8]. These

P. Siarry et al. (Eds.): WCNA 2023, LNEE 1361, pp. 134–146, 2025.
https://doi.org/10.1007/978-981-96-2409-6_14

techniques typically include the following categories: model-based sensor fault diagnosis methods, knowledge-based sensor fault diagnosis methods, data-based sensor fault diagnosis methods, and so on [9].

The model-based approach was first proposed by Beardsh from the Massachusetts Institute of Technology in 1971 [10]. It involves creating detailed mathematical models of sensors to achieve fault diagnosis. However, a thorough understanding of the internal mechanisms of the sensors is required to implement the model-based approach for diagnosing sensor faults. Mehranbod et al. used probabilistic data based on a multi-sensor model [11]. However, this approach has limited capabilities in identifying sensor defects and can only be used to detect biases, drifts, and noise [12]. In addition, machine learning methods such as principal component analysis and artificial neural networks have also been used for data reconstruction in the presence of missing data. However, these algorithms may have limitations when dealing with large-scale long-term continuous data loss scenarios [13].

Deep learning can adaptively extract abstract features from a large amount of data and train in an end-to-end manner, effectively learning the complex mapping relationships between the data, which can partially address the aforementioned issues [14]. Deep learning has been widely used in sensor fault diagnosis due to its powerful nonlinear mapping capabilities, generalization abilities, and learning and fault-tolerance capabilities. Neural networks and support vector machines have demonstrated strong competitiveness in handling classification applications and have been widely applied in various sensor fields [15–17]. Researchers have already applied these techniques to fault detection in pressure sensors, achieving satisfactory results in terms of detection accuracy and efficiency, far surpassing manual detection [18]. FAN et al. proposed a dense connected convolutional network (DenseNet) for reconstructing real building responses under environmental excitations [19]. NI et al. applied autoencoder-related theories to data compression and reconstruction [20].

Although convolutional neural networks (CNN) have shown good performance in fault diagnosis, they may face challenges such as long training times and lower accuracy on small sample datasets [21].

In this paper, the CNN fault diagnosis method in infrared sensor fault diagnosis is improved based on transfer learning; the improved model and the pre-improvement comparative study is carried out to verify the effect of the improved model. The research results can provide technical support for improving the efficiency of infrared sensor fault diagnosis in substations, which is of great significance for the construction of intelligent substations.

2   Type Style and Fonts

Wherever Times is specified, Times Roman or Times New Roman may be used. If neither is available on your word processor, please use the font closest in appearance to Times. Avoid using bit-mapped fonts if possible. True-Type 1 or Open Type fonts are preferred. Please embed symbol fonts, as well, for math, etc.

2.1 Color Deviation

The infrared sensor may experience faults due to the aging of insulation in the line outlet section caused by high temperature conditions and the effects of high-temperature stress and temperature cycling stress. This can lead to color deviation in the infrared sensor. The specific manifestations are shown in Fig. 1.

Fig. 1. Color shift fault

2.2 Halo

When a strong electromagnetic pulse affects the sensor, it can temporarily malfunction or break down, causing the sensor to capture a halo effect. The specific manifestations are shown in Fig. 2.

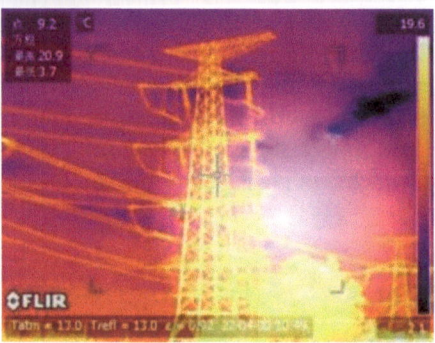

Fig. 2. Halo fault

2.3 Floating Image

Due to interference, when a strong electromagnetic pulse affects the internal components of the sensor, it can cause an increase in internal noise or the appearance of new interference signals, leading to data acquisition faults in the sensor. The specific manifestations are shown in Fig. 3.

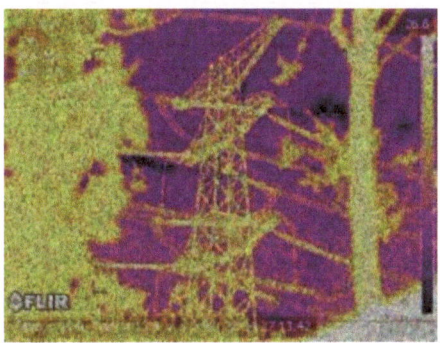

Fig. 3. Floating flower fault

2.4 Blurring

Long-term exposure to vibration stress can cause continuous strain in the sensor material, resulting in blurry image capture. The specific manifestations are shown in Fig. 4.

Fig. 4. Blurring fault

3 The Principle of the Proposed Method for Infrared Sensor Diagnostic

This paper uses the convolutional neural network model ResNet, which is trained on the ImageNet dataset; then migration learning is added to improve the training, retaining the weights that are trained on large data samples, and the fully connected layer is trained on a small sample dataset of infrared sensor faults; the final classification results are obtained. The flow of this paper is shown in Fig. 5.

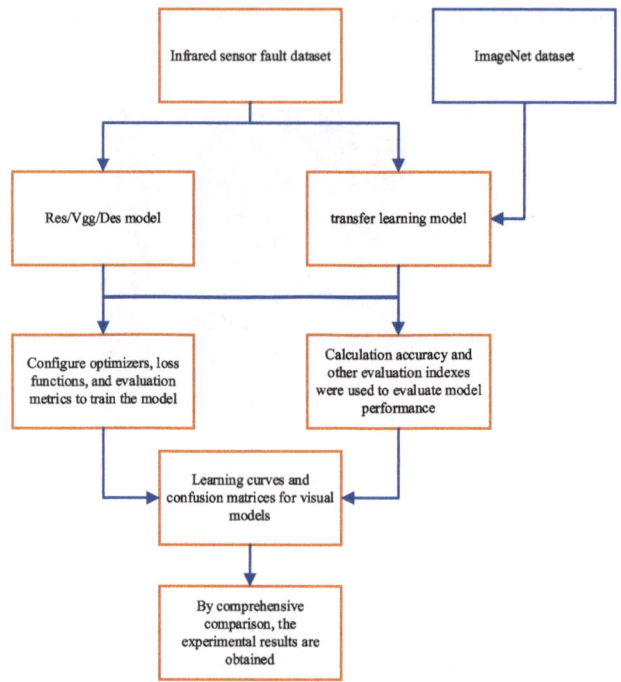

Fig. 5. Infrared sensor diagnostic process

3.1   Training Model Weights Based on ImageNet Dataset

First, we train the model by ImageNet dataset, and the specific training process is as follows:

- Choose the Res-Net model to classify the Infrared sensor fault dataset;
- Train model with ImageNet Database;
- Fix the weight of the model in order to do transfer learning below.

We use the Res-Net model adopting the convolution structure and the residual unit structure. Among them, the convolution structure is as follows:

In the two-dimensional convolutional layer, a two-dimensional input array and a two-dimensional kernel array output a two-dimensional array through the convolution operation. The convolution operation involves sliding the window of the convolution kernel at regular intervals, multiplying the elements of the convolution kernel at each position by the corresponding element of the input, and then summing (sometimes this computation is referred to as the multiply-accumulate operation), and storing this result at the corresponding position of the output. For an image, the convolution kernel slides over each region of the image in turn, from the very beginning of the image, from left to right and from top to bottom, at a spacing of one pixel or a specified number of pixels.

The convolution kernel size ($f \times f$) we choose can also be varied, e.g., $1 \times 5 \times 5$, etc., at which point the fill size needs to be adjusted according to the size of the convolution kernel. Generally, the convolution kernel size is taken as an odd number (because we want the convolution kernel to have a center for easy processing of the output).

In addition, before processing the convolution operation, sometimes you have to fill in fixed data (such as 0, etc.) around the input data, and the purpose of using the filling is to adjust the size of the output so that the output dimension is the same as the input dimension. If the dimensions are not adjusted, after many layers of convolution, the output dimensions will become very small. So, in order to minimize the loss of, edge information caused by the convolution operation, we need to perform padding.

Eventually, after the convolution operation, the output dimension of the image can be expressed as follows:

$$w_{out} = \frac{w + 2 \times PaddingSize - f}{s} + 1 \tag{1}$$

$$h_{out} = \frac{h + 2 \times PaddingSize - f}{s} + 1 \tag{2}$$

f denotes the size of the convolution kernel, s denotes the step size, w denotes the width of the image, h denotes the height of the image, Padding Size denotes the size of the surrounding padding, w_{out}, h_{out} which denotes the width and height of the image after the convolution operation. Second, the structure of the residual module is as follows:

The stacked residual module structure of Res-Net improves gradient propagation, speeds up model training to some extent, and significantly improves model performance over current networks. At the same time, Res-Net is highly scalable and can simply be incorporated into training with other network models to improve performance.

In order to increase feature propagation and speed up model training, we constantly uses feature skip connections at the model's nearest neighbor layer. The main feature of skip connections is that the weights of the layers are directly connected, allowing the model to retain some of the output feature maps from the previous layers. Skip connections are used to create residual units, which then feed the feature maps from the previous layer directly into the next convolutional layer, as shown in Fig. 6.

The above figure shows the residual module in the residual model, and in order to cope with the gradient vanishing in the gradient descent method, the input features are added directly to the output term so that the derivation of the input term is not too small.

Specifically, for the kth residual module in the network, the feature extraction process can be expressed as:

$$y_k = x + f(x) \tag{3}$$

$$x_{k+1} = a(y_k) \tag{4}$$

where x_k is the input to the kth residual structure, f is a function of the convolutional layer and the output of the model features, and a is an activation function, typically a ReLU function, x_k representing the jump connections in the residual module. x_{k+1} is the output of the kth residual structure and the input of the k + 1th residual structure.

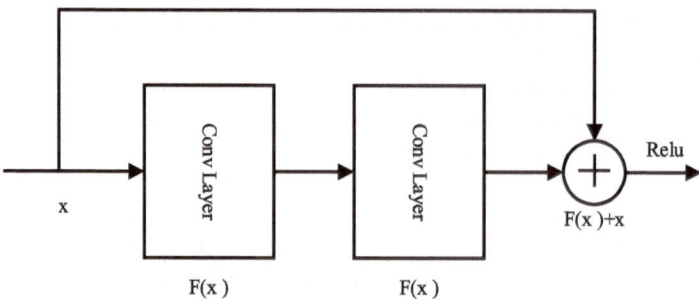

Fig. 6. Residual module

3.2 Transfer Learning Based on Infrared Sensor Failure Dataset

In order to improve the accuracy of the result, we add the transfer learning in the training of the model Transfer learning is used to describe how one activity affects another task in a training approach designed to accomplish knowledge transfer between comparable problems, it solves a problem by applying a currently solved task model to another similar task. The goal of transfer learning is to learn existing knowledge to improve the model's ability to learn new information. The basic idea is to utilize the similarity between previously acquired knowledge and new knowledge. The basic procedure is to utilize the weights of a specific model previously trained on some dataset and migrate to achieve feature extraction on an existing dataset. Utilizing transfer learning on a new model can effectively improve the accuracy of model classification while reducing the training time of the model.

On the task of classification of infrared sensor fault datasets we have, a large amount of labeled data is required to train a high-performance model for a particular task. However there are only a small number of infrared sensor fault datasets available we have. Therefore, it is possible to efficiently achieve the classification performance of the model on a new task by migrating similar models to be trained on a large sample dataset.

The specific training process of the model after improvement based on transfer learning we use to train is shown in Fig. 7. The deep transfer learning network model built by fusing the above convolutional neural network with transfer learning and sharing the weight parameters of the convolutional part trained in the ImageNet image classification task.

4   Experimental Results

This section contains the experimental results of classifying infrared sensor defect images using transfer learning on a small sample dataset. A convolutional neural network based on transfer learning is employed, with the ResNet152 model as the primary model, and VGG19 and DenseNet169 as control groups. Due to constraints in experimental equipment, weight files obtained from training on ImageNet data were sourced from the PyTorch official website. Therefore, the experimental results can be primarily categorized into two classes: 1. Comparative results of deep convolutional neural network

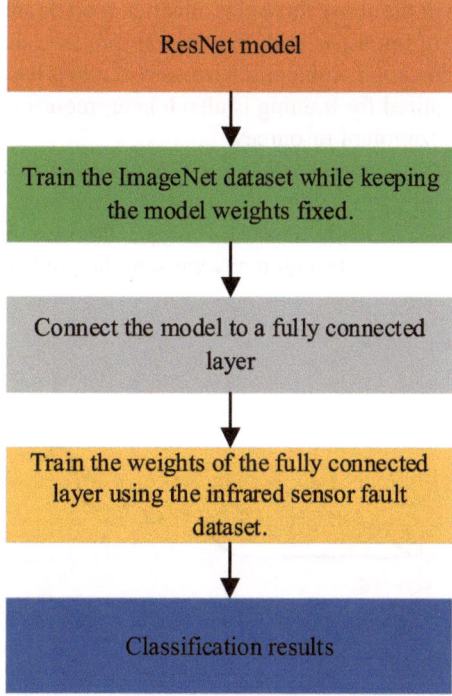

Fig. 7. Training process of transfer learning model

models without transfer learning. 2. Classification results of deep transfer convolutional neural network models in the presence of transfer learning. These two categories of results will be presented and briefly analyzed.

4.1 Analysis of Classification Results Without Transfer Learning

Table 1 presents the classification performance of various models without transfer learning on the small sample dataset of infrared sensor defects. The worst average accuracy is only 60.8%, while the best average accuracy is only 77.8%. From the table, it can be observed that directly using the original deep convolutional neural network models for classification on this small sample dataset does not yield particularly good results. Such classification accuracy is difficult to satisfy for classification tasks.

Table 1. Classification performance of various deep convolutional neural network models on a small infrared sensor dataset

Model	Vgg19	ResNet152	Densenet169
Accuracy	74.6%	60.8%	77.8%

The training curves of the above three classification models are shown in Fig. 8. With the increasing number of iterations with the training process, the model is difficult to converge, and as a classification problem, such a correct rate is hardly satisfactory. At the same time, the time required for training is also longer, requiring 35 training sessions, which is not a good utilization of resources.

Confusion matrix is a commonly used analysis table in machine learning for evaluating the predicted results of classification models. It summarizes the predictions of the model against the true labels in a matrix format. The rows of the confusion matrix represent the true values, while the columns represent the predicted values.

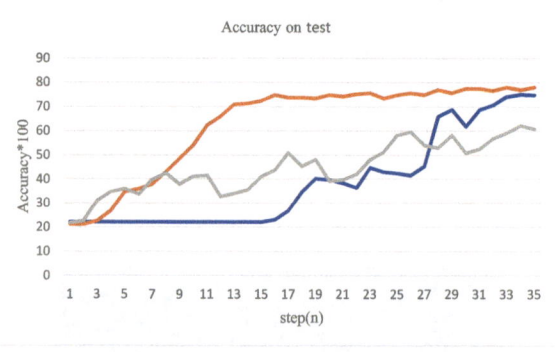

Fig. 8. Accuracy of three models on the training set

Based on the confusion matrix in Fig. 9, the model exhibits good classification performance on images that are prone to blurriness, color deviation and noise, achieving accuracy rates of 97.7%, 98.9%, 100% respectively after training. However, it performs poorly on images representing halo and normal features, with accuracy rates of only 50.6% and 84.6% respectively. This indicates that the model lacks the ability to extract features from images with halo characteristics. In the subsequent classification results of the transfer learning convolutional neural network model experiment, special attention will be given to the classification results of these categories.

4.2   Analysis of Classification Results with Transfer Learning

On the small sample dataset of infrared sensor images, Table 2 presents the classification performance of three deep transfer convolutional neural networks. By comparing Table 2 and Table 2, it can be observed that transfer learning significantly enhances the classification performance of each model on the small sample dataset. The largest gain is 25.4% points, while the smallest gain is an increase of 5.2% points. Among the three models, the ResNet152 model exhibits the highest classification performance after transfer learning. In conclusion, it can be seen that transfer learning helps facilitate training on small sample datasets, and models trained on the ImageNet dataset have excellent feature extraction capabilities.

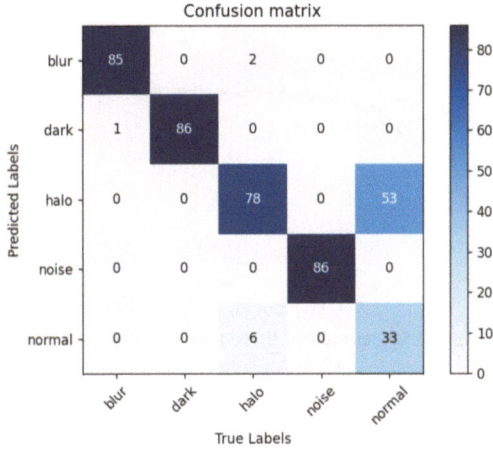

Fig. 9. Confusion matrix of ResNet152 on the test set

Table 2. Fault diagnosis results after improvement

Model	Vgg19	ResNet152	Densenet169
Accuracy	92.8%	94.2%	94.2%

Figures 10 and 11 represent the correctness of the different models on the training and test sets. It can be seen that compared to direct training, the model has basically converged in roughly 10 training sessions, which greatly reduces the training time compared to the previous difficult convergence. At the same time, a high correctness rate is maintained, basically over 90% in all cases.

The confusion matrix in Fig. 12 illustrates that the model's classification performance on the five different categories of images has significantly improved after training. The noise category has the highest classification accuracy, reaching a perfect 100% classification rate, while the halo category has the lowest accuracy at 85.8%. It is important to pay special attention to the classification performance of the normal category. Without transfer learning, it had the poorest performance, but with transfer learning, its classification performance improved significantly. The chart shows a notable improvement in the accuracy of both the halo and normal categories. The accuracy of the normal category increased by 26%, demonstrating the highest gain. It can be observed that the features extracted through transfer learning are particularly advantageous in recognizing typical categories.

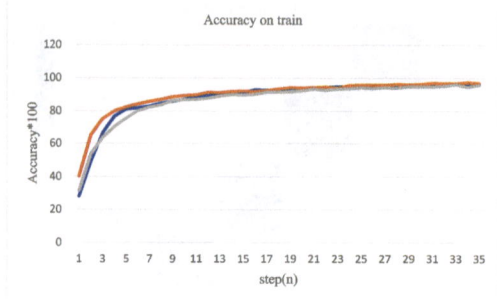

Fig. 10. Accuracy of three models on the train set

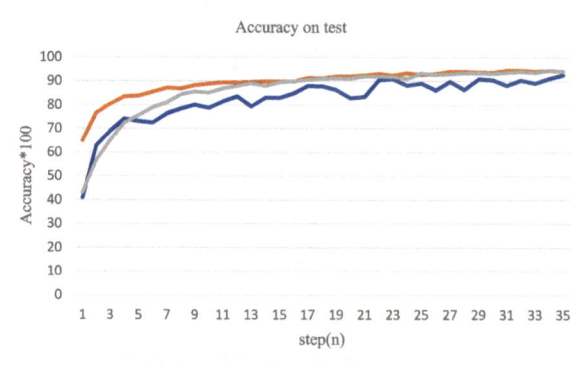

Fig. 11. Accuracy of three models on the test set

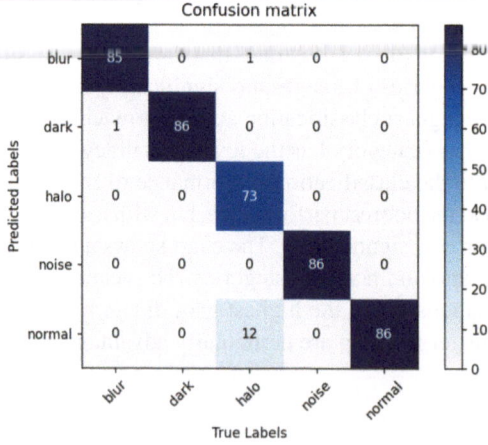

Fig. 12. Confusion matrix of resnet152 on the test set

5   Conclusion

This article proposes a transfer learning-based method for infrared sensor fault diagnosis, which combines transfer learning with convolutional neural networks. The method involves pretraining the model on the ImageNet dataset, fixing the weights, and then fine-tuning the model on the infrared sensor fault dataset. The specific conclusions obtained are as follows:

1) The convolutional structure can effectively identify features in infrared images and effectively recognize faults, but it performs poorly in the case of halo faults.
2) The ResNet model based on residual modules has a significant advantage in classification accuracy compared to other models such as VGG and DES models:
3) Compared to the model before transfer learning improvement, the improved model significantly reduces training time and also shows a substantial increase in classification accuracy.

Acknowledgment. This research was supported by State Grid Corporation of China: Research and application of reliability improvement technology of substation sensors and sensor network considering the influence of multi-physics coupling (5700-202232441A-2-0-ZN). Yingyi Liu is the corresponding author.

References

1. Huo, W., Li, W., Sun, C.: Research on fuel cell fault diagnosis based on genetic algorithm optimization of support vector machine. Energies **15**(6) (2022)
2. Yu Q., Wan C.: A model-based sensor fault diagnosis scheme for batteries in electric vehicles. Energies **14**(4) (2021)
3. Phan, A.T., Bui, V.C.: Data compensation with Gaussian processes regression: application in smart building's sensor network. Energies **15**(23) (2022)
4. Iliev, A., Kyurkchiev, N., Markov, S.: On the approximation of the step function by some sigmoid functions. Math. Comput. Simul **133**, 223–234 (2017)
5. Karras, T., Laine, S., Aila, T.: A style-based generator architecture for generative adversarial networks. In: Proceedings of the IEEE/CVF Conference on Computer Vision and Pattern Recognition, pp. 4401–4410 (2019)
6. Corsi, C.: History highlights and future trends of infrared sensors. J. Mod. Opt. **57**(18), 1663–1686 (2010)
7. Zhou, Z.H.: Machine Learning, pp. 15–64. Springer, Singapore (2021). https://doi.org/10.1007/978-981-15-1967-3
8. Hu, J., et al.: Robustness of deep equilibrium architectures to changes in the measurement model. In: ICASSP 2023–2023 International Conference on Acoustics, Speech and Signal Processing (ICASSP). IEEE (2023)
9. Huang, G., Liu, Z., Van Der Maaten, L., et al.: Densely connected convolutional networks. In: Proceedings of the IEEE Conference on Computer Vision and Pattern Recognition, pp. 4700–4708 (2017)
10. Kshatri, S.S., Singh, D.: Convolutional neural network in medical image analysis: a review. Arch. Comput. Methods Eng. **30**(4) (2023)

11. Malibari, A.A., Obayya, M., Gaddah, A.: Artificial hummingbird algorithm with transfer-learning-based mitotic nuclei classification on histopathologic breast cancer images. Bioengineering **10**(1) (2023)
12. Shao, B., Li, Q., Jiang, X.: A survey of DCGAN based unsupervised decoding and image generation. Int. J. Comput. Appl. **178**(23) (2019)
13. Sensor research; report summarizes sensor research study findings from national Tsing Hua university (IR sensor based on low bandgap organic photodiode with up-converting phosphor). Technol. Bus. J. (2015)
14. Narzary, D., Veluvolu, K.C.: Multiple sensor fault detection using index-based method. Sensors **22**(20) (2022)
15. Nham, D.-H.-N., Trinh, M.-N., Nguyen, V.-D..: An EffcientNet-encoder U-net joint residual refinement module with Tversky–Kahneman Baroni–Urbani–Buser loss for biomedical image Segmentation. Biomed. Signal Process. Control **83** (2023)
16. Yu, F., Xiu, X., Li, Y.: A survey on deep transfer learning and beyond. Mathematics **10**(19) (2022)
17. Smith, B.: An approach to graphs of linear forms (2014). arXiv:2105.02824
18. Chen, X., Yang, R., Xue, Y., Huang, M., Ferrero, R., Wang, Z.: Deep transfer learning for bearing fault diagnosis: a systematic review since 2016. IEEE Trans. Instrum. Meas. **72**, 1–21 (2023). Article no. 3508221. https://doi.org/10.1109/TIM.2023.3244237
19. Zeng, N., Wu, P., Wang, Z., Li, H., Liu, W., Liu, X.: A small-sized object detection oriented multi-scale feature fusion approach with application to defect detection. IEEE Trans. Instrum. Meas. **71**, 1–14, Article no. 3507014 (2022). https://doi.org/10.1109/TIM.2022.3153997
20. Lin, A., Chen, B., Xu, J., Zhang, Z., Lu, G., Zhang, D.: DS-TransUNet: dual swin transformer u-net for medical image segmentation. IEEE Trans. Instrum. Meas. **71**, 1–15, Article no. 4005615 (2022). https://doi.org/10.1109/TIM.2022.3178991
21. Zhang, B., Chen, S., Gao, S., Gao, Z., Wang, D., Zhang, X.: Combination balance correction of grinding disk based on improved quantum genetic algorithm. IEEE Trans. Instrum. Meas. **72**, 1–12, Article no. 1000112 (2023). https://doi.org/10.1109/TIM.2022.3227990

Graph Neural Network Knowledge Graph Recommendation Model Integrating Deep Domain Information and Important Domain Information

Xuelian Zhang[✉]

Computer Science Department, Sichuan University Jinjiang College, Meishan, China
zhangxuelian@scujj.edu.cn

Abstract. The importance between source and target nodes is often overlooked in higher-order information, which can introduce excessive noise in complex network scenarios, thereby affecting recommendation performance. To address this issue, we propose a graph neural network-based knowledge graph recommendation model that integrates deep domain information and important domain information, aiming to enhance recommendation accuracy and diversity. Initially, we pre-train the knowledge graph using graph embedding techniques to obtain structural information. Subsequently, we employ a graph convolutional network to delve deeper into the semantic information of the knowledge graph, capturing rich structural and semantic details from both depth and importance perspectives. Finally, we compute the interaction probability between users and items using the inner product of the enhanced vectors and user vectors, facilitating recommendations. Experiments on the Last-FM, Book-Crossing, and MovieLens-20M datasets yield AUC and F1 scores of 83.2% and 75.4%, 74.9% and 66.8%, 97.9% and 93.1%, respectively. Additionally, Recall@50 scores are 34.5%, 11.2%, and 35.0%. Our model outperforms RippleNet, KGCN, LKGR, and other models, indicating that integrating knowledge graph recommendation models with meta-graph neighborhoods effectively improves recommendation performance.

Keywords: knowledge graph · recommendation system · graph neural network · Deep Search Sampling · importance sampling

1 Introduction

The key to recommendation systems lies in mining and extracting features from massive information. However, traditional recommendation systems solely rely on historical interaction data between users and items as input, limiting their effectiveness due to the sparsity of interaction information and the cold-start problem. The core issue is the insufficiency of information. Therefore, introducing more auxiliary information as input to the model to compensate for the lack of interaction data is a simple and effective solution.

P. Siarry et al. (Eds.): WCNA 2023, LNEE 1361, pp. 147–161, 2025.
https://doi.org/10.1007/978-981-96-2409-6_15

In recent years, the Knowledge Graph (KG) has been widely used as auxiliary information in recommendation systems, attracting keen attention from scholars. Initially, the Knowledge Graph was mainly used to enhance the performance of search systems by building rich relational networks among massive entities, providing users with deeper information exploration capabilities. With further research, the Knowledge Graph has been cleverly integrated into recommendation systems, effectively enhancing users' understanding and cognition of item relevance, thereby significantly improving recommendation accuracy.

Compared with other additional information, the application of the Knowledge Graph in recommendation systems offers numerous distinct advantages. Firstly, it can introduce more semantic relationships, deeply exploring users' personalized preferences and enhancing the precision of recommendations. Secondly, the Knowledge Graph provides diverse connection paths, contributing to the broad coverage of recommendation results, avoiding monotony, and increasing diversity. Thirdly, by effectively linking users' interaction history with recommendation results, the Knowledge Graph improves the interpretability of recommendations, enhancing users' acceptance and trust in the recommended results, and ultimately improving user satisfaction.

Integrating research findings from both domestic and international sources, we can broadly categorize recommendation systems based on knowledge graphs into the following four types:

- Embedding-based approach: This approach employs knowledge graph node embedding techniques to learn low-dimensional vector representations of entities within the graph. These representations can serve as prior information for items. By mapping entities to a low-dimensional vector space, it effectively encodes and represents the semantic features of entities. DeepWalk [1], a pioneering network embedding method, treats network nodes as words in natural language processing and generates random walks to derive network embeddings using NLP models like Skip-Gram.

- Path-based approach: This approach views the knowledge graph as a heterogeneous information network and introduces meta-paths to precisely characterize the similarity between users and items. For instance, PER [2] mines associative relationships by analyzing paths between users and entities. Hu [3] et al. propose the use of meta-graphs instead of meta-paths to provide richer connectivity information. However, this approach has efficiency issues as it requires manual design of meta-paths, unable to automatically infer implicit path patterns. Especially with large-scale data, millions of paths may be generated, significantly increasing the complexity of data storage and model training.

- Graph neural network-based models: The core idea of these models is to aggregate the features of a node itself and its adjacent nodes through stacked graph convolutional layers, thereby generating high-order feature representations for central nodes. Representative models include Knowledge Graph Convolutional Networks (KGCN) [4], Knowledge Graph Attention Network Recommendation (KGAT) [5], and Collaborative Knowledge-Aware Attention Network Recommendation (CKAN) [6].

- Reinforcement learning-based models: These models aim to capture user preference information in the knowledge graph by designing reinforcement learning agents combined with path finding strategy networks. However, the training process of the strategy network is often complex and may face convergence difficulties or the challenge of finding optimal solutions. Typical representative models include the KGPolicy model [7] for knowledge graph-based policy recommendations and the PGPR model [8] for policy-guided path reasoning recommendations.

To overcome the limitations of the above methods, this paper argues that developing a knowledge graph recommendation model that combines simplicity and expressive-ness is crucial. Inspired by embedding and path research, combining depth search layer-by-layer sampling with importance neighbor sampling within a single graph convolutional layer has the potential to achieve this goal. Therefore, this paper proposes a method based on a knowledge graph network that integrates embedding and path, namely the DSINS model (Knowledge graph network that integrates depth search layer by layer sampling and importance neighbor sampling).

The contributions of this work are summarized as follows:

- This paper proposes a recommendation method for knowledge graph networks that combines embedding and path.
- This paper integrates depth search node sampling with importance sampling strategies.
- The application of NCKN to three real-world scenarios of music recommendation, book recommendation, and movie recommendation demonstrates that compared with mainstream algorithms such as RippleNet and KGCN, DSINS achieves accuracy improvements of 1.77% and 0.31% in CTR prediction, respectively, and also exhibits superior performance in top-k prediction.

2   Problem Definition

Within this model, a user set $U = \{u1, u2, ..., um\}$ and an item set $V = \{v1, v2, ..., vn\}$ are provided, where m and n represent the total number of users and items, respectively. The user-item interaction matrix is defined as $Y \in Rm \times n$, with $Yij = 1$ indicating the occurrence of an interaction between user i and item j (such as browsing, clicking, favoring, or purchasing), and $Yij = 0$ otherwise.

Moreover, a knowledge graph $G = \{(E, R)\}$ containing structured information is utilized to represent entities and the relationships among them. Here, E and R denote the set of entities and the set of relationships within G, respectively.

The problem addressed in this study is defined as follows:

Given the user-item interaction matrix Y and the knowledge graph G, the objective is to predict the probability of a user u potentially being interested in an item v that they have not interacted with previously. Furthermore, a top-N recommendation list should be returned for a specific user, sorted in descending order of the predicted probabilities.

3　Method

Traditional methods of utilizing knowledge graphs for recommendation primarily include path-based and embedding based approaches, both of which have their limitations. In this study, we propose a DSINS model (as illustrated in Fig. 1) that combines embedding-based and path-based methods to fully leverage their respective advantages, enabling the capture of richer semantic and structural information from the knowledge graph. Specifically, we first employ graph embedding techniques to pretrain the knowledge graph and obtain structural information about the network. Subsequently, we utilize graph convolutional networks to conduct deep semantic information mining on the knowledge graph, thereby extracting rich structural and semantic information from both depth and importance perspectives.

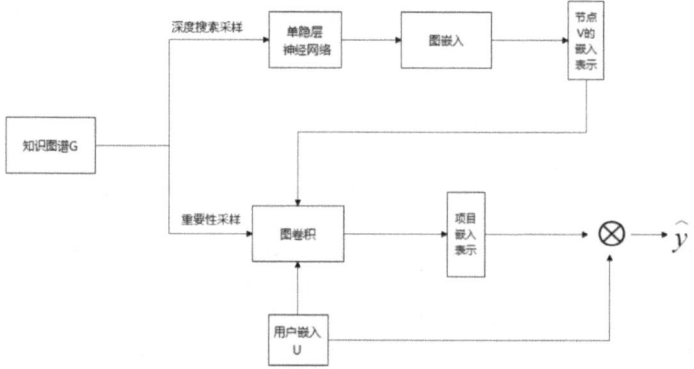

Fig. 1. The Integrated framework of DSINS

Given a specific knowledge graph and user historical interaction information, this model adopts two strategies for learning the knowledge graph: deep search layer-by-layer sampling and importance-based neighborhood sampling.

3.1　Deep Search Layer-by-Layer Sampling

In the knowledge graph, a deep search strategy is employed to conduct layer-by-layer sampling of nodes. For instance, considering the interaction (u, v) between user u and item v, we first locate item v within the knowledge graph. Then, starting from the item node v, we sample nodes in the graph by traversing from the inside out, layer by layer, to obtain multiple node sequences. The depth information of the nodes is controlled by regulating the number of layers traversed. Unlike traditional random walk methods, we specify a global direction during the traversal process without restricting it to local directions. For example, in Fig. 2 below, we begin the traversal from node v (layer 0). The next node can be any random node in layer 1, and the subsequent traversal will target nodes in layer 2, rather than staying within layer 1. This process continues until the designated exploration depth N (here, 4) is reached.

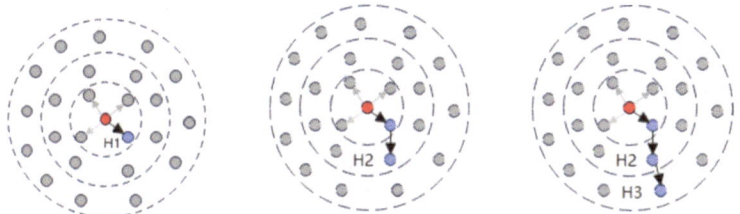

Fig. 2. Navigate the sampling nodes layer by layer

After obtaining multiple traversal paths, a single-hidden-layer neural network is employed to train the set of sampled node sequences, as depicted in Fig. 3. The input layer receives the node sequences, where "1" indicates the presence of a node at the corresponding position, and "0" represents the node to be predicted. In the training process, information of other nodes connected to the current node is predicted based on the current node. After convergence through training on multiple similar paths, the weight matrix of the hidden layer is taken as the embedding matrix vD for the nodes.

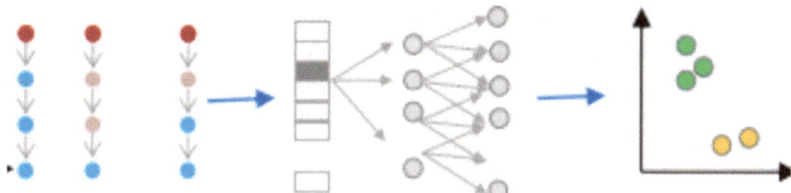

Fig. 3. Graph embedding model

3.2 Graph Embedding

In this study, we adopt Word2vec for graph embedding. Word2vec primarily comprises two models: Continuous Bag of Words (CBOW) and Skip-gram [9]. Comparatively, Skip-gram is more commonly used as it predicts the probability of context based on the current word. Consequently, during the training process, Skip-gram performs more predictions and adjustments, leading to more accurate word embedding.

3.3 Importance-Based Neighborhood Sampling

Employing the aforementioned deep search method for node information sampling may lead to the loss of important node information while introducing excessive noise, making it difficult to fully grasp the potential preferences of users. To address this issue, this paper proposes a method that combines the deep search sampling strategy with importance-based neighborhood sampling during the sampling process, aiming to more accurately capture the potential preferences of users and items. This approach not only reduces the uncertainty associated with random sampling but also provides a guide for model training, accelerating the training process.

The specific steps of the importance-based neighbor sampling algorithm are as follows:

Step1: Calculate the closeness between all nodes in the network and the central node. A higher value indicates a closer relationship between the node and the central node.
Setp2: Sort the closeness values and select the top k values. The nodes corresponding to these k values are the target nodes for sampling.

The metrics used to calculate the closeness of relationships include the CN coefficient (common neighbors) [10], the Adamic-Adar coefficient [11], the RA coefficient (resource allocation coefficient) [12], and others.

The CN coefficient measures the closeness of the relationship between nodes by comparing the number of common neighbors. The more common neighbors there are, the closer the relationship between the nodes. The specific calculation formula is as follows:

$$sim(x, y) = |\Gamma(x) \cap \Gamma(y)| \qquad (1)$$

The Resource Allocation (RA) index algorithm introduces the consideration of degree values on the basis of calculating the number of common neighbors to distinguish the different impacts of common neighbors on the strength of relationships. It holds that the higher the degree of a common neighbor, the more valuable it is. The specific calculation formula is as follows:

$$S_{xy} = \sum_{k \in \Gamma(x) \cap \Gamma(y)} \frac{1}{k(z)} \qquad (2)$$

The Adamic-Adar (AA) index algorithm applies a logarithmic transformation to the degree calculation in the RA coefficient. This is done to account for the possibility that extremely high degrees of certain nodes may interfere with the calculation of relationship closeness. For instance, if a node is connected to most other nodes, it will become a common neighbor for many nodes, and its high degree value can overshadow the influence of other common neighbors. In such cases, the size of the degree can actually hinder the accurate calculation of relationship closeness. Therefore, by applying a logarithmic transformation to the degree, we can reduce the absolute value of the degree, effectively considering its influence without overemphasizing it. The specific calculation formula is as follows:

$$S_{xy} = \sum_{k \in \Gamma(x) \cap \Gamma(y)} \frac{1}{\log k(z)} \qquad (3)$$

The Jaccard similarity index [13] was proposed by Jaccard himself. This index addresses the issue that scientists may have common neighbors because each scientist has many neighbors. However, these common neighbors are not necessarily closely related to each other. The specific definition is as follows:

$$sim(x, y) = \frac{|\Gamma(x) \cap \Gamma(y)|}{|\Gamma(x) \cup var\Gamma(y)|} \qquad (4)$$

In the above formula, Sxy represents the closeness between the neighborhood node y and the central node x. $\Gamma(x)$ represents the set of first-order neighbor nodes of node x, and k(i) represents the degree value of node i.

3.4 Neighborhood Aggregation

In this paper, a graph convolutional network is employed to learn the embedded representations of nodes in the local network of item v. We define the attention factor of user u towards relation r, and function f represents the attention score of user u towards relation r. The specific formula is as follows:

$$\alpha_r^u = f(u, r) = u \cdot r \tag{5}$$

Normalization:

$$\tilde{\alpha}_r^u = \frac{exp(\alpha_r^u)}{\sum_{e \in N(v)} exp(\alpha_r^u)} \tag{6}$$

The attention factors are utilized to aggregate the neighbor information of node v, denoted as $V_{T(v)}^u$, as follows:

$$V_{T(v)}^u = \sum_{e \in T(v)} \tilde{\alpha}_r^u e \tag{7}$$

Finally, the information of node v itself is fused with its neighbor information to obtain a new representation of the node. Here, σ represents a nonlinear activation function.

$$V_k^u = \sigma(W(v + v_{T(v)}^u) + b) \tag{8}$$

3.5 Prediction

For user u and item v, the predicted score output by the model is denoted as, and the true prediction score is denoted as y_{uv}. The cross-entropy loss function is denoted as $J(y_{uv})$, and its calculation form is shown in Eq. (9). The cross-entropy loss function J can reflect the gap between the predicted score and the true score, thereby more accurately assessing the performance of the model.

$$J(y_{uv}, \hat{y}_{uv}) = y_{uv} \lg \hat{y}_{uv} + (1 - y_{uv}) \lg(1 - \hat{y}_{uv}) \tag{9}$$

The loss function Loss is calculated as shown in Eq. (10), where P+ represents positive samples and P− represents negative samples.

$$Loss = \sum_{u \in U} \left(\sum_{v \in \{v | (u,v) \in P^+\}} J - \sum_{v \in \{v | (u,v) \in P^-\}} J \right) \tag{10}$$

4 Experiment

4.1 Introduction to Experimental Environment

This experiment was implemented using the deep learning framework TensorFlow based on Python. TensorFlow is an excellent machine learning platform that adopts a dataflow graph and can run on CPUs or GPUs, facilitating deployment after the dataflow graph is established.

4.2 Introduction to Datasets

In this paper, the CRT and Top-K tasks were conducted on three public datasets: Last.FM (https://grouplens.org/data-sets/hetrec-2011/), Book-Crossing (http://www2.informati kuni-freiburg.de/cziegler/BX/), and MovieLens-20M (Movie). These three datasets differ in size and sparsity. Last.FM contains data on music listening by users of the Last.FM online music system, including 1,872 users' listening data for 3,846 songs. Listens with a count greater than 1 are considered positive samples, and the rest are negative samples. Book-Crossing is a dataset collected from a book community containing reader ratings. MovieLens-20M is an open-source dataset based on user ratings of movies. It contains user feedback information from a movie website with ratings ranging from 1 to 5. In this paper, explicit feedback is converted into implicit feedback, where ratings above 3 are considered positive samples, and the rest are negative samples. Table 1 lists the specific statistics of each dataset.

4.3 Evaluation Metrics

To intuitively demonstrate the effectiveness of the proposed improvement strategies in this paper, we employ the commonly used metric AUC in KGCN and recommendation system CRT tasks for validation. Additionally, in the Top-K recommendation task, the metric F1 is utilized for verification.

AUC: It is used to evaluate the performance of a recommendation system in distinguishing between items that a user likes and dislikes. Let 'a' represent the item liked by the user and 'b' represent the item disliked by the user. Each comparison involves scoring 'a' and 'b' by the recommendation system. 'm' represents the total number of comparisons, 'm' prime represents the number of times when the score of 'a' is higher than that of 'b', and 'm' double prime represents the number of times when the scores of 'a' and 'b' are equal. The AUC is calculated as shown in Eq. (11).

$$AUC = \frac{m' + 0.5m'}{m} \tag{11}$$

The precision of the model's recommended items is calculated as in Eq. (12), where R(u) represents the list of recommended items generated by the system for user 'u' based on the training set, and T(u) represents the list of recommended items generated for user 'u' based on the test set.

$$Precision = \frac{\sum_{u \in U} |R(u) \cap T(u)|}{\sum_{u \in U} |R(u)|} \tag{12}$$

$$Recall = \frac{\sum_{u \in U} |R(u) \cap T(u)|}{\sum_{u \in U} |T(u)|} \tag{13}$$

F1: It is a weighted combination of Precision and Recall, and the value of F1 better reflects the performance of the model.

$$F1 = \frac{2 \times \text{Pr} \, ecision \times \text{Re} call}{\text{Pr} \, ecision + \text{Re} call} \tag{14}$$

Programming for this paper was conducted in the environment of pytorch 1.3.0, and the parameters of all comparison algorithms were adjusted. The learning rate was adjusted within the range of $[10 - 3, 5 \times 10 - 3, 10 - 2, 5 \times 10 - 2]$, and the dimension size of embeddings was adjusted within [8, 16, 32, 64, 128, 256].

4.4 Introduction to Comparison Models

To validate the effectiveness of the proposed DSINS model in this paper, the following models were compared.

BPRMF [14] (Bayesian Personalized Ranking optimizations for Matrix Factorization):

Bayesian matrix factorization employs a Bayesian pairwise loss function to optimize matrix factorization algorithms.

CKE [15] (Collaborative Knowledge base Embedding): This is a model based on entity embedding, which jointly trains knowledge graphs with collaborative filtering and utilizes feature embeddings of multimodal information.

PER [2]: Leveraging the relational heterogeneity in item knowledge graphs, meta-paths are introduced to represent the connectivity between users and items in different relational paths, and items are recommended based on path similarity.

RippleNet [16] : The RippleNet model is the first preference propagation based model. This model analogizes the high-order propagation of user preference information in the knowledge graph to ripple diffusion, thereby enriching the high-order representation of users.

KGCN [4]: This introduces traditional graph convolutional network models into knowledge graphs and treats the combination of users and relationships in the knowledge graph as an important factor in aggregating neighborhood information in the knowledge graph, thereby enriching the high-order representation of item entities in the knowledge graph.

4.5 Introduction to Experimental Parameters

Given that implicit feedback can provide richer interaction content, which is beneficial for alleviating cold start issues, this paper first converts explicit feedback to implicit feedback during data preprocessing. Here, 1 represents positive ratings by users, while 0 represents negative samples randomly sampled from the set of never-interacted items. The interaction data in Last.FM and Book-Crossing are sparse, hence no threshold is set. However, for MovieLens-20M, the positive rating threshold is set to 4. In the deep search layer-by-layer sampling strategy, the exploration depth is set to 8, and the search frequency is set to 8 for the Last.FM dataset, while it is set to 32 for both Book-Crossing and MovieLens-20M datasets.

4.6   Analysis of Experimental Results

In this section, we conducted performance comparison experiments with baseline models, ablation experiments, and model parameter sensitivity analysis experiments.

1) performance comparison experiments

Performance comparison between the proposed model and the comparison models. The experimental results are presented in Table 1 below.

Table 1. Prediction results in CTR task based on AUC metric

Model	Last.FM		Book-Crossing		Movielens-20M	
	AUC	F1	AUC	F1	AUC	F1
BPRMF	0.752	0.718	0.669	0.617	0.962	0.916
CKE	0.781	0.691	0.684	0.628	0.929	0.877
PER	0.658	0.628	0.608	0.569	0.841	0.801
RippleNet	0.797	0.714	0.729	0.651	0.978	0.927
KGCN	0.813	0.731	0.691	0.639	0.976	0.929
DSINS	0.832	0.754	0.749	0.668	0.979	0.931

In the CTR task, we first selected the three datasets of Last.FM, Book-Crossing, and MovieLens-20M to compare and evaluate the proposed model D3INS with the aforementioned five baseline models. The prediction results based on the AUC metric in the CTR task are presented in Table 1, while the prediction results in the top-k scenario are shown in Figs. 4, 5 and 6, respectively.

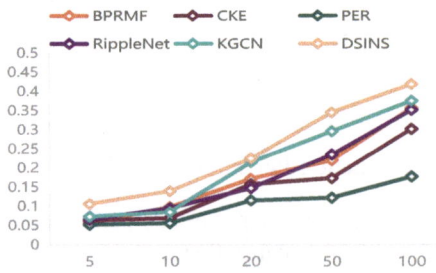

Fig. 4. Prediction results of Last.FM in top-k

Based on the experimental results, the following conclusions can be drawn:

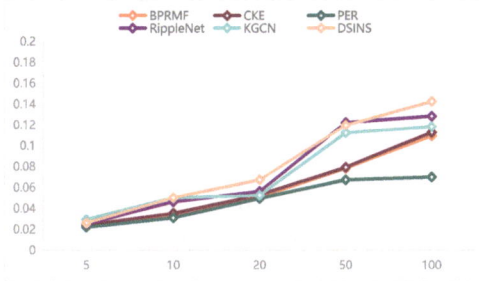

Fig. 5. Prediction results of Book-Crossing in top-k

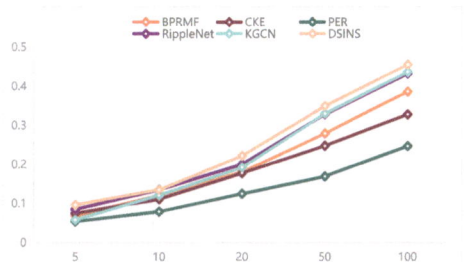

Fig. 6. Prediction results of MovieLens in top-k

The performance of the four models, namely BPRMF, CKE, PER, and RippleNet, is relatively low. BPRMF solely utilizes the information inherent in the dataset without introducing auxiliary information. Although PER incorporates auxiliary information through a knowledge graph, this method heavily relies on manually defined meta-paths. CKE fails to consider high-order information from the knowledge graph and, in practical operations, does not incorporate crucial information such as text and vision, resulting in suboptimal performance. RippleNet asymmetrically generates user embeddings and item embeddings, leading to poor performance. KGCN solely explores knowledge graph-related information on the item-side embeddings, neglecting user-side embedding-related information from the knowledge graph, thus resulting in subpar performance.

Upon observation, it is evident that the movie dataset exhibits superior performance across all methods compared to the music and book datasets. This may be attributed to the movie dataset's inherent richness in interactive behaviors and relational links, providing more abundant information that enables recommendation models to learn latent feature representations more accurately.

Compared to all other methods, the proposed DSINS model demonstrates the best performance across the three datasets. Specifically, in CTR prediction, DSINS significantly outperforms mainstream algorithms such as RippleNet and KGCN on the Last.FM and Book-Crossing datasets. On the Movielens-20M dataset, DSINS achieves approximately 1.77% and 1.64% improvements in AUC and F1, respectively, compared to the

BPRMF model, and approximately 0.31% and 0.22% improvements compared to the KGCN model.

2) Ablation experiments

The ablation study of the model aims to validate the effectiveness of its two crucial components. The experimental results, as shown in Table 2, indicate that DSINS_1 solely utilizes graph embedding techniques and deep sampling methods to learn structural information from the knowledge graph. In contrast, DSINS_2 employs importance-based neighborhood sampling to capture vital neighbor information from the knowledge graph.

Table 2. Results of ablation experiments

Model	Last.FM		Book-Crossing		Movielens-20M	
	ACU	F1	ACU	F1	ACU	F1
DSINS_1	0.643	0.506	0.712	0.531	0.793	0.615
DSINS_2	0.691	0.594	0.598	0.603	0.896	0.804
DSINS	0.767	0.695	0.839	0.681	0.921	0.859

Based on the experimental results above, the following conclusions can be drawn:

DSINS1 exhibits the worst performance across all datasets, demonstrating that relying solely on structural information from the knowledge graph for recommendation using deep sampling based on meta-path embeddings is insufficient.

DSINS2 outperforms DSINS1 in various metrics, indicating that the graph convolutional network based on importance sampling is more effective and has a significant impact on recommendation results.

A comparison between DSINS1 and DSINS2 reveals a significant improvement in performance for the latter. This suggests that integrating semantic information with the wide-area approach can effectively enhance the performance of a single method.

The optimal performance of DSINS demonstrates that the fusion of multiple methods based on graph neural networks is more advantageous than a single meta-path-based approach in utilizing the knowledge graph for recommendation.

3) Sensitivity analysis experiments on model parameters

Impact of exploration depth L in the deep-area exploration method. As shown in the Fig. 7, increasing L within a certain range can significantly improve model performance. However, further increasing L beyond a certain threshold has little effect on model efficiency and may even lead to a decline. This verifies that enhancing the deep-area information of each node is beneficial to the model, but deeper exploration is not always better. This is because when L is too large, overfitting issues may arise.

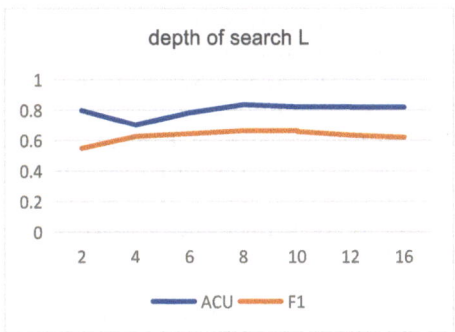

Fig. 7. Experimental results for search depth L in deep search

5   Conclusions

The DSINS model proposed in this paper combines embedding-based and path-based methods, aiming to fully exploit the semantic and structural information within the knowledge graph. Through the pre-training of graph embedding techniques, the model successfully captures the structural characteristics of the network. Furthermore, by utilizing graph convolutional networks to conduct deep semantic mining of the knowledge graph, it not only explores information in depth but also considers the importance of the information, thus obtaining a richer and more comprehensive representation. Experimental results show that the DSINS model outperforms other baseline models in movie and music recommendation scenarios, fully demonstrating its effectiveness and practicality. However, as with any research work, there are some directions worthy of further exploration and improvement in this paper.

Firstly, current recommendation systems primarily rely on users' static preferences for recommendations, which may limit the system's understanding of users' real-time interests to a certain extent. Therefore, future research can focus on how to incorporate users' dynamic preferences into recommendations to more accurately capture their changing needs and provide more personalized services. Secondly, for the processing of large-scale knowledge graphs, most current methods adopt the approach of sampling to construct subgraphs, which is effective but relatively time-consuming. To improve processing efficiency, future research can further explore domain-specific construction and randomly generated sampling strategies, aiming to reduce computational costs and increase processing speed while maintaining information integrity.

Acknowledgment. This work is supported by Youth Fund Project of Sichuan University Jinjiang College (QNJJ-2022-A03) and Network and Data Security Key Laboratory of Sichuan Province, University of Electronic Science and Technology of China Chengdu (No. NDS2023-3).

References

1. Hu, J., Xu, Z., Liu, L., et al.: Network embedding method incorporating multi-granularity community information. J. Comput. Appl. **42**(03), 663–670 (2022)

2. Yu, X., Ren, X., Sun, Y.Z., et al.: Personalized entity recommendation: a heterogeneous information network approach. In: WSDM 2014: Proceedings of 7th ACM International Conference on Web Search and Data Mining, New York, pp. 283–292. ACM (2014)
3. Hu, B.B, Shi, C., Zhao, X.W., et al.: Leveraging meta-path based context for top-n recommendation with a neural co-attention model. In: KDD 2018: Proceedings of the 24th ACM SIGKDD International Conference on Knowledge Discovery & Data Mining, New York, pp. 1531–1540. ACM (2018)
4. Wang, H.W., Zhao, M., Xie, X., et al.: Knowledge graph convolutional networks for recommender systems. In: WWW 2019: Proceedings of the 28th World Wide Web Conference, New York, pp. 3307–3313. ACM (2019)
5. Wang, X., He, X.N., Cao, Y.X., Liu, M., Chua, T.S.: KGAT: know-ledge graph attention network for recommendation. ar Xiv pre-print ar Xiv: 1905.07854 (2019)
6. Xu, Z., Liu, H., Li, J., et al.: CKGAT: collaborative knowledge-aware graph attention network for top-N recommendation. Appl. Sci. **12**(3), 1669–1669 (2022)
7. Wang, X., Xu, Y., He, X., et al.: Reinforced negative sampling over knowledge graph for recommendation. In: Proceedings of the Web Conference 2020, pp. 99–109 (2020)
8. Xian, Y.K., Fu, Z.H., Muthukrishnan, S., de Melo, G., Zhang, Y.F.: Reinforcement knowledge graph reasoning for explainable recom-mendation. ar Xiv preprint arXiv:1906.05237 (2019)
9. Dai, Y.: Automatic keyword extraction algorithm based on word embedding and multi-feature fusion. Guangdong Polytechnic Normal University (2019). https://doi.org/10.27729/d.cnki. ggdjs.-2019.000006
10. Lü, L., Zhou, T.: Link prediction in complex networks: a survey. Physica A **390**(6), 1150–1170 (2011)
11. Adamic, L.A., Adar, E.: How to search a social network. Soc. Netw. **27**(3), 187–203 (2005)
12. Zhou, T., Lü, L., Zhang, Y.: Predicting missing links via local information. Eur. Phys. J. B. **71**(5), 623–630 (2009)
13. Zhang, C.: Research on link prediction algorithm based on compactness and node contribution. Yanshan University (2017)
14. Rendle, S., Freudenthaler, C., Gantner, Z., et al.: BPR: Bayesian personalized ranking from implicit feedback. In: UAI 2009: Proceedings of the 25th Conference on Uncertainty in Artificial Intelligence. Montreal: AUAI, pp. 452–461 (2009)
15. Zhang, F., Yuan, N.J., Lian, D., et al.: Collaborative knowledge base embedding for recommender systems. In: KDD 2016: Proceedings of the 22nd ACM SIGKDD International Conference on Knowledge Discovery and Data Mining, New York. ACM, pp. 353–362 (2016)
16. Wang, H.W., Zhang, F.Z., Wang, J.L., et al.: RippleNet: propagating user preferences on the knowledge graph for recommender systems. In: CIKM 2018: Proceedings of the 27th ACM International Conference on Information and Knowledge Management, New York, pp. 417–426. ACM (2018)

An Approach to Microenvironment-Based Particle Swarm Optimization Algorithm

Tingting Chen[1,2], Zhenya Zhang[1,2(✉)], Ping Wang[1,2], and Hongmei Cheng[1,2]

[1] Anhui Province Key Laboratory of Intelligent Building and Building Energy Saving, Anhui Jianzhu University, Hefei, China
`chenting2@stu.ahjzu.edu.cn`, `zzychm@ustc.edu.cn`,
`wangping@ahjzu.edu.cn`, `hmcheng@mail.ustc.edu.cn`
[2] School of Electronics and Information Engineering, Anhui Jianzhu University, Hefei, China

Abstract. The Insect Intelligent Building (I^2B) platform is a distributed information processing system based on a microenvironment network. To solve optimization problems in microenvironments using the Particle Swarm Optimization (PSO) algorithm, this paper designs the information exchange strategy of the PSO algorithm in microenvironments. It optimizes the particle velocity and position updating mechanism to propose the Microenvironment-based Particle Swarm Optimization (MPSO) algorithm. In the microenvironment network, the information exchange strategy is used for collaborative computation among microenvironments, which can simultaneously solve a given optimization problem, thus generating multiple candidate optimal solutions and finally determining the optimal solution. The experimental results indicate that the MPSO algorithm proposed in this paper can solve the optimization problem more efficiently.

Keywords: insect intelligent building · building unit · information process unit · microenvironment · particle swarm optimization algorithm

1 Introduction

Architecture is a human-made environment created using material and technical means, combining scientific laws, feng shui concepts, and aesthetic principles to meet the needs of social life. It covers buildings and structures [1]. The building can be divided into several spatial units with different functions [2], in which the facilities and equipment are connected through complex pipelines and skillfully laid out in the corresponding spatial units according to their respective functional characteristics [3]. The building units follow the principles of distribution and localization to maintain their functions [4]. Buildings can be divided into adjacent spatial units based on their structure, forming a network of branches based on spatial adjacencies. Building facilities consist of facility units distributed within these spatial units, with each unit constructing a network of facility units based on spatial adjacency [5]. Therefore, a building can be regarded as a complex network system composed of a series of interrelated spatial and facility units [6]. The Insect Intelligent Building (I^2B) platform assigns an information processing unit

© The Author(s) 2025
P. Siarry et al. (Eds.): WCNA 2023, LNEE 1361, pp. 162–169, 2025.
https://doi.org/10.1007/978-981-96-2409-6_16

to each building unit to enable independent information processing [7]. Optimization is often necessary for critical issues in building units, such as intelligent sensor failure detection and people counting [8]. The Particle Swarm Optimization (PSO) algorithm can be applied to building functional maintenance due to its computational simplicity and efficiency [9, 10]. As many optimization problems in building units involve multi-objectives [11], the Niche Particle Swarm Optimization (NPSO) algorithm is used to solve multi-objective optimization problems [12, 13], which divides the search space into different regions and uniformly distributes particles to form subpopulations in each area. Equations (1) and (2) describe the velocity and position of particles in the PSO algorithm.

$$v^{(t+1)} = \omega \times v^{(t)} + c_1 \times r_1 \times \left(p\text{Best} - x^{(t)} \right) + c_2 \times r_2 \times \left(g\text{Best} - x^{(t)} \right) \quad (1)$$

$$x^{(t+1)} = x^{(t)} + v^{(t+1)} \quad (2)$$

Equations (1) and (2), at time t, $v(t)$ denotes the velocity, $x(t)$ denotes the position, ω denotes the inertia factor, c_1, c_2 denote the individual and the global learning factor, respectively, r_1 and r_2 denote the random numbers, $pBest$ denotes the individual best position of the particle, and $gBest$ denotes the global best position of the whole particle.

In the I^2B platform, if the information processing unit of the building unit is considered a microenvironment, spatially adjacent microenvironments can exchange information. All microenvironments in the building form a microenvironment network. When using PSO algorithms to solve optimization problems in microenvironment networks, one has to face the following: (1) When the computing power of the information processing unit is weak, too many particles can lead to slower convergence. Conversely, too few particles can converge quickly, but the accuracy of the solution cannot be guaranteed; and (2) The PSO algorithm is stochastic and usually requires multiple attempts to approach the optimal solution, increasing the computational power or time demand. In fact, the optimization problem can be disassembled, where each microenvironment uses only a small number of particles to run PSO instances. The necessary results are shared through collaborative computation in the network, which reduces the arithmetic requirements for individual microenvironments and allows them to jointly approximate the optimal optimization problem solution, thus obtaining multiple near-optimal solutions more quickly. The main works in this paper are:

(1) According to the working principle of the PSO algorithm, the information exchange strategy of the PSO algorithm and the updating mechanism of particle velocity and position in the microenvironment network are studied, and the MPSO algorithm is proposed.
(2) Evaluate the effectiveness of the MPSO algorithm by using a benchmarking function.

Section 2 proposes the MPSO algorithm. Section 3 tests the MPSO algorithm using standard test functions. Section 4 concludes with a summary and outlook.

2 Microenvironment-Based Particle Swarm Optimization Algorithm

A building is composed of multiple building units with different functions. The computing environment within a building unit is defined as a microenvironment equipped with sensing and processing devices responsible for collecting and processing data within the unit; these microenvironments communicate with each other. Adjacent building units that can communicate and share data are called neighbor microenvironments. So, these microenvironments form a network of microenvironments that only allow communication between directly connected neighbors (see Fig. 1).

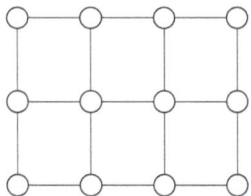

Fig. 1. Microenvironment network topology.

Figure 1 shows microenvironment nodes as dots and their adjacency as edges. The microenvironment nodes are arranged in a 3 × 4 form, totaling 12 nodes.

In the MPSO algorithm, each microenvironment approximates the optimal solution based on the number of particles assigned. During each iteration, the local computation results are compared with those of the neighboring microenvironments to update the local global optimal solution and gradually converge to the optimal solution. The historical optimal positions of particles are kept in each microenvironment. The setup follows: (1) Each microenvironment contains multiple particles; (2) The global best position of a microenvironment is denoted as $gBest_L$; and (3) If a microenvironment has m neighboring microenvironments, the global best position of its ith neighboring microenvironment is denoted as $gBest_{Li}$. In a given microenvironment, position x(t) and velocity v(t) at time t are determined by Eqs. (4) and (2), where $gBest_A$ is defined by Eq. (3), which represents the data exchange mechanism used by the MPSO algorithm.

$$gBest_A = \min\left(gBest_L, gBest_{L_1}, gBest_{L_2}, \ldots\ldots, gBest_{L_m}\right) \tag{3}$$

$$v^{(t+1)} = \omega \times v^{(t)} + c_1 \times r_1 \times \left(pBest - x^{(t)}\right) + c_2 \times r_2 \times \left(gBest_A - x^{(t)}\right) \tag{4}$$

Algorithm 1 outlines the steps for solving the optimization problem in a microenvironment.

Algorithm 1: Localized MPSO Algorithm

Input: network topology *topo*, number of particles *n*, c_1, c_2, ω, maximum number of iterations *maxN*, fitness function *fun*

Output: $gBest_L$

1. Initialization settings

2. Compute and save the local global best position $gBest_L$

3. Begin the main loop:

4. Obtain the global optimal position of each neighboring microenvironment $gBest_{Li}$, $i=1,2...m$.

5. Calculate $gBest_A$ according to Equation (3)

6. Update v and x according to Equation (4) and (2)

7. Calculate *fun(x)* for each local particle

8. Update *pBest* and $gBest_L$ of each local particle

9. Save $gBest_L$ for this iteration

10. The main loop ends when the condition is met.

Algorithm 1, the fitness value is calculated using the corresponding fitness function *fun*, which measures the superiority of the position. The smaller the fitness value, the better the position. Step 4 involves obtaining data from neighboring microenvironments, divided into (1) the synchronous MPSO algorithm, which obtains data from neighboring microenvironments for a specified number of iterations at each iteration; and (2) the asynchronous MPSO algorithm obtains the latest iteration data of the neighboring microenvironments at each iteration. In a microenvironment network, population particles are evenly distributed to each microenvironment. Let *gBest* be the globally optimal position of all particles in the microenvironment network at time *t*. For any microenvironments, $fun(gBest_L) \geq fun(gBest_A) \geq fun(gBest)$. Algorithm 1 is used to update the velocity and position of the particles so that the particles in this microenvironment converge. Furthermore, although *gBest* may not appear in the particle swarms of each microenvironment, any two microenvironments are reachable in the microenvironment network by a finite number of steps. *gBest* can be passed to the particle swarms in each microenvironment after a finite number of iterations based on the connectivity properties between microenvironments and Algorithm 1. It means that the optimal position of the particle swarm within each microenvironment in the microenvironment network is finitely reachable from *gBest*. The slight difference in the local optimal value of the particle swarm in each microenvironment at the end of Algorithm 1 demonstrates this.

3 Experimental Results

The evaluation used three benchmark functions [14] (see Table 1). The experiments used 12 computers with Intel Core i7 processors and 16 GB of RAM, Linux Fedora 6.4.14-200.fc38.x86_64 system. The MariaDB 10.5 database was used to store the information locally. The first experiment uses the standard PSO algorithm with 240 particles, $\omega = 0.8$, and c1 and c2 2.0. The experiments were repeated 12 times with two iteration termination criteria: (1) 200 iterations; and (2) 200 iterations or the expected precision value reached 1e−05. Next, the MPSO algorithm, the microenvironment network topology is shown in Fig. 2, which adopts the spatial distribution structure of the Key Laboratory of Intelligent Building and Building Energy Saving in Anhui Province, with 12 microenvironments, each containing 20 particles (total 240 particles), the iteration termination criteria and ω, c1, c2 are the same as the standard PSO algorithm and consider both synchronous and asynchronous MPSO algorithms.

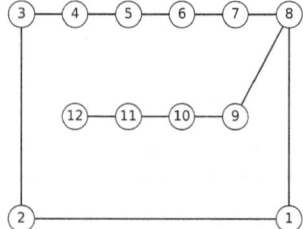

Fig. 2. Network topology for 12 microenvironments.

Table 1. Benchmark Function

Benchmark Function	Definition	Search Scope
Sphere	$$f(x) = \sum_{i=1}^{n} x_i^2$$	[−100, 100]
Rosenbrock	$$f(x) = \sum_{i=1}^{n-1} \left[100\left(x_{i+1} - x_i^2\right)^2 + (x_i - 1)^2 \right]$$	[−100, 100]
Ackley	$$f(x) = -20e^{-0.2\sqrt{\frac{1}{n}\sum_{i=1}^{n} x_i^2}} - e^{\frac{1}{n}\sum_{i=1}^{n} \cos(2\pi x_i)} + 20 + e$$	[−100, 100]

Tables 2, 3, 4, 5, 6 and 7 show the performance of the PSO and MPSO algorithms, including the maximum, minimum, and median values. Both the PSO and the synchronous MPSO algorithms achieve convergence in terms of accuracy. However, the synchronous MPSO algorithm exhibits higher stability and accuracy, especially the Ackley function. Regarding time, the MPSO algorithm significantly reduces execution time compared to The PSO algorithm. Regarding iterations, the synchronous MPSO

Table 2. Performance of the PSO algorithm ($maxN = 200$)

Benchmark Function	Time (Seconds)				Accuracy		
	Max	Min	Median value	Total Time	Max	Min	Median value
Sphere	9	7	8	96	2.70e−08	1.43e−11	1.08e−09
Rosenbrock	8	6	7	88	3.91e−05	1.61e−09	1.76e−07
Ackley	10	8	9	108	8.77e−04	2.93e−06	5.65e−05

Table 3. Performance of the PSO algorithm ($maxN = 200$, *desired accuracy* = 1e−05)

Benchmark Function	Time (Seconds)				Accuracy			Iterations		
	Max	Min	Median value	Total Time	Max	Min	Median value	Max	Min	Median value
Sphere	4	2	3	36	9.93e−06	1.05e−06	3.70e−06	109	44	78
Rosenbrock	7	4	5	62	9.01e−05	1.04e−06	3.62e−05	200	94	140
Ackley	10	8	9	105	5.44e−04	6.71e−06	5.32e−05	200	179	200

Table 4. Performance of the synchronous MPSO algorithm ($maxN = 200$)

Benchmark Function	Time (Seconds)			Accuracy		
	Max	Min	Median value	Max	Min	Median value
Sphere	15	9	12	2.77e−10	1.90e−11	2.77e−10
Rosenbrock	16	10	13	2.69e−06	2.69e−06	2.69e−06
Ackley	16	10	13	5.22e−06	5.22e−06	5.22e−06

Table 5. Performance of the synchronous MPSO algorithm ($maxN = 200$, *desired accuracy* = 1e−05)

Benchmark Function	Time (Seconds)			Accuracy			Iterations		
	Max	Min	Median value	Max	Min	Median value	Max	Min	Median value
Sphere	15	2	9.5	6.91e−06	6.91e−06	6.91e−06	60	55	58
Rosenbrock	11	6	8.5	3.52e−06	3.52e−06	3.52e−06	137	132	135
Ackley	15	10	12.5	9.29e−06	9.29e−06	9.29e−06	185	180	183

algorithm has better convergence characteristics than the PSO algorithm. However, the asynchronous MPSO algorithm requires more iterations to achieve the desired accuracy.

Table 6. Performance of the Asynchronous MPSO algorithm (*maxN* =200)

Benchmark Function	Time (Seconds)			Accuracy		
	Max	Min	Median value	Max	Min	Median value
Sphere	11	8	9	5.22e−08	1.76e−10	5.71e−09
Rosenbrock	12	7	8.5	8.44e−01	2.37e−05	2.14e−03
Ackley	12	7	9	5.54e−04	1.49e−05	2.81e−05

Table 7. Performance of the Asynchronous MPSO algorithm (*maxN* = 200, *desired accuracy* = 1e−05)

Benchmark Function	Time (Seconds)			Accuracy			Iterations		
	Max	Min	Median value	Max	Min	Median value	Max	Min	Median value
Sphere	7	3	6	7.29e−06	1.29e−06	4.50e−06	178	88	116
Rosenbrock	11	6	8	2.96e−03	1.45e−06	1.54e−05	200	143	200
Ackley	13	7	9	1.72e−04	5.79e−06	2.31e−05	200	179	200

In summary, the synchronous MPSO algorithm has a more stable performance, while the asynchronous MPSO algorithm has a shorter computation time.

4   Conclusion

Aiming at the information processing process in buildings, which is naturally distributed, this paper proposes the MPSO algorithm. Using the lab microenvironment network structure and benchmark functions, the experiments verified the effectiveness of the MPSO algorithm and its faster convergence in solving the optimization problem compared to the standard PSO algorithm, and the MPSO algorithm's use of a distributed architecture to achieve a more stable solution. In addition, each microenvironment performs information processing tasks within a building unit, which, from an edge computing perspective, can be understood as a manifestation of the potential for edge computing within a building. The MPSO algorithm can manifest in forms such as distributed systems or edge computing systems. However, its adaptation is not limited to the I^2B platform. Fast problem-solving can be achieved using similar computational mechanisms in computing environments with weak computational capabilities.

The MPSO algorithm is expected to be evaluated in real-world applications for optimization problems in building operations and maintenance. Additionally, it is crucial to explore and verify the most effective communication mechanisms for coordinating and cooperating tasks when multiple microenvironments must work together to achieve a common goal.

References

1. Zhang, Z.H.: A Study on the integration of interior design and architectural design. China Décor. **5**, 116–117 (2020). (in Chinese)
2. Wang, W.: Design of public building space in smart city based on big data. J. Environ. Public Health **2022**, 1–10, Article no. 4733901 (2022)
3. Diao, P.H., Shih, N.J.: BIM-based AR maintenance system (BARMS) as an intelligent instruction platform for complex plumbing facilities. Appl. Sci. **9**(8), 1592 (2019)
4. Xu, J., Li, D., Gu, W., Chen, Y.: UAV-assisted task offloading for IoT in smart buildings and environment via deep reinforcement learning. Build. Environ. **222**, 109218 (2022)
5. Zhao, T.Y., Guan, X.L., Chen, Y.F., Hua, P.M.: Research on information model of architectural spatial units for distributed architecture. Build. Sci. **39**(8), 233–240 (2023). (in Chinese)
6. Zhang, Z.Y., Luo, L.C., Wang, Y., Wang, P., Cheng, H.M.: A distributed particle swarm optimization algorithm based on CPN network. J. Anhui Jianzhu Univ. **29**(6), 27–34 (2021). (in Chinese)
7. Zhang, Z.Y., Fang, B., Wang, P., Cheng, H.M.: A local area network-based insect intelligent building platform. Int. J. Pattern Recognit Artif Intell. **37**(2), 1–16 (2023)
8. Ahmad, N., Egan, M., Gorce, J.M. , Dibangoye, J.S., Le Mouël, F.: Codesigned communication and data analytics for condition-based maintenance in smart buildings. IEEE Internet Things J. **10**(18), 15847–15856 (2023)
9. Kennedy, J., Eberhart, R.: Particle swarm optimization. In: Proceedings of ICNN'95 - International Conference on Neural Networks, Perth, WA, Australia, vol. 4, pp. 1942–1948 (1995)
10. Huang, Y., Zhang, J., Mo, Y., Lu, S., Ma, J.: A hybrid optimization approach for residential energy management. IEEE Access **8**, 225201–225209 (2020)
11. Yu, M.G., Pavlak, G.S.: Extracting interpretable building control rules from multi-objective model predictive control data sets. Energy **240** (2022)
12. Brits, R., Engelbrecht, A.P., van den Bergh, F.: A niching particle swarm optimizer. In: Proceedings of the 4th Asia-Pacific Conference on Simulated Evolution and Learning, vol. 2, pp. 692–696 (2002)
13. Zhuang, Y., Huang, Y., Liu, W.: Integrating sensor ontologies with niching multi-objective particle swarm optimization algorithm. Sensors **23**(11), Article no. 5069 (2023)
14. Stacey, A.: Particle swarm optimization with mutation. In: 2003 Congress on Evolutionary Computation, pp. 1425–1430. IEEE (2003)

A Verifiable Ciphertext Retrieval Scheme for Smart Grids

Xiuqing Lin, Jueyu Chen$^{(\boxtimes)}$, Yupeng Qin, Zhou Yang, and Jinjin Li

Measurement Center of Guangxi Power Grid Co., Ltd., NanNing, Guangxi, China
jueyuchen@qq.com, yang_z.sy@gx.csg.cn

Abstract. With the expanding volume of smart grid data, cloud storage has emerged as a cost-effective scheme for their maintenance. However, storing plaintext data in the cloud may lead to a significant risk of privacy leakage of web data. Existing cryptography-based systems are often inefficient because data must be decrypted to access it. To address these challenges, a new approach for verifiable digital text discovery is proposed. The program allows recovery operations to be performed while ensuring data confidentiality and availability. This new scheme involves: Generating root hashes using Merkle trees and leverage Blockchain guarantees the trustworthiness of search results. In addition, a multi-cloud heterogeneous distributed system storage system is designed to store user data, which has the ability to repair damaged data. This feature not only improves data security, but also ensures data availability. Finally, the security analysis and simulation results verify the security and efficiency of the protocol.

Keywords: Smart grid · Merkle tree · Searchable encryption · Data confidentiality · Data availability

1 Introduction

The electricity sector has experienced considerable development in recent years, but traditional grid systems have inherent flaws such as limited interactivity, low intelligence and insufficient security, making it difficult to meet real demand. New smart grid systems are gradually replacing traditional grids, improving their stability, reliability and efficiency [1, 2]. In smart grid operating scenarios, environmental data is collected by detection devices on a regular or irregular basis. Uploading network data to cloud storage for centralized management and unified planning by power companies helps reduce storage costs [3].

Deploying cloud storage for power grid data management can effectively address the challenges of big data processing, but it also brings potential security risks. Smart grids need to process a large amount of sensitive user information, such as personal grid data and usage habits. Data privacy protection has become an important obstacle to the development of smart grids [1]. It is common practice to upload data to a cloud server before encryption, but this limits the availability of the data for operations such as search and computing. Downloading ciphertext files from cloud storage and then decrypting

© The Author(s) 2025
P. Siarry et al. (Eds.): WCNA 2023, LNEE 1361, pp. 170–179, 2025.
https://doi.org/10.1007/978-981-96-2409-6_17

and performing search operations not only consumes a lot of bandwidth resources, but also is inefficient [4, 5]. In addition, cloud data is vulnerable to serious threats such as hacking, natural disasters, and equipment failures. Cloud storage service providers may also selectively delete inaccessible duplicate data without notifying the data owner, often for data redundancy reasons [6]. Therefore, ensuring the confidentiality, integrity, and availability of data on cloud storage is a key issue we explore in this article.

Searchable Encryption (SE) is a technology that enables encrypted data to be searched, allowing users to query specific information within ciphertext datasets without revealing the content. SE algorithms can greatly improve data availability and are commonly used in fields like healthcare and finance [4]. However, current SE algorithms cannot ensure the integrity of search results, and in the event that cloud storage is compromised, these systems fail. To solve this problem, we suggest a novel approach to searchable encryption scheme specifically designed for power grid data uploaded to the cloud, permitting data searches without decryption. Upon receiving the ciphertext, the scheme uses the Merkle tree root to verify the authenticity of retrieval results. By utilizing the open-source Tahoe-LAFS, we have created a multi-cloud heterogeneous distributed storage cluster to secure user data. Even if some cloud nodes are compromised, user data can still be recovered. This proposed scheme enhances the reliability of search results while protecting data confidentiality and availability, making it applicable to various scenarios.

The main contributions of this paper are as follows:

- We propose a novel searchable encryption scheme based on the decisional bilinear Diffie-Hellman (DBDH) hard problem. The new scheme guarantees that multiple encryption results for the same keyword are different, thereby resisting keyword guessing attacks (KGA), and ensures that indexes and trapdoors do not disclose any secret information. We construct a Merkle hash tree (MHT) and store the root hash value on the blockchain. When utilizing the proposed SE scheme for data retrieval, the results are verified to ensure their credibility, leveraging the blockchain's non-tampering characteristics.
- We design a multi-cloud heterogeneous distributed storage cluster using Tahoe-LAFS to manage grid data. Even if some cloud nodes are compromised, user data can still be recovered, significantly enhancing data availability.
- We analyze the security of the new SE algorithm and then assess its performance, demonstrating its superior efficiency.

2 Related Work

The majority of contemporary smart grid data comprises time series data, structured in chronological order, and is characterized by frequent updates and a fixed data format. However, traditional symmetric searchable encryption schemes are deemed unsuitable for securing smart grid data [7]. In 2013, Wen et al. proposed a range query scheme based on encrypted metering data, a with the objective of addressing privacy concerns within financial audits of smart grids [8]. In 2014, Li et al. introduced a searchable encryption scheme enabling multi-keyword range queries. This approach aimed to address information privacy concerns for both buyers and sellers in smart grid energy auctions, along with facilitating encrypted retrieval of energy data [9]. In 2019, Uwizeye et al. developed a connection keyword searchable public key encryption scheme for smart grid applications, with the goal of minimizing communication costs. They proved the security of the scheme under the random oracle model [10]. In 2021, Tur et al. incorporated chaos cryptography and searchable encryption technology into communication protocols among power systems. This integration aimed to provide real-time defense against network attacks and keyword guessing attacks [11].

In addition, data integrity in smart grids is also very important. It can ensure that grid data is not tampered with or damaged during transmission and storage. Ibrahim et al. introduced a verifiable single-keyword ciphertext retrieval scheme, which relies on a prefix tree constructed using keyed hash functions and bitmap technology to ensure the integrity of power grid data. However, the prefix tree in this method takes up a lot of space [12]. Sun et al. proposed a searchable integrity verification scheme using bilinear accumulators. It constructs a completeness proof through polynomial coprime conditions and realizes the verification of multi-keyword searches [13]. Ping et al. introduced an integrity verification method designed using algebraic signature and elliptic curve cryptography technology. This method efficiently ensures data integrity while keeping computational and communication costs low [14].

However, each of the mentioned schemes has its own set of constraints. This paper seeks to tackle the confidentiality, integrity, and availability concerns regarding smart grid data through the introduction of a verifiable ciphertext retrieval scheme. This scheme integrates Merkle tree and blockchain technology with the Tahoe-LAFS system to achieve secure storage and retrieval of power grid data.

3   System Model

This article's scheme mainly consists of four entities: the data owner (DO), the heterogeneous storage clusters of Tahoe-LAFS (heterogeneous storage clusters), the blockchain (Blockchain), and the data user (DU). The scheme's system model is depicted in Fig. 1.

(1) Data owner. The data owner holds the raw power grid data and continually gathers newly produced data. Its key duties involve computing security indices, creating ciphertexts for grid data, and uploading them to diverse storage clusters. Furthermore, to uphold the integrity of the power grid data, the data owner constructs a Merkle tree for the set of ciphertexts and stores it on the cloud server. Following this, it acquires the associated root node hash value and uploads it to the blockchain.

(2) Tahoe-LAFS heterogeneous storage cluster. The Tahoe-LAFS [15] heterogeneous storage cluster consists of cloud servers offered by various service providers. Its primary responsibility is to store the grid data ciphertext, the security index and the Merkle tree uploaded by the data owner, receive the data user's trapdoor, execute retrieval operations, and transmit the retrieval results to the data user.

(3) Blockchain. It is primarily employed to store the root hash value of the Merkle tree associated with the power grid data ciphertext uploaded by the data owner. The root hash value corresponding to the ciphertext is transmitted to the user based on the power grid data number provided by the user. This enables the user to verify the integrity of the data.

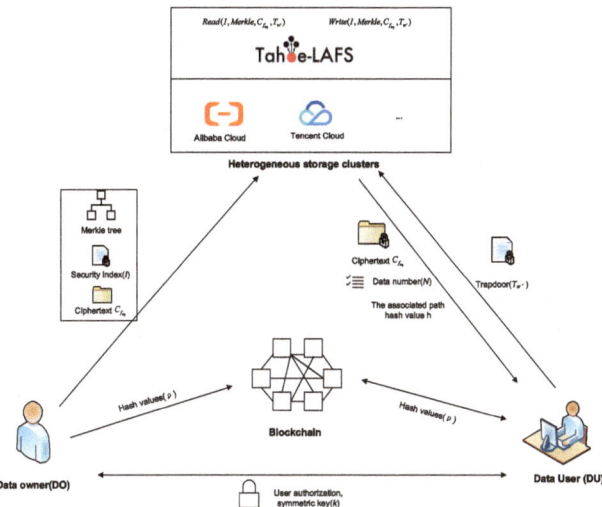

Fig. 1. System Model

(4) User. Its primary task involves computing trapdoors and uploading them to the heterogeneous storage cluster for retrieval purposes. Users retrieve grid data, verify its integrity, and decrypt it.

4 Proposed Scheme

This article primarily encompasses five stages: system initialization, ciphertext data generation, trapdoor generation, search, verification and decryption. The details are as follows:

4.1 System Initialization

$Setup(1^\lambda) \rightarrow params$. Given security parameters λ to generate public parameters $params$. Among them, G, G_T are two cyclic groups with prime order p, g is the generator of the group G, and they satisfie the bilinear mapping $e: G \times G \rightarrow G_T$. Randomly select two elements $a, b \in Z_p$, and calculate $g_1 = g^a$, $g_2 = g^b$. Finally, two collision-resistant hash functions are selected, that is, $H_1 : \{0, 1\}^* \rightarrow G$, $H_2 : \{0, 1\}^* \rightarrow G$, then the system public parameters are $params = \{g, g_1, g_2, p, a, G, G_T, e, H_1, H_2\}$.

4.2 Ciphertext Generation

$Encrypt(f_{m_i}, w, k) \rightarrow (C_{f_{m_i}}, I, \rho)$. This stage mainly includes three steps: data encryption, security index generation and verification data generation.

1) *Data encryption*

The data owner uses the symmetric encryption key k to encrypt the plaintext document f_{m_i} to obtain the corresponding ciphertext $C_{f_{m_i}}$.

2) *Security index generation*

The data owner first selects a random number $r \to Z_P$, and then performs a hash operation $H_1(w)$ on the keyword w, and the security index is $I = \{e(g_1, g_2)^r N, g^r, H_1(w)^r\}$, Among them, N is the file number containing the keyword w. The data owner uploads the encrypted ciphertext C_{fm_i} to the Tahoe-LAFS heterogeneous storage cluster for storage operations, and uploads the security index I to the Tahoe-LAFS heterogeneous storage cluster for query operation.

The power grid data is encrypted before transmission to the heterogeneous storage cluster, rendering the Advanced Encryption Standard (AES) unnecessary. Moreover, Reed-Solomon Codes (RS) erasure coding is employed for the power grid data, meaning that each cloud server stores the encoded block of C_{fm_i}. The utilization of erasure coding ensures high fault tolerance in power grid data storage, enabling robust recoverability in case of failures.

3) *Verification data generation*

The data owner acquires the Merkle hash tree of the ciphertext data through the following operations, as depicted in Fig. 2. The specific steps are as follows:

- Create leaf nodes for each ciphertext file C_{fm_i} and compute their corresponding hash values, denoted as $h_i = H_2(C_{fm_i})$.
- Pair adjacent leaf nodes and compute their combined hash values, denoted as $h_{i,i+1} = H_2(H_2(C_{fm_i})||H_2(C_{fm_{i+1}}))$.
- If there are more than two nodes remaining, continue pairing and hashing them until only one node, the root node h_{root}, remains.

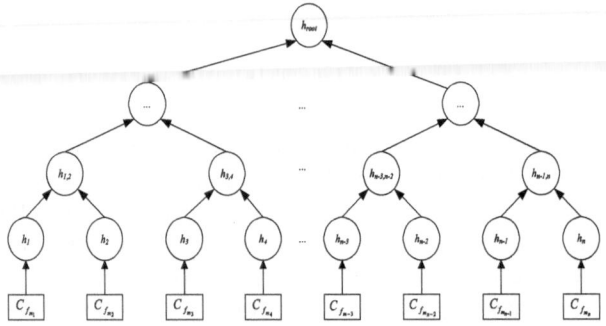

Fig. 2. Merkle hash tree

Let $\rho = h_{root}$ as verification data for cloud storage data and upload it to the blockchain for integrity verification. Upload the hash values of the remaining nodes of the Merkle hash tree (excluding the root hash value h_{root}) to the Tahoe-LAFS heterogeneous storage cluster for storage.

4.3 Trapdoor Generation

$Trapdoor(w') \rightarrow T_{w'}$. The user calculates the trapdoor $T_{w'}$ for a specific query keyword w' as follows: first, conduct a hash operation on the query keyword w' to obtain $H_1(w')$; then select a random number $t \rightarrow Z_P$ and compute $T_{w'} = \{g_2^a H_1(w')^t, g^t\}$.

4.4 Search

$search(I, T_{w'}) \rightarrow N$. Input trapdoor $T_{w'}$ and security index I, and output related file number N. The calculation process is as the Eq. (1).

$$
\begin{aligned}
e(g_1, g_2)^r N & \frac{e(g^t, H_1(w)^r)}{e(g_2^a H(w')^t, g^r)} \\
= \, & e(g_1, g_2)^r N \frac{e(g^t, H_1(w)^r)}{e(g_2^a, g^r)e(H(w')^t, g^r)} \\
= \, & e(g_1, g_2)^r N \frac{e(g, H_1(w))^{tr}}{e(g_2, g)^{ar}e(H(w'), g)^{tr}} \\
= \, & e(g_1, g_2)^r N \frac{e(g, H_1(w))^{tr}}{e(g_2, g^a)^r e(H(w'), g)^{tr}} \\
= \, & e(g_1, g_2)^r N \frac{e(g, H_1(w))^{tr}}{e(g_2, g_1)^r e(H(w'), g)^{tr}} \\
= \, & N \frac{e(g, H_1(w))^{tr}}{e(H(w'), g)^{tr}}
\end{aligned}
\tag{1}
$$

If $w = w'$, the final result is the corresponding file number N.

The Tahoe-LAFS heterogeneous storage cluster transmits to the user the file number N, the corresponding ciphertext $C_{f_{m_i}}$, and the hash values of all nodes between the Merkle hash tree root node and $C_{f_{m_i}}$ (excluding the root node hash value h_{root}).

4.5 Verification and Decryption

$Verify(C_{f_{m_i}}, \rho) = 1 \, or \, 0$. The user executes this algorithm to verify whether each power grid data ciphertext $C_{f_{m_i}}$, retrieved from the Tahoe-LAFS heterogeneous storage cluster, has been tampered with by the cloud server. The user recalculates the Merkle hash tree based on the hash value on the obtained path to quickly obtain the root hash value A.

The user recalculates the Merkle hash tree based on the hash values obtained along the path to quickly obtain the root hash value $\rho' = h'_{root}$. Then, determine whether $\rho' = \rho$ is true. If they are equal, it indicates that the data $C_{f_{m_i}}$ has not been tampered with. The user acquires the symmetric key k through a secure channel to decrypt and access the plaintext information of the ciphertext data. Otherwise, it indicates that $C_{f_{m_i}}$ has been tampered with, and consequently, the user will reject the returned result.

5 Security Analysis

5.1 Confidentiality

The scheme proposed in this article effectively ensures the security of keywords. Keyword trapdoors are randomly encrypted, thus meeting the IND-KGA security standards. In addition, the security index is constructed using the quintuple construction method in the deterministic bilinear Diffie-Hellman (DBDH) difficult problem. Therefore, the security of ciphertext can be reduced to the difficult problem of decisional bilinear Diffie-Hellman hypothesis.

5.2 Credibility

The Merkle hash tree is constructed based on the ciphertext of the power grid data uploaded to the heterogeneous storage cluster, and the resulting root node hash value is uploaded to the blockchain as the verification value. The immutability of the blockchain ensures the authenticity of this verification value. Moreover, the user utilizes the verification value to authenticate the received retrieval results, thus guaranteeing that the outcomes provided by the cloud server remain untampered with and enhancing the credibility of the data.

6 Experimental Analysis

In this study, the experimental environment is a 64-bit Windows operating system, equipped with Intel(R) Core (TM) i7-4790 CPU @ 3.60 GHz and 16 GB of memory. JAVA language is selected as the main programming tool, and the encryption function comes from the JPBC function library.

The experiment in this section compares the scheme proposed in this article with two schemes from the literature [16, 17], evaluating trapdoor generation time, index generation time, and keyword retrieval time. Analysis of Fig. 3, Fig. 4, and Fig. 5 reveals that as the number of keywords increases, the running time of each algorithm also increases proportionally. During the index generation stage, the proposed scheme demonstrates significant efficiency advantages, particularly in comparison to the scheme presented in literature [16], and exhibits a marginal improvement over the scheme outlined in literature [17]. The primary reason is that the index calculation in this scheme necessitates only one bilinear calculation and one hash calculation, as depicted in Fig. 3. In the trapdoor generation stage, this scheme exhibits superior performance compared to the other two schemes, with its advantage becoming more pronounced as the number of keywords increases, as illustrated in Fig. 4. In the keyword retrieval stage, this scheme requires only three bilinear pairing calculations. Its computational overhead is smaller compared to the other two schemes, as illustrated in Fig. 5.

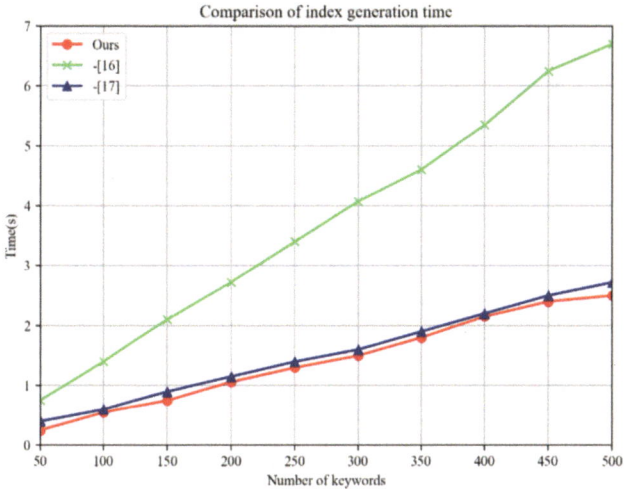

Fig. 3. Index generation time

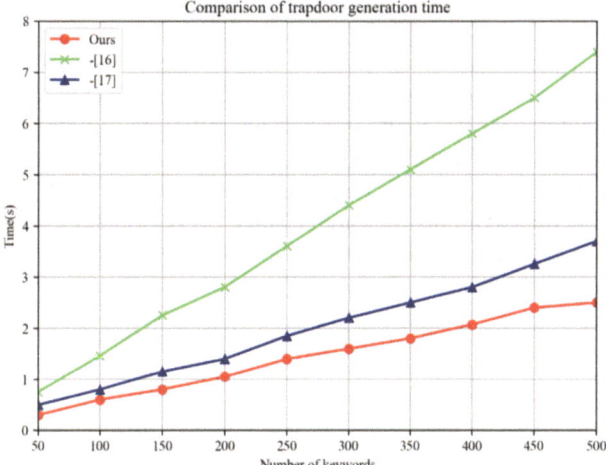

Fig. 4. Trapdoor generation time

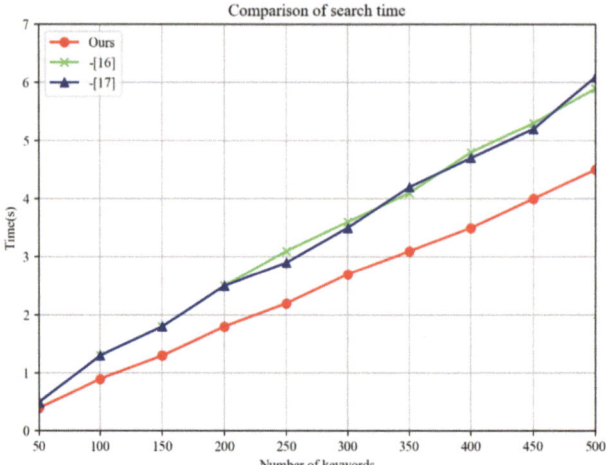

Fig. 5. Keyword search time

7   Conclusion

This study proposes a verifiable ciphertext retrieval scheme tailored to the smart grid scenario, primarily addressing the following three key issues: (1) Utilizing searchable encryption technology addresses the issue of low efficiency in smart grid ciphertext data retrieval. (2) Leveraging the Merkle tree root hash value and the tamper-resistant properties of the blockchain prevents malicious cloud servers from tampering with data, thereby enhancing the credibility of search results. (3) Building upon the Tahoe-LAFS system, a multi-cloud heterogeneous distributed storage cluster is implemented to store user data and enhance data availability.

References

1. Butt, O.M., Zulqarnain, M., Butt, T.M.: Recent advancement in smart grid technology: future prospects in the electrical power network. Ain Shams Eng. J. **12**(1), 687–695 (2021)
2. Khalid, H., Shobole, A.: Existing developments in adaptive smart grid protection: a review. Electr. Power Syst. Res. **191**, 106901 (2021)
3. Kim, Y., Hakak, S., Ghorbani, A.: Smart grid security: attacks and defence techniques. IET Smart Grid (2022)
4. Andola, N., Gahlot, R., Yadav, V.K., Venkatesan, S., Verma, S.: Searchable encryption on the cloud: a survey. J. Supercomput. **78**(7), 9952–9984 (2022)
5. Tahir, S., Ruj, S., Rahulamathavan, Y., Rajarajan, M., Glackin, C.: A new secure and lightweight searchable encryption scheme over encrypted cloud data. IEEE Trans. Emerg. Top. Comput. **7**(4), 530–544 (2017)
6. Zhang, C., Xu, Y., Hu, Y., Wu, J., Ren, J., Zhang, Y.: A blockchain-based multi-cloud storage data auditing scheme to locate faults. IEEE Trans. Cloud Comput. **10**(4), 2252–2263 (2021)
7. Li, J., Niu, X., Sun, J.S.: A practical searchable symmetric encryption scheme for smart grid data. In: ICC 2019–2019 IEEE International Conference on Communications (ICC), pp. 1–6. IEEE (2019)

8. Wen, M., Lu, R., Zhang, K., Lei, J., Liang, X., Shen, X.: PaRQ: a privacy-preserving range query scheme over encrypted metering data for smart grid. IEEE Trans. Emerg. Top. Comput. **1**(1), 178–191 (2013)
9. Lu, R., Li, H., Yang, Y., Wen, M., Luo, H.: EMRQ: an efficient multi-keyword range query scheme in smart grid auction market. KSII Trans. Internet Inf. Syst. (2014)
10. Uwizeye, E., Wang, J., Cheng, Z., Li, F.: Certificateless public key encryption with conjunctive keyword search and its application to cloud-based reliable smart grid system. Ann. Telecommun. **74**, 435–449 (2019)
11. Tur, M.R., Ogras, H.: Transmission of frequency balance instructions and secure data sharing based on chaos encryption in smart grid-based energy systems applications. IEEE Access **9**, 27323–27332 (2021)
12. Ibrahim, I.M., El-Din, S.H.N., Elgohary, R., Faheem, H., Mostafa, M.G.: A robust owner-to-user data sharing framework in honest but curious cloud environments. In: 2013 8th International Conference on Computer Engineering & Systems (ICCES), pp. 51–56. IEEE (2013)
13. Sun, W., Liu, X., Lou, W., Hou, Y.T., Li, H.: Catch you if you lie to me: efficient verifiable conjunctive keyword search over large dynamic encrypted cloud data. In: 2015 IEEE Conference on Computer Communications (INFOCOM), pp. 2110–2118. IEEE (2015)
14. Ping, Y., Zhan, Y., Lu, K., Wang, B.: Public data integrity verification scheme for secure cloud storage. Information **11**(9), 409 (2020)
15. Selimi, M., Freitag, F.: Tahoe-LAFS distributed storage service in community network clouds. In: 2014 IEEE Fourth International Conference on Big Data and Cloud Computing, pp. 17–24. IEEE (2014)
16. Huang, Q., Li, H.: An efficient public-key searchable encryption scheme secure against inside keyword guessing attacks. Inf. Sci. **403**, 1–14 (2017)
17. Wu, L., Chen, B., Zeadally, S., et al.: An efficient and secure searchable public key encryption scheme with privacy protection for cloud storage. Soft. Comput. **22**, 7685–7696 (2018)

Panoptic Semantic Mapping Method for Tomato Growing Environment Based on K-Net and OctoMap

Junxiong Zhang[1(✉)], Yu Zhang[1], Jinyi Xie[1], Xiajun Zheng[1], Fan Zhang[2],
Weijie Rao[1], and Jiayang Guo[1]

[1] College of Engineering, China Agricultural University, Beijing, China
cau2007@cau.edu.cn
[2] Faculty of Mechanical and Electrical Engineering, Kunming University of Science and
Technology, Kunming, China
20240035@kust.edu.cn

Abstract. In modern greenhouses, complicated tasks and unstructured environments generate the imperious demand for advanced semantic information about each object at work scenes. A significant problem that mainstream methods intend to resolve is that the refinement and understanding of environmental information cannot efficiently cover the entire task in real time. Therefore, this paper proposes a panoptic semantic mapping method to identify each object that is supposed to be concerned in greenhouses. This method builds grid maps with advanced semantic information based on RGB and depth images. For the agricultural task with tomato as the working object, the categories of various objects in the grid map are divided into four groups: fruits, pedicels, stems and obstacles. This method consists of three steps: semantic segmentation from RGB images with K-Net, reconstruction of point cloud data based on depth images and semantic masks and transformation of the point cloud data into OctoMap. Experimental results show the semantic segmentation algorithm reaches a mean precision of semantic segmentation of 93.83%, a mean IoU of 88.39% and an average accuracy of 98.28%. Meanwhile, the refresh frequency of publishing point cloud data with advanced semantic information holds steady at 2 Hz with the resolution of 8 mm.

Keywords: panoptic mapping · tomato · greenhouse · K-Net · OctoMap

1 Introduction

Nowadays, tomato becomes one of the most popular greenhouse crops, with global production tripling over the past four decades [1]. Aiming at efficient tomato production and standardized management, facility cultivation has become the mainstream instead of open field cultivation. In recent years, with the continuous rise in labor costs, agricultural robots have become a research focus in modern agricultural facility cultivation. Providing information about work objectives and environments is crucial for guiding agricultural robots to accomplish tasks. Today, technologies like artificial intelligence, big data and

P. Siarry et al. (Eds.): WCNA 2023, LNEE 1361, pp. 180–193, 2025.
https://doi.org/10.1007/978-981-96-2409-6_18

the Internet of Things have found their way into practice, and have started to converge [2]. Digital technologies are supposed to support a more profound comprehension of the interconnections within the agricultural production system [3]. The greenhouse is a complex and dynamic environment with varying shapes, sizes, positions, and orientations of objects [4]. Nowadays, recognizing the work environment in unstructured and complex greenhouses is one of the serious challenges facing the development of agricultural robots.

To enhance the ability of robots to recognize and perceive objects in three-dimensional space, Tao [5] optimized a classifier for categorizing point cloud data into apples, branches and leaves. Zhou [6] improved a double convolutional chain fast R-CNN algorithm for detecting tomato flowers, fruits, and stems. Tian [7] proposed an improved YOLO-V3 model to detect apples in different stages of growth in orchards with complex backgrounds. Xiong [8] proposed a method to rapidly detect green mangoes and estimate the number of mangoes in orchards by using UAV vision. Feng [9] utilized a Mask-RCNN model to recognize the main stem of a plant. You [10] introduced an algorithm to generate the marked skeleton by using the topological and geometric priors of Upright Fruiting Offshoot (UFO) trees. Li [11] proposed a method based on UAVs for locating branch picking points in complex natural environments and guiding cinnamon picking thorough using an improved YOLOv5s model and an improved DeepLabv3+ model. Aiming at picking point recognition of ripe tomatoes in complex environments, Rong [12] proposed a picking point recognition algorithm based on the connection of tomato fruit, calyx, and stem. There is a growing demand for efficient and precise agricultural approaches that can provide high-level semantic information.

The research mentioned above presents some reasonable methods to extract useful information from the unstructured agricultural environment. Two serious and pressing problems remain: (1) the refinement and understanding of environmental information is completed before the implementation of agricultural tasks and is not maintained throughout the process; (2) the quality of immediate data collected by sensors would cause dramatic fluctuations in subsequent work. To solve these problems, technologies such as SLAM (Simultaneous Localization and Mapping), for example, are applied to the cognitive environment task of agricultural robots based on the digital twin concept. Zhang [13] proposed a semantic SLAM system based on object-level entities to construct semantic mapping. Gené-Mola [14] proposed an apple detection and 3D location method based on the combination of case segmentation algorithm and structure-from-motion. Liu [15] proposed a Multi-view Self-Constructing Graph Convolutional Networks (MSCGNet) which achieved a competitive result with much fewer parameters compared to pure-CNN-based networks. End-to-end networks TPM and TPMv2 are proposed to simulate tomato bunch pose [16, 17]. In summary, though many studies have proposed effective methods for localized scene construction, real-time processing speed, and robustness still need to be improved. Roggiolani [18] proposed a novel automatic postprocessing, which solved the problem of crop field joint semantics, plant instance and leaf instance segmentation in RGB data. Tomatoes in greenhouses are taken as the research objects in this paper. Based on the combined use of deep learning algorithms and point cloud grid maps, the advanced semantic map of the local scene in greenhouses is constructed.

To achieve the above functions, this paper finished the following work: (1) A semantic segmentation model based on K-Net is used to accurately identify tomato fruits, pedicels, stems and obstacles and is compared with three other typical semantic segmentation algorithms using accuracy and precision as evaluated metrics; (2) Based on OctoMap, a local scene reconstruction framework is built for advanced semantic information extraction of agricultural greenhouse environments.

This paper is structured as follows: Sect. 2 presents the acquired dataset and the methodology pipeline which includes a description of the semantic segmentation network and the systematic framework for creating grid maps of localized scenes; Sect. 3 shows the results of filed experiments and the performances of each part in our panoptic mapping method; Sect. 4 discusses the effects and problems reflected in field experiments; the conclusion is presented in Sect. 5.

2   Materials and Methods

2.1   Data Acquisition

Tomatoes' images are captured with an Azure Kinect DK camera (Microsoft, Redmond, America), which can get RGB-D images with vary resolutions. We selected the mode of 2048×1536 pixels. As shown in Fig. 1, 173 images of tomatoes are acquired in HongFu Agricultural Tomato Production Park in Daxing District (116.283601 E, 39.603538 N), Beijing, China. This greenhouse follows the Dutch standard planting and management mode. A variety of tomatoes called 'red pearl' is the mainstream selection in this greenhouse. It grows in clusters and its fruits are big, grow dense, stick to each other and get squeezed. The RGB-D camera is 400–600 mm away from tomato plants. To ensure the diversity of samples, those images include the conditions of two growth stages (mature and immature) and three periods of one day (morning, noon and afternoon). Thus, sampled images of tomato growing environment included various environment construct situations, different growth stages and various poses of tomato plants. In addition, all images were enhanced with rotation and brightness adjustment, resulting in 500 images. Labelme software was used to manually mark the polygon area and generate label files.

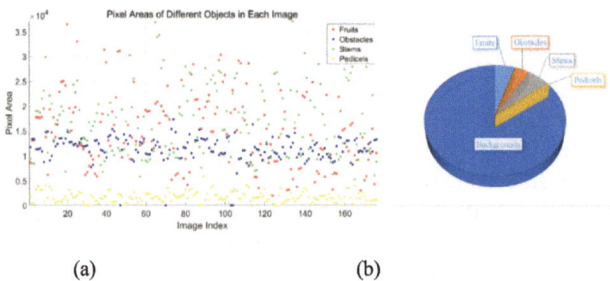

(a)   (b)

Fig. 1. Dataset details. (a) The pixel area of each type of object in each image. (b) The proportion of pixel area occupied by various objects in the dataset.

2.2   Overview of Semantic Mapping Method

Generally, these methods obtain results through semantic segmentation algorithms. The major improvement of these methods focuses on recognition accuracy and instantaneity. Moreover, instance segmentation results also hold significant guidance for subsequent specific tasks in agricultural work. Combining semantic and instance segmentation, panoptic segmentation is a more comprehensive selection for providing an objective understanding of the scene. The whole method is implemented with the ROS (Robot Operating System), which provides a structured communications layer above the host operating systems of a heterogeneous compute cluster. Through the utilization of ROS and SLAM, the accuracy and instantaneity of the semantic map are ensured. As shown in Fig. 2, the semantic mapping framework is divided into the following three steps: (1) Detect tomato fruits, pedicels, stems and obstacles from RGB images with K-Net; (2) Reconstruct point cloud data based on depth images and semantic masks; (3) Transform the point cloud data into OctoMap.

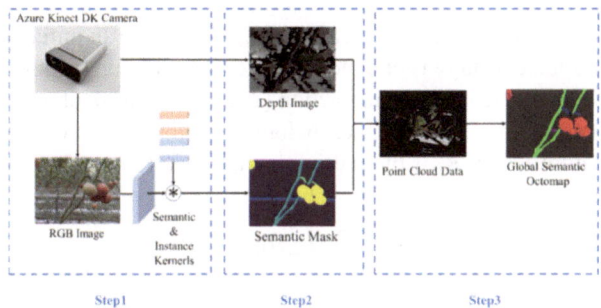

Fig. 2.   Overall framework of semantic mapping method.

2.3   Panoptic Segmentation Algorithm

To detect each object in the greenhouse, an approach of panoptic segmentation is chosen. The panoptic segmentation model in step (1) is deployed based on MMSegmentation. The light weight and easy deployment of this framework help to improve the performance of the whole system. In the greenhouse environment filled with intricate interferences, it is anticipated that precise semantic segmentation can be achieved through visual algorithms. Simultaneously, the effective segmentation of target individuals is a prerequisite for the automated completion of greenhouse tomato planting tasks, including picking. It is expected that a panoptic segmentation network takes into account both requirements. Therefore, K-Net is chosen and expected to reach good performance in unstructured and complex greenhouses.

K-Net is a simple, and effective framework which considers various segmentation tasks through a unified perspective of kernels [19]. According to the experiments' results on the challenging COCO dataset, K-Net obtained new state-of-the-art single-model performance on panoptic and semantic segmentation. The structural characteristics and experimental performance of K-Net network show that the network is suitable for panoptic segmentation in greenhouse.

The segmentation task essentially assigns each pixel into a predefined meaningful group. For panoptic segmentation, in our mission, we set four semantic groups: Fruits, Pedicels, Stems and Obstacles. The RGB images captured in the greenhouse serve as inputs to the K-Net model, which outputs the classification results for each pixel.

The core idea of K-Net is to learn the convolution kernel, which divides pixels into N meaningful groups. So that semantic segmentation, instance segmentation and panoptic segmentation can be achieved by the same formulation. Given a feature map $F \in R^{B \times C \times H \times W}$ and N kernels $K \in R^{B \times N \times C}$, K-Net produces N mask predictions by performing convolution of F and K

$$M = K * F \tag{1}$$

with $M \in R^{B \times N \times H \times W}$.

K-Net uses K randomly initialized Kernels K_0 to produce an initial mask prediction M_0. This kernel update head f_S is obtained in 3 steps: (1) Group feature assembling; (2) Adaptive kernel update; (3) Kernel interaction [20]. As the mask of each kernel in M_{i-1} essentially defines whether or not a pixel belongs to the kernel's related group, we can assemble the feature F^k for K_{i-1} by multiplying the feature map F with the M_{i-1} as

$$F^K = \sum_u^H \sum_v^W M_{i-1}(u, v) \cdot F(u, v) \tag{2}$$

The kernel update head then updates the kernels using the obtained F^K to improve the representation ability of kernels. Specifically, we first conduct element-wise multiplication between F^K and K_{i-1} as

$$F^G = \phi_1\left(F^K\right) \otimes \phi_2(K_{i-1}) \tag{3}$$

where ϕ_1 and ϕ_2 are linear transformations. Then the head learns two gates, G^F and G^K, which adapt the contribution from F^K and K_{i-1} to the updated kernel $\tilde{K}$, respectively.

As shown in Fig. 3, this process can be done iteratively, as finer partitioning usually reduces noise in group features, resulting in more discriminating cores. This process is formulated as

$$K_S, M_S = f_S(M_{S-1}, K_{S-1}, F) \tag{4}$$

2.4   Point Cloud Reconstruction and Post-processing

This paper uses the raw depth images and semantic segmentation results for point cloud reconstruction to turn original data into an advanced semantic map that guides greenhouse robots on their tasks. Within the process that matches semantic information and

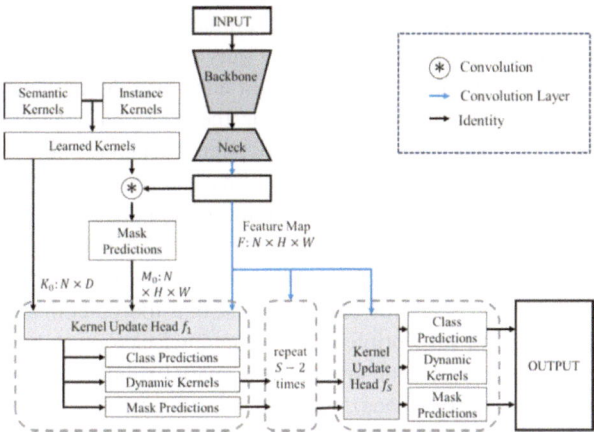

Fig. 3. K-Net for panoptic segmentation

the point cloud data, fruits, pedicel, stems and obstacles are presented as red, blue, green, and purple respectively. After obtaining the point cloud data with semantic information, outlier filtering, and distance filtering are used to remove noise to improve point cloud quality. For the process to distinguish abnormal values effectively, the number of neighboring points is set to 20 and the standard deviation multiplier threshold is set to 2.0. As shown in Fig. 4, statistical outliers are removed from the point cloud data of each semantic group using multiple of standard deviation as an index.

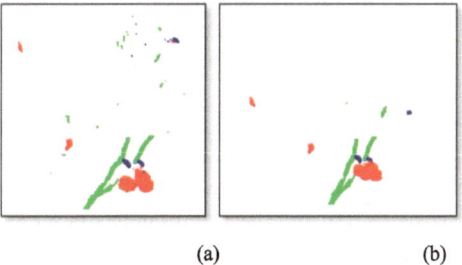

(a)   (b)

Fig. 4. Processed Point Cloud Data visualizations. (a) The visual result after reconstruction with semantic information. (b) The visual result after filtering process.

2.5   Semantic OctoMap Generation

Most robotics applications require a probabilistic representation, modeling of free, occupied and unmapped regions, and additional efficiencies regarding runtime and memory usage. In robotic systems, people often have to deal with sensor noise and temporary or permanent changes in the environment. In this case, a separate occupancy label is not enough. Therefore, occupancy has to be modelled probabilistically, for example by applying occupancy grid mapping.

Therefore, for the third step of the method, a mapping approach based on octrees is adopted. This approach explicitly represents not only occupied space, free and unknown areas, but also advanced semantic information of each grid. Meanwhile, this approach effectively updates the presentation and models the data consistently, while keeping memory requirements to a minimum. A series of experimental results on real robots and open real world data sets demonstrate the reliability and practicability of the proposed method [21]. The capability to continuously monitor and update the probability of possession in real-time renders this method highly suitable for the unstructured greenhouse environment, which poses challenges in comprehensive observation. In addition, the proposed approach compensates for the drawback that the outcomes of the semantic segmentation algorithm are susceptible to environmental factors.

An octree is a hierarchical data structure for spatial subdivision in 3D [22]. Each node in the octree represents the space contained in a cubic volume. The volume is recursively subdivided into eight sub-volumes until a given minimum voxel size is reached. The minimum voxel size determines the resolution of the octree. Since an octree is a hierarchical data structure, the tree can be cut at any level for a coarser breakdown if the internal nodes are maintained accordingly.

3   Experiments and Results

3.1   Training of Models

The models are trained and tested using the Pytorch framework with an Intel Core i7-12650H processor and NVIDIA GeForce RTX4050 graphics card. The resolution of the input image is 640×480. These models are trained using transfer learning methods. In addition, mosaic data enhancement is used in image preprocessing of detection tasks, rotation, contraction and random cropping of original images are performed, which greatly increases the size of data set.

The batch size of the above three models is set to 4, the maximum number of iterations is set to 40,000, and the number of iterations when the model converges is taken as the actual number of iterations of the model. The initial learning rate is set to 0.01, the momentum parameter is set to 0.9, and the weight decay is set to 0.0005. The learning rate is controlled by PolyLR method.

3.2   Evaluation Indicators

For convenience of explanation, assume the following: $k + 1$ classes (from L_0 to L_k, which contain an empty class or background), p_{ij} represents the number of pixels that belong to class i but are predicted as class j; p_{ii} represents the real quantity, and p_{ij} p_{ji} represent a false positive and false negative, respectively.

Pixel accuracy is the ratio of correctly labeled pixels to the total number of pixels. This metric compares all semantic groups obtained by pixel-by-pixel model segmentation with standard objects obtained by manual labeling. The average accuracy (aAcc) is then calculated from these comparisons. The mean pixel accuracy (mAcc) is the average proportion of correctly classified pixels in each category. This involves calculating the precision of each class individually and then taking the average of those values. The mean intersection over union (mIoU) measures the ratio of the intersection to the union of two sets: the underlying truth value and the prediction split. This ratio can be expressed as the sum of true positives divided by the sum of true positives, false negatives, and false positives. The mean recall (mRecall) measures the proportion of truly positive samples that the model correctly predicts to be positive. It quantifies the ratio of correctly labeled positive pixels to the total number of true pixels. The aAcc, mAcc, mIoU, mPrecision and mRecall formulas are shown in Eq. (5), (6), (7), (8) and (9), respectively:

$$aAcc = \frac{\sum_{i=0}^{k} p_{ii}}{\sum_{i=0}^{k} \sum_{j=0}^{k} p_{ij}} \tag{5}$$

$$mAcc = \frac{1}{k+1} \sum_{i=0}^{k} \frac{p_{ii}}{\sum_{j=0}^{k} p_{ij}} \tag{6}$$

$$mIoU = \frac{1}{k+1} \sum_{i=0}^{k} \frac{p_{ii}}{\sum_{j=0}^{k} p_{ij} + \sum_{j=0}^{k} p_{ji} - p_{ii}} \tag{7}$$

$$mPrecision = \frac{1}{k+1} \sum_{i=0}^{k} \frac{p_{ii}}{\sum_{j=0}^{k} p_{ji}} \tag{8}$$

$$mRecall = \frac{1}{k+1} \sum_{i=0}^{k} \frac{p_{ii}}{p_{ii} + p_{ji}} \tag{9}$$

3.3 Comparison of Semantic Segmentation Models

Table 1 shows the experimental results of DeepLabv3+, PSP-Net, K-Net and U-Net. Through the comparison and analysis of DeepLabv3+, PSP-Net and U-Net models, it is proved that K-Net is correct in choosing the semantic segmentation of tomato greenhouse image. While DeepLabv3+ shows commendable overall accuracy, its emphasis on average accuracy and recall metrics can hurt accuracy levels. Although PSP-Net shows strong accuracy performance, it is slightly behind in terms of average accuracy. While U-Net achieves a respectable level of accuracy, it faces challenges in terms of average accuracy. In contrast, K-Net achieves coordination by ensuring the highest average accuracy without compromising competitive level accuracy. This precision-focused approach has important implications in the agricultural environment, where misclassification can severely impact the decision-making process.

In conclusion, the advantage in average accuracy of K-Net emphasizes its ability to maintain a high level of accuracy in agricultural Settings while minimizing misclassification to a large extent. This performance is exactly what the semantic mapping method proposed in this paper needs. The ongoing need to generate and maintain OctoMap requires semantic segmentation network with stable high-precision output and low noise level. This preference is consistent with the overall goal of enhancing the fidelity and reliability of the generated map.

Table 1. Evaluation index results using different semantic segmentation models.

Models	aAcc (%)	mAcc (%)	mIoU (%)	mPrecision (%)	mRecall (%)
DeepLabv3+	98.01	**94.00**	87.72	92.40	**94.00**
PSP-Net	97.85	91.97	86.44	92.71	91.97
K-Net	**98.28**	93.24	**88.39**	**93.83**	93.24
U-Net	95.89	85.56	78.32	89.48	85.56

3.4   Field Experiment of Semantic Mapping Method

In order to verify the working performance of the semantic mapping method, field experiment was conducted in a standardized modern greenhouse provided by HongFu Agricultural Tomato Production Park. The greenhouse environment is shown in Fig. 5. Meanwhile, the real-time mapping effect is also displayed.

(a)   (b)

Fig. 5. Filed experiments in a standardized modern greenhouse. (a) Field experiments. (b) The visual result of the semantic mapping method.

In addition, as shown in Fig. 6, the collaborative work between the semantic mapping method and the robotic manipulator is completed. The joint work with the manipulator control system under the ROS framework was an important test for the real-time performance of the semantic mapping method. The robot manipulator's transform information updated at a frequency of about 30 Hz, while the camera's original point cloud was released at a frequency of 3 Hz.

To achieve semantic mapping, each frame of point cloud data must correspond uniquely to the coordinate system data of the robot manipulator. Otherwise, as shown in Fig. 7, grid maps will refer to incorrect transformation relationships. The yellow rectangle represents the grid map generated from the point cloud data and the transformation of the previous frame. Similarly, the blue rectangle represents the grid map generated from the same point cloud data and the transformation of the current frame. This means that a set of point cloud data will be misused multiple times. Therefore, after matching each frame of point cloud data with the corresponding time-stamped transform information, a point cloud data with no information is published to empty the stranded content of the topic.

Fig. 6. The entire system for collaborative experiments.

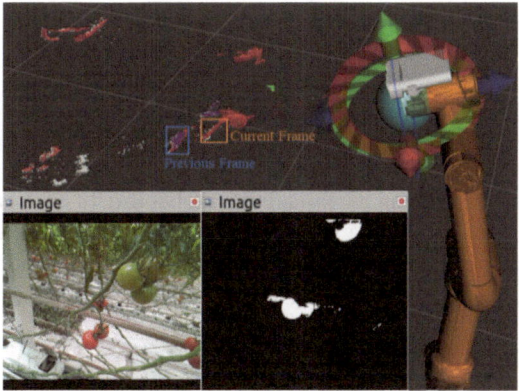

Fig. 7. The grid map with incorrect transformation relationships.

To enhance real-time performance, we have implemented a dual-thread working mode. As depicted in Fig. 8, the semantic mapping method performs two-dimensional image segmentation and publishes point cloud data with semantic information simultaneously. Line_profiler is used to measure the time spent on each step. Following this processing, the semantic mapping method's refresh frequency in the actual working environment stabilizes at 2 Hz. In this experiment, different map resolutions were tested on the premise of ensuring real-time performance. The experimental results show that the resolution can be refined to 8mm while maintaining the semantic map in real-time.

4   Discussions

In this section, the focus shifts to two optimization directions identified through the field experiment, along with an exploration of the potential applications of the semantic mapping method. The primary optimization goal is to reduce noise in semantic maps, as shown in Fig. 9. The noise is caused by two main sources. Firstly, inaccuracies in the mask arise from the semantic segmentation algorithm during the segmentation of

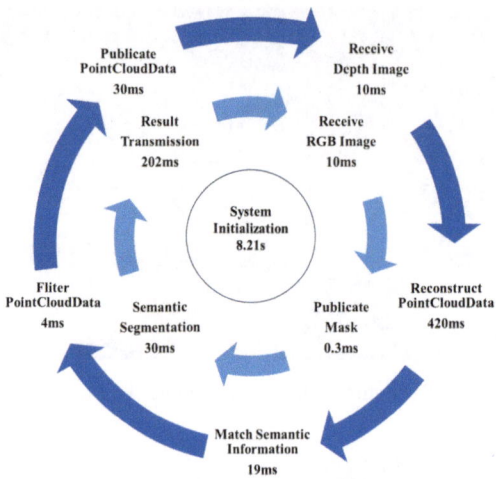

Fig. 8. The average time of each process per frame.

two-dimensional images. Secondly, noise in the original depth image captured by the camera contributes to the overall noise level. Although OctoMap offers some noise immunity by gradually updating the possession probability of each grid, the update rate of generated grids lags behind the rate at which new point cloud data generates new grids. As a result, the generated grid map can integrate new point cloud data in real time but encounters challenges in maintaining the generated portion in real time, particularly at finer resolutions.

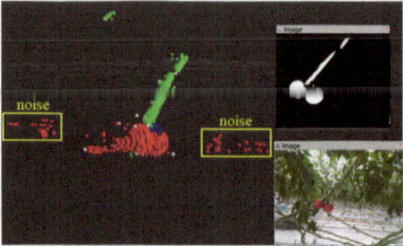

Fig. 9. The grid map disturbed by noise.

This brings us to the second optimization focus, centering on the refresh rate of maintaining the map. Two strategies are considered: one involves reducing resolution, albeit at the cost of sacrificing physical shape information in the semantic map. The alternative method, inspired by ORB-SLAM, employs keyframes and key points to decrease the amount of processed information. Despite the proposal by Cheng [23] for SG-SLAM, which combines ORB-SLAM with dense mapping to generate semantic instance maps, the map accuracy in their method falls short of the requirements for this task.

In terms of the prospective applications of the proposed semantic mapping method, its relevance extends to the execution of exploratory agricultural tasks utilizing robot manipulators and other specialized equipment. This innovative approach not only adeptly distinguishes between known and unexplored regions but also opens avenues for the seamless development of strategies across a spectrum of agricultural activities. By harnessing grid maps enriched with high-level semantic information, the method becomes a versatile tool for enhancing decision-making processes in tasks such as crop monitoring, precision farming, and autonomous navigation within agricultural environments. This not only optimizes operational efficiency but also lays the foundation for the advancement of intelligent and adaptive agricultural practices in a dynamic and evolving landscape.

5 Conclusions

This study proposes a semantic mapping method based on the K-Net model and OctoMap to accurately identify tomato fruits, pedicels, stems and obstacles in an unstructured greenhouse environment. The experimental results demonstrate a mean precision of semantic segmentation of 93.83%, a mean IoU of 88.39%, an average accuracy of 98.28%, and a stable refresh frequency of publishing point cloud data at 2Hz. In summary, the panoptic semantic mapping method based on K-Net and OctoMap can accurately distinguish groups of objects in greenhouse scenes, generate grid maps with a resolution of 8mm, and update the map in real-time based on subsequent point cloud data. Future work will involve collecting more images containing complex planting scenarios to train the semantic segmentation model and improve precision. Additionally, the framework for maintaining grid maps will be enhanced by utilizing the concept of key points and keyframes to increase efficiency.

Acknowledgment. This research is supported by Beijing Innovation Consortium of Agriculture Research System (grant number: BAIC12-2024). We are most grateful to HongFu Agricultural Tomato Production Park for giving us access to their farms.

References

1. Zhang, J., et al.: Greenhouse tomato detection and pose classification algorithm based on improved YOLOv5. Comput. Electron. Agric. **216**, 108519 (2024)
2. Pylianidis, C., Osinga, S., Athanasiadis, I.N.: Introducing digital twins to agriculture. Comput. Electron. Agric. **184**, 105942 (2021)
3. Nasirahmadi, A., Hensel, O.: Toward the next generation of digitalization in agriculture based on digital twin paradigm. Sensors **22**(2), 498 (2022)
4. Ye, L., Duan, J., Yang, Z., Zou, X., Chen, M., Zhang, S.: Collision-free motion planning for the litchi-picking robot. Comput. Electron. Agric. **185**, 106151 (2021)
5. Tao, Y., Zhou, J.: Automatic apple recognition based on the fusion of color and 3D feature for robotic fruit picking. Comput. Electron. Agric. **142**, 388–396 (2017)
6. Zhou, Y., Xu, T., Deng, H., Miao, T.: Recognition method of tomato key organs based on dual convolution Fast R-CNN. J. Shenyang Agric. Univ **49**(01), 65–74 (2018)

7. Tian, Y., Yang, G., Wang, Z., Wang, H., Li, E., Liang, Z.: Apple detection during different growth stages in orchards using the improved YOLO-V3 model. Comput. Electron. Agric. **157**, 417–426 (2019)

8. Xiong, J., et al.: Visual detection of green mangoes by an unmanned aerial vehicle in orchards based on a deep learning method. Biosyst. Eng. **194**, 261–272 (2020).

9. Feng, Q., Cheng, W., Zhang, W., Wang, B.: Visual tracking method of tomato plant mainstems for robotic harvesting. In: 2021 IEEE 11th Annual International Conference on CYBER Technology in Automation, Control, and Intelligent Systems (CYBER), pp. 886–890. IEEE, July 2021

10. You, A., Grimm, C., Silwal, A., Davidson, J.R.: Semantics-guided skeletonization of upright fruiting offshoot trees for robotic pruning. Comput. Electron. Agric. **192**, 106622 (2022)

11. Li, D., et al.: A novel approach for the 3D localization of branch picking points based on deep learning applied to longan harvesting UAVs. Comput. Electron. Agric. **199**, 107191. (2022)

12. Rong, Q., Hu, C., Hu, X., Xu, M.: Picking point recognition for ripe tomatoes using semantic segmentation and morphological processing. Comput. Electron. Agric. **210**, 107923 (2023)

13. Zhang, L., Wei, L., Shen, P., Wei, W., Zhu, G., Song, J.: Semantic SLAM based on object detection and improved octomap. IEEE Access **6**, 75545–75559 (2018)

14. Gené-Mola, J., et al.: Fruit detection and 3D location using instance segmentation neural networks and structure-from-motion photogrammetry. Comput. Electron. Agric. **169**, 105165 (2020)

15. Liu, Q., Kampffmeyer, M.C., Jenssen, R., Salberg, A.B.: Multi-view self-constructing graph convolutional networks with adaptive class weighting loss for semantic segmentation. In: Proceedings of the IEEE/CVF Conference on Computer Vision and Pattern Recognition Workshops, pp. 44–45 (2020)

16. Zhang, F., Gao, J., Zhou, H., Zhang, J., Zou, K., Yuan, T.: Three-dimensional pose detection method based on keypoints detection network for tomato bunch. Comput. Electron. Agric. **195**, 106824 (2022)

17. Zhang, F., et al.: TPMv2: an end-to-end tomato pose method based on 3D key points detection. Comput. Electron. Agric. **210**, 107878 (2023)

18. Roggiolani, G., Sodano, M., Guadagnino, T., Magistri, F., Behley, J., Stachniss, C.: Hierarchical approach for joint semantic, plant instance, and leaf instance segmentation in the agricultural domain. In: 2023 IEEE International Conference on Robotics and Automation (ICRA), pp. 9601–9607. IEEE, May 2023

19. Zhang, W., Pang, J., Chen, K., Loy, C.C.: K-Net: Towards unified image segmentation. Adv. Neural. Inf. Process. Syst. **34**, 10326–10338 (2021)

20. Schön, M., Buchholz, M., Dietmayer, K.: RT-K-Net: revisiting K-Net for real-time panoptic segmentation. arXiv preprint arXiv:2305.01255 (2023)

21. Hornung, A., Wurm, K.M., Bennewitz, M., Stachniss, C., Burgard, W.: OctoMap: an efficient probabilistic 3D mapping framework based on octrees. Auton. Robot. **34**, 189–206 (2013)

22. Meagher, D.: Geometric modeling using octree encoding. Comput. Graph. Image Process. **19**(2), 129–147 (1982)

23. Cheng, S., Sun, C., Zhang, S., Zhang, D.: SG-SLAM: a real-time RGB-D visual SLAM toward dynamic scenes with semantic and geometric information. IEEE Trans. Instrum. Meas. **72**, 1–12 (2022)

An Adversarial Attack Method for Multivariate Time Series Classification Based on AdvGAN

Yubo Wang[1], Hui He[2], Peng Zhang[1], Yuanchi Ma[1], Zhongxiang Lei[1], and Zhendong Niu[1(✉)]

[1] School of Computer Science and Technology, Beijing Institute of Technology, Beijing, China
{wangyubo,pengzhang_cs,yma,npcleilei,zniu}@bit.edu.cn
[2] School of Medical Technology, Beijing Institute of Technology, Beijing, China
hehui617@bit.edu.cn

Abstract. Considering the complexity of time series data and real-world applications, multivariate time series classification models are vulnerable to adversarial attacks. Although existing white-box attack strategies have made progress in generating adversarial samples, they rely on access to the target model's parameters, training data, and gradients. Therefore, we apply AdvGAN framework for multivariate time series classification. AdvGAN is designed as a framework based on Generative Adversarial Networks (GANs), encompassing a generator, discriminator. The generator creates multivariate perturbations, and the perturbations combine with original data to form adversarial samples. The discriminator assesses the authenticity of these samples. These samples are then used to evaluate the security of the target model. We conducts experiments across three University of East Anglia (UEA) and University of California Riverside (UCR) datasets, employing the Multivariate Long Short Term Memory Fully Convolutional Network (MLSTM_FCN) as the target model for adversarial attack testing. The results indicate that our designed attack method effectively enhances the success rate of adversarial attacks while maintaining a similar level of Mean Squared Error (MSE) between the generated adversarial samples and the original samples.

Keywords: Adversarial Attack · Multivariate Time Series Classification · Generative Adversarial Network

1 Introduction

In the field of artificial intelligence, the rapid evolution of technologies has positioned deep learning model as extensively applied in tasks such as the classification of multivariate time series [1]. The applications span the prediction of stock market movements [2], and the detection of irregularities in electronic health records [3].

Conversely, the application of these models has highlighted the adversarial sample attacks that prey on the susceptibilities of deep neural networks. These attacks are common threat. And the attack capitalize on the heightened sensitivity of neural networks to inputs. Adversarial samples [4] are generated by making slight alterations to the original samples, which, when input into the target model, lead to erroneous predictions. The

P. Siarry et al. (Eds.): WCNA 2023, LNEE 1361, pp. 194–202, 2025.
https://doi.org/10.1007/978-981-96-2409-6_19

consequences of such mispredictions are particularly severe in fields such as finance [5] and healthcare [6].

Adversarial attack methods currently in existence are broadly divided into two categories: black-box [7] and white-box [8]. Within the white-box paradigm, the generation of adversarial samples is facilitated by access to the model's architecture, parameters, and training datasets, as seen with techniques such as the Fast Gradient Sign Method (FGSM) [9] and Projected Gradient Descent (PGD) [10]. Nonetheless, the majority of these attack methodologies are tailored to image classification models and do not consider the intricate features inherent in multivariate time series classification data, nor do they address the subtlety of adversarial samples. The paper encapsulates the extant issues and challenges as: White-box attacks presuppose that the structure and parameters of the model under attack are laid bare, a scenario that frequently eludes practicality in real-world scenarios.

Considering the issues previously discussed, this paper design an adversarial attack network tailored for multivariate time series data classification models, based on Adv-GAN. Specifically, this network acknowledges the impracticality of acquiring details such as the target model's training data, architecture, and parameters in real-world scenarios, opting for an attack strategy that constructs adversarial samples based solely on the target model's inputs and outputs. To counter the vulnerability of adversarial samples being readily detectable, the GAN framework [11] is adopted, capitalizing on its proficiency in generating samples that closely resemble authentic ones. The generator then employs these features along with the target labels to fabricate perturbations for the multivariate time series. By incorporating this perturbations, the multivariate time series classification data is transformed into adversarial samples, which are authenticated for verisimilitude by a discriminator.

The principal contributions of this study are encapsulated in the following points:

- For the classification task of multivariate time series, we design and implement an AdvGAN model and we use KL divergence to optimize the adversarial examples.
- The efficacy of AdvGAN has been comprehensively verified across three multivariate time series datasets. The experimental outcomes demonstrate that AdvGAN has realized an elevated rate of attack success.

2   Related Work

Due to the successful application of adversarial attacks in the field of image recognition, researchers have also explored extending these methods to time series classification models. Fawaz et al. [12] applied the BIM and FGSM attack methods to the univariate time series classification datasets in UCR [13], demonstrating the vulnerability of time series classification models. Pialla et al. [14] generated smooth adversarial examples by maximizing the KL divergence between the predicted distributions of the original and adversarial samples. Harord et al. [15] proposed using a distilled model to imitate the classification behavior of the target model under attack, and then utilized an adversarial transformation network to attack the target model. Yang et al. [16] proposed a black-box attack method called TSadv for univariate time series data that does not require gradient information, generating adversarial examples through a Differential Evolution (DE) algorithm. Ding et al. [17] identified positions with significant influence on classification results through tree search to generate adversarial examples.

Significant strides have been made in the field of adversarial attacks, particularly with white-box and black-box attacks [18]. In image classification tasks, Xiao et al. [19] proposed the AdvGAN model, which leverages the generator of AdvGAN to create adversarial perturbations that are applied to the original data to produce adversarial examples capable of deceiving the target model. And AdvGAN also utilizes the advantages of GAN to generate realistic adversarial samples. Wang et al. [20] showed that by leveraging key features within images, more deceptive adversarial samples can be created. Multivariate time series data are characterized by multiple variables within time series that possess unique temporal patterns and structures. These patterns and structures have a significant impact on the model's predictive decision-making process because they reflect the intrinsic dynamics and the distinguishing features between classes.

3   Methodology

3.1   Problem Description

Given a dataset $D_{data} = \{X^{(i)}, y^{(i)}\}_{i=1}^N$ consisting of N samples, where $X^{(i)} \in R^{M \times T}$ represents multivariate time series data with M variables and T time steps, and $y^{(i)}$ denotes the category label of that multivariate time series data, the goal of multivariate time series classification is to train a classifier f that maps the multivariate time series classification data $X^{(i)}$ to the label $y^{(i)}$.

The objective of adversarial attacks is to craft adversarial examples from original samples in such a way that they lead the target model to make incorrect classification predictions. In the context of adversarial attacks on multivariate time series classification tasks, given an original sample $X^{(i)} \in R^{M \times T}$, the adversarial attack involves optimizing a perturbation $\delta \in R^{M \times T}$ generated by a generator G, so that the adversarial example $Y + \delta$ deceives the target model into classifying it into a target category t. The formulation can be represented as follows:

$$\min L(f(X + \delta), t) \tag{1}$$

This formulation seeks to find the perturbation δ that minimizes the loss function L, which in turn maximizes the likelihood of the model making an incorrect classification, subject to the constraint that $f(X + \delta)$ is classified as the target category t.

3.2   Attack Restriction

In this study, attackers are unable to access details such as the target model's architecture, parameters, or the training data it was built upon. They must devise their strategies based only on the observable inputs and outputs of the model. We begin by segmenting the dataset D_{data} into a training subset D_{train} and a testing subset D_{test} in a 1:1 ratio. Beyond this primary bifurcation, the testing subset D_{test} is further divided. It is apportioned into two equivalently-sized datasets that mirror the proportion of each class: the adversarial training set $D_{advtrain}$ and the adversarial testing set $D_{advtest}$. The adversarial training set $D_{advtrain}$ serves as the foundation for the training of our adversarial attack model. On the other hand, the adversarial testing set $D_{advtest}$ is dedicated to the assessment of the efficacy of the adversarial samples crafted by the attack model.

3.3 Model

We borrow from AdvGAN and apply it to the multivariate temporal classification task. By doing so, attackers can create targeted and stealthy adversarial samples that mislead the model, thus increasing the

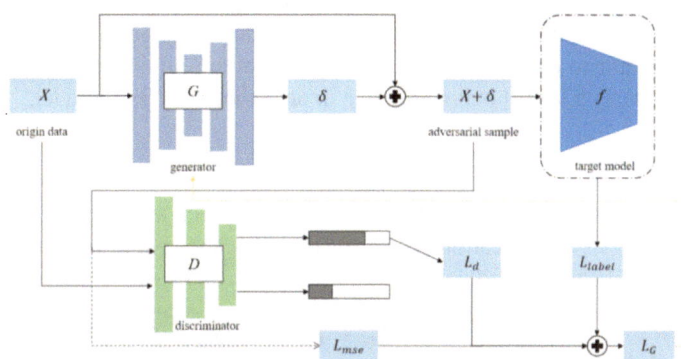

Fig. 1. AdvGAN for Multivariate time series

attack success rate. The generator within the AdvGAN framework is tasked with producing multivariate time series perturbations. This perturbations is integrated with the original input samples to construct adversarial samples designed to challenge the classification model. A discriminator evaluates the verisimilitude of these adversarial samples, ensuring they closely mimic genuine data. The comprehensive workflow of this approach is depicted in Fig. 1.

The original sample is fed into the generator G to produce a perturbation $\delta = G(X)$, which is then combined with the original sample to form the adversarial sample $X_{adv} = X + \delta$. Both the adversarial sample X_{adv} and the original sample X are combined to form a mixed dataset X_{mix}. The corresponding labels for this mixed data are denoted as y_{mix}. This dataset X_{mix} is then sent into the discriminator D. The objective of the discriminator D is to differentiate between the generated adversarial samples X_{adv} and the original samples X. It also guides the generator in producing more realistic adversarial samples.

The generator's goal is to create perturbations that, when incorporated, form adversarial samples capable of launching a successful attack on the target model while closely resembling real samples. In this paper, the binary cross-entropy (BCE) loss for the generator, denoted as L_d, can be expressed as follows:

$$L_d = -[y_{real} \log D(X_{adv}) + (1 - y_{real}) \log(1 - D(X_{adv}))] \qquad (2)$$

And y_{real} represents the label for real samples as perceived by the discriminator.

To ensure that adversarial samples successfully deceive the target model into classifying them into the target category t, we design a loss function L_{label} that measures the distance between the output of the adversarial sample when input into the target model f and the target category t. This loss function is crucial for guiding the generation of adversarial samples that can achieve the desired misclassification. We use p_t represents the probability distribution of target category t such that the probability value is 1 at the position of the target class. Given that KL represents the Kullback-Leibler (KL) divergence between the classification result of the adversarial sample input into the target model f and the target category probability distribution.

$$L_{label} = KL(f(X_{adv}), p_t) = \sum f(X_{adv}) \log(f(X_{adv})/p_t) \tag{3}$$

To restrict the magnitude of the perturbation generated, a mean squared error (MSE) loss L_{mse} is utilized. This loss function penalizes larger perturbations, encouraging the generator to create adversarial samples that are minimally different from the original samples while still being effective in deceiving the target model.

$$L_{mse} = MSE(X_{adv}, X) \tag{4}$$

The complete loss function for the adversarial attack in the context of multivariate time series classification typically combines above several components to achieve the desired behavior of the adversarial samples, can be expressed as follows:

$$L_G = L_d + \alpha L_{label} + \beta L_{mse} \tag{5}$$

where α, β are hyperparameters that control the relative importance of the label loss and the MSE loss, respectively.

4   Experiments and Results

In this experiment, we utilize the MLSTM_FCN as the target mode f, which consists of LSTM block and fully convolutional block with filters numbers set to 128, 256, 128, and kernel sizes of 7, 5 and 3, respectively. Considering the excellent performance of MLP [21] on multivariate time series datasets, a 5-layer MLP is employed as the generator with the adversarial framework, and a 4 layer MLP is used as the discriminator.

4.1   Dataset and Baseline

This paper selects FGSM adversarial attack methods for comparison. FGSM is a white-box attack technique that leverages the gradients of the target model to craft adversarial samples. It is an efficient method that adds a small, carefully calculated perturbation to the original samples to mislead the model's predictions. In order to control the MSE easily, we do not clip the adversarial samples.

In our experiment, we have curated a selection of three datasets: BasicMotions, Libras, and NATOPS. These datasets encapsulate a range of fields, including human motion analysis, sign language identification, and gesture recognition, offering a multi-faceted scenario to evaluate the performance of our AdvGAN model. And we standardize the dataset. We perform label encoding on the categories of each dataset, setting the target category to 0, which means that the classification target for each adversarial sample is set to 0. Our experiment are implemented in Python and executed on a server equipped with four NVIDIA GeForce RTX 4090 GPUs.

4.2 Evaluation Methodology

To ascertain the success rate of adversarial attacks on the test set, a selection process for the samples to be attacked is imperative. Initially, we assess whether the target model's prediction y' for an original sample x aligns with its true label y. Should y equate to y', we proceed to verify if the label y corresponds to the target category t. If the classification is accurate and pertains to the target category, an attack is deemed unnecessary. Similarly, if y does not equal y', the original sample is also exempt from attack. Subsequently, adversarial samples are crafted from the original samples that fulfill the aforementioned conditions and are subsequently fed into the target model to evaluate the efficacy of the attack. This study employs a duo of assessment metrics to gauge the performance of the adversarial attack methodologies.

- Attack Success Rate (ASR): The Attack Success Rate is a metric used to calculate the proportion of samples that have been successfully attacked out of the total adversarial samples. And $N_{success}$ is the number of adversarial samples that successfully mislead the target model into target class. N_{attack} is the total number of original samples used to generate adversarial samples. It is represented as follows:

$$ASR = N_{success} / N_{attack} \tag{6}$$

- Mean Squared Error (MSE): The Mean Squared Error (MSE) is utilized to denote the magnitude of the perturbation between the original samples and the adversarial samples that successfully mislead the target model on the test set. The MSE is computed as (4).

  In this study, the Mean Squared Error (MSE) serves as a metric for the size of the perturbation introduced to craft adversarial samples. Our objective is to attain a high Attack Success Rate (ASR) while maintaining MSE for these samples.

4.3 Preparation for Experiment

The objective of this experiment is to conduct adversarial attacks on a multivariate time series classification model known as MLSTM_FCN. To ensure that our attack is meaningful, the target model f must first be capable of accurately classifying the original samples. Therefore, we use D_{train} as train dataset to train the target model f. After training, we test the target model using D_{test} as a test dataset.

Table 1. The result of the target model MLSTM_FCN

Dataset	ACC (%)
BasicMotions	95.00
Libras	90.56
NATOPS	96.11

From Table 1, we can find that MLSTM_FCN model has high accuracy on these three datasets. And we think it makes sense to target a model with a high accuracy rate as an attack. Therefore, we think that it is reasonable to use MLSTM_FCN as the target model.

Table 2. The result of attack

Dataset	AdvGAN		FSGM	
	ASR (%)	MSE	ASR (%)	MSE
BasicMotions	78.57	0.0158	21.43	0.0169
Libras	82.72	0.2623	65.43	0.2704
NATOPS	57.75	0.0402	38.03	0.0400

4.4 Result

We conducted experiments on three datasets from the UEA and UCR archive. In our experiments, we use $D_{advtrain}$ as train dataset and $D_{advtest}$ as test dataset. We recorded the mean squared error (MSE) between the adversarial samples and the original samples. We maintain the mean square error (MSE) between the adversarial samples of successful attacks generated by different methods. The main metric we focus on is the Attack Success Rate (ASR), which provides insight into the efficacy of each method in launching adversarial attacks.

Table 2 shows the results of our experiments on three datasets. In our experiments, different α, β parameters are used for different datasets. From the table, it can be observed that, while maintaining the MSE, AdvGAN outperforms FGSM in terms of attack rates on all three datasets. By utilizing GAN, AdvGAN is able to produce high quality adversarial samples in the context of multivariate time series data.

5 Conclusion

To summarize our findings, we apply AdvGAN to multivariate time series data. In the experiments, MSE and ASR are used as metrics and experimentally validated on three UEA and UCR datasets. Experiments demonstrate that AdvGAN can effectively generate adversarial samples. And we think that the generated adversarial samples can be used to test the security of the target model.

References

1. Foumani, N.M., Miller, L., Tan, C.W., et al.: Deep learning for time series classification and extrinsic regression: a current survey. ACM Comput. Surv. (2023)

2. Zhan, X., Li, Y., Li, R., Gu, X., Habimana, O., Wang, H.: Stock price prediction using time convolution long short-term memory network. In: Liu, W., Giunchiglia, F., Yang, B. (eds.) KSEM 2018, Part I. LNCS, vol. 11061, pp. 461–468. Springer, Cham. https://doi.org/10.1007/978-3-319-99365-2_41

3. Che, Z., Cheng, Y., Zhai, S., et al.: Boosting deep learning risk prediction with generative adversarial networks for electronic health records. In: 2017 IEEE International Conference on Data Mining (ICDM), pp. 787–792. IEEE (2017)

4. Szegedy, C., Zaremba, W., Sutskever, I., et al.: Intriguing properties of neural networks. arXiv preprint arXiv:1312.6199 (2013)

5. Xie, Y., Wang, D., Chen, P.Y., et al.: A word is worth a thousand dollars: Adversarial attack on tweets fools stock predictions. arXiv preprint arXiv:2205.01094 (2022)

6. Ma, X., Niu, Y., Gu, L., et al.: Understanding adversarial attacks on deep learning based medical image analysis systems. Pattern Recognit. **110**, 107332 (2021)

7. Mahmood, K., Mahmood, R., Rathbun, E., et al.: Back in black: a comparative evaluation of recent state-of-the-art black-box attacks. IEEE Access **10**, 998–1019 (2021)

8. Meng, L., Lin, C.T., Jung, T.P., Wu, D.: White-box target attack for EEG-based BCI regression problems. In: Gedeon, T., Wong, K., Lee, M. (eds.) ICONIP 2019, Part I. LNCS, vol. 11953, pp. 476–488. Springer, Cham (2019). https://doi.org/10.1007/978-3-030-36708-4_39

9. Goodfellow, I.J., Shlens, J., Szegedy, C.: Explaining and harnessing adversarial examples. arXiv preprint arXiv:1412.6572 (2014)

10. Madry, A., Makelov, A., Schmidt, L., et al.: Towards deep learning models resistant to adversarial attacks. arXiv preprint arXiv:1706.06083 (2017)

11. Gui, J., Sun, Z., Wen, Y., et al.: A review on generative adversarial networks: algorithms, theory, and applications. IEEE Trans. Knowl. Data Eng. **35**(4), 3313–3332 (2021)

12. Fawaz, H.I., Forestier, G., Weber, J., et al.: Adversarial attacks on deep neural networks for time series classification. In: 2019 International Joint Conference on Neural Networks (IJCNN), pp. 1–8. IEEE (2019)

13. Dau, H.A., Bagnall, A., Kamgar, K., et al.: The UCR time series archive. IEEE/CAA J. Autom. Sin. **6**(6), 1293–1305 (2019)

14. Pialla, G., et al.: Smooth Perturbations for time series adversarial attacks. In: Gama, J., Li, T., Yu, Y., Chen, E., Zheng, Y., Teng, F. (eds.) PAKDD 2022, pp. 485–496. LNCS, vol. 13280. Springer, Cham (2022). https://doi.org/10.1007/978-3-031-05933-9_38

15. Harford, S., Karim, F., Darabi, H.: Adversarial attacks on multivariate time series. arXiv preprint arXiv:2004.00410 (2020)

16. Yang, W., Yuan, J., Wang, X., et al.: TSadv: black-box adversarial attack on time series with local perturbations. Eng. Appl. Artif. Intell. **114**, 105218 (2022)

17. Ding, D., Zhang, M., Feng, F., et al.: Black-box adversarial attack on time series classification. In: Proceedings of the AAAI Conference on Artificial Intelligence, vol. 37, no. 6, pp. 7358–7368 (2023)

18. Li, Y., Cheng, M., Hsieh, C.J., et al.: A review of adversarial attack and defense for classification methods. Am. Stat. **76**(4), 329–345 (2022)

19. Xiao, C., Li, B., Zhu, J.Y., et al.: Generating adversarial examples with adversarial networks. arXiv preprint arXiv:1801.02610 (2018)

20. Wang, Z., Guo, H., Zhang, Z., et al.: Feature importance-aware transferable adversarial attacks. In: Proceedings of the IEEE/CVF International Conference on Computer Vision, pp. 7639–7648 (2021)

21. Yi, K., Zhang, Q., Fan, W., et al.: Frequency-domain MLPs are more effective learners in time series forecasting In: Advances in Neural Information Processing Systems, vol. 36 (2024)

Recognition and Calculation of Fish Rafts in Mariculture on the Basis of Artificial Intelligence

Weibo Zhang[1], Li Zhang[2], Yaozhao Zhong[1], Peitu Lin[3], and Feng Zhang[1(✉)]

[1] Fuzhou Institute of Oceanography, Miniiang University, Fuzhou, China
zhongyz@mju.edu.cn, zhang168feng@gmail.com
[2] Fujian Academy of Environmental Sciences, Fujian Technology Center of Emission Storage, Fuzhou, China
[3] Fujian Key Laboratory of Autonomous Controllable Software, Linewell Software Co., Ltd., Quanzhou, China
lpeitu@linewell.com

Abstract. Mariculture is crucial for grain security and sustainable development, but identifying and counting for the different types of mariculture remains a challenge. In this research, the data of satellite remote sensing, unmanned aerial vehicle (UAV) and shore-based camera is adopted to collect high-definition pictures, and the deep learning algorithm is adopted to detect and position the marine aquaculture areas. Then the classification model of convolutional neural network is adopted to identify the aquaculture species, and the OpenCV model is adopted to assess the area of aquaculture area. The approach obviously improves the efficiency and accuracy of marine aquaculture type identification and area assessment, reduces the subjectivity and error rate of manual statistics, has higher robustness and applicability, can further optimize the algorithm in the future, and is expanded to be applied to the scope on ocean environment monitoring, fishery management and the like.

Keywords: artificial intelligence · aquaculture facilities · YOLO · OpenCV · ocean environment monitoring

1 Introduction

Mariculture plays an important role in guaranteeing global grain security and promoting sustainable economic development. With the increasing demand for ocean resources, it becomes increasingly crucial to accurately identify and count the different types of mariculture. However, this mission is often challenging due to a great deal of work and subjective judgment. Therefore, there is a need to develop efficient, accurate, and automated solutions to respond this challenge.

The goal of this paper is to put forward an innovative approach on the basis of satellite images and deep learning techniques to solve the problem of marine aquaculture type identification and calculation. Our research is motivated by the limitations of existing

© The Author(s) 2025
P. Siarry et al. (Eds.): WCNA 2023, LNEE 1361, pp. 203–210, 2025.
https://doi.org/10.1007/978-981-96-2409-6_20

approaches and the need for sustainable mariculture management. This new approach is designed to increase productivity, reduce human error and revolutionize the mariculture industry. The challenges of mariculture and existing identification approaches are briefly introduced. Then, we introduce the proposed approaches in detail, containing data acquisition, image processing, object detection and classification on the basis of deep learning and area calculation. We also discuss the potential applications of the designed methodology and future research directions. The structure of this paper is as follows: the second section introduces the research background, the third section describes the methodology, the fourth section presents the discussion of the research results, and the fifth section is the conclusion.

2 Background

2.1 Application Progress of Artificial Intelligence in Marine Aquaculture Area Identification

In the past few decades, mariculture research has made significant progress. Early research paid attention to understanding the impact of mariculture on the environment [1, 2] Adopting remote sensing techniques to detect and monitor mariculture areas and explore their influence on ecosystems [3], scholars provided a comprehensive overview of the present status, challenges, and prospects of the sustainable development of mariculture [4]. These researches laid the foundation for sustainable mariculture management. However, most of these researches rely on traditional artificial identification approaches, which are inefficient and lack accuracy. With the rapid development of artificial intelligence and deep learning on the scope of image processing, scholars are beginning to apply these techniques to mariculture, to improve the efficiency of identification and monitoring [5, 6].

2.2 Limitations of Existing Approaches and Innovation Needs

Although there have been some researches adopting remote sensing and artificial intelligence techniques, most of them are specific to particular farming types or areas [7] and there is no simultaneous classification and area calculation of farming types. In addition, the robustness and adaptability of these approaches need to be improved when they are applied in large-scale multi-culture areas. The limitations of existing approaches point to the need for a comprehensive, automated and efficient approach to the identification and calculation of mariculture. This approach should be highly accurate, robust and widely applicable to provide effective support for sustainable mariculture management.

2.3 Innovations

Different from the existing research, we put forward a comprehensive approach on the basis of multi-source satellite images and deep learning algorithm to realize the accurate identification and calculation of mariculture types. Our innovations contain:

1) Multi-source data fusion: the data of high-definition satellite pictures, unmanned aerial vehicle images and shore-based camera is adopted to cover a larger range and improve image quality.
2) Advanced target detection and classification: adopt proven YOLO algorithm [8] and convolutional neural network classifier [9] to realize efficient and accurate detection and classification.
3) Innovation of area calculation: We put forward a novel area calculation approach by combining computer vision and culture characteristics.
4) Strong robustness and applicability: the approach is verified to have strong robustness and can adapt to various marine culture environments and types.

3 Method

3.1 Technology Roadmapping

Figure 1 displays the technical roadmap for this experiment, with detailed content to be elaborated below.

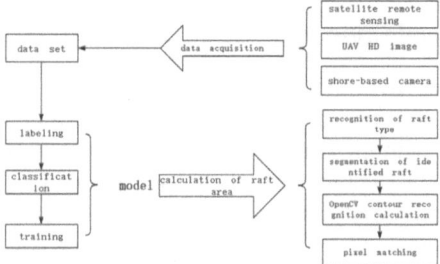

Fig. 1. Technology roadmap

3.2 Data

To acquire high-quality training data sets, we collected data from multiple sources (Fig. 2 for data collection areas): 1) Satellite images: We adopted high-definition satellite pictures covering multiple mariculture areas. These images are abundant in visual information and can be adopted to train the model. In order to improve the image quality, we preprocess the image, including clipping, denoising and correction. 2) UAV Images: We adopt UAVs to take high-definition pictures of specific areas with a variety of mariculture types. The UAV image has high-definition and coverage, which can provide more training samples for the model. 3) Shore-based camera data: we install high-definition cameras in coastal areas to continuously take images of the mariculture area. These images provide more detailed farming information.

We combined the data sets from these sources to create a large data set. Each image is annotated to mark the mariculture area. We adopt the tool (this article uses the LabelImg tool) to manually label the image to guarantee that the labeling is accurate. After labeling, we split the dataset into a training set, a verification set, and a test set.

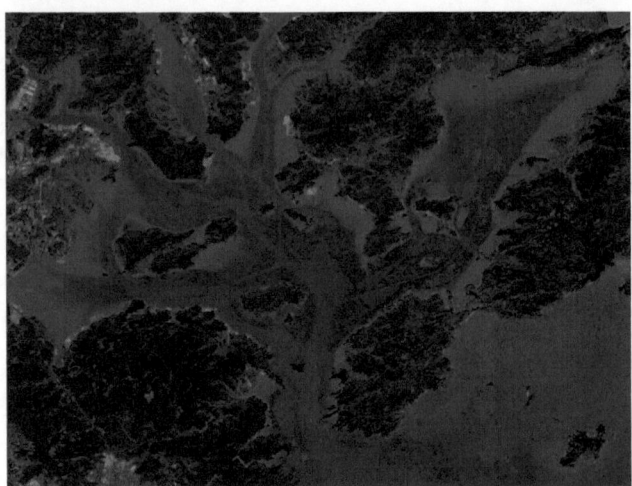

Fig. 2. High-precision remote sensing image

3.3 Model Training and Optimization

Adopt the YOLO series algorithm (YOLOv8) to detect the target in the mariculture area. YOLO algorithm is a real-time target detection algorithm. Its key idea is to divide the image into meshes, and predict the bounding box and corresponding target probability in each mesh. YOLOv8 improves detection accuracy while ensuring real-time performance, making it more advantageous compared to other target detection methods. YOLOv8 is the latest improved version of the YOLO series, which adopts a more complex network architecture and optimization technology, and correspondingly improves detection capabilities. YOLOv8 system can then be the method applied to accurately locate target objects in aquaculture settings while also exponentially maximizing the efficiency and effectiveness of the supervision and management processes (Fig. 3).

For high-speed applications, YOLOv8 is greatly useful, giving real-time object detection with high accuracy. Hence, because YOLOv8 inherits the feature fusion from the multi-scale feature map, YOLOv8 can capture target information at different scales and improve its detection accuracy and robustness. YOLOv8 adapts to one single detection to recognize many targets in an image, where it led to the minimization of excessive computations over other methods of traditional object detection, thus leading to the improved detection efficiency. Simple architecture design of YOLOv8 makes it suitable for the deployment of the network in various systems; the network requires fewer parameters, and it has an ultra-fast training and inference speed as well. Subsequently, YOLOv8 is being fine-tuned on a large-scale dataset, which gives rise to strong generalization ability and robustness across a complicated case, thus making it handy for any complex detection task. Also, we need to train on object detection tasks via YOLOv8 algorithm. This algorithm was chosen because it can find multiple targets in a single test with high accuracy. We train the YOLO model on the training dataset and optimize the model adopting techniques such as batch normalization, sample dropping and learning rate scheduling. The detected mariculture region images are input to a pre-trained Convolutional Neural

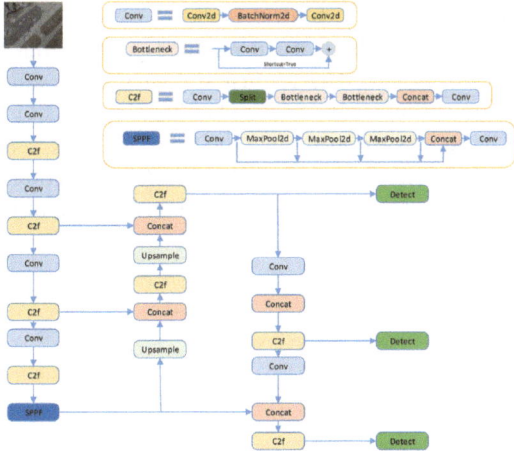

Fig. 3. Schematic diagram of improved YOLOv8 neural network structure

Network (CNN) classifier [11]. The CNN classifier can identify the aquaculture type. We adopt transfer learning techniques, adopting pre-trained VGG or ResNet models, to fine tune our farming datasets. And calculating the area of the detected breeding area by means of adopting the OpenCV library.

OpenCV is an open source computer vision library, which provides numerous advanced image processing and computer vision algorithms. We adopt OpenCV for accurate area calculations. The hash tables and machine learning algorithms provided in OpenCV are effective in identifying and processing objects in images. Adopting the computer vision ability of OpenCV, we designed a set of area calculation model, which can accurately calculate the area of mariculture area in accordance with the shape and size of fish raft. OpenCV can accurately calculate the number of pixels occupied by a fish raft in accordance with its shape and size characteristics. Then, we convert the number of pixels into the actual area of the fish raft in accordance with the conversion relationship between the pixels and the actual area.

4   Results and Discussion

In this research, the YOLOv8 model and OpenCV library are adopted to detect the target and calculate the area of mariculture area, and a series of satisfactory results are acquired. Through the training of YOLOv8 model, we acquire the classification model and segmentation model for mariculture area. The experimental results illustrate that the YOLO model can recognize the target contour accurately and separate the target object completely. However, for the sake of the simple characteristics of the raft itself, the model will mistakenly identify similar objects in the sea as the raft, which will affect the subsequent area statistics. We have further verified the accuracy of the model. By calculating the ratio of the number of samples correctly detected and classified to the total number of samples, the accuracy of the model is approximately 93%, which illustrates that deep learning is suitable for extraction of marine aquaculture areas.

We analyzed the recall rate (Fig. 4-left) and the F1 score (Fig. 4-right) of the model, and found that the model had a high recall rate and good overall performance while maintaining a high accuracy rate. The confusion matrix is the situation analysis table that summarizes the prediction results of the classification model in machine learning. The records in the data set are summed up in the form of matrix in accordance with the two criteria of the true classification and the classification prediction of the classification model. The chaos matrix of the model is illustrated on the right of Fig. 5.

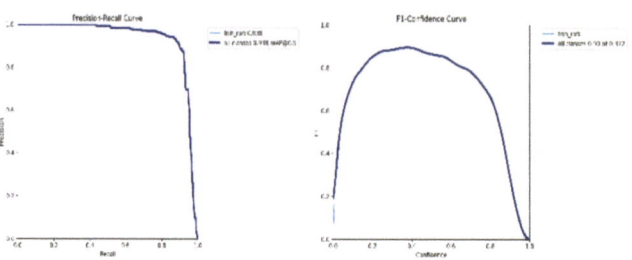

Fig. 4. The recall rate (left) and the F1 score (right) of the model

However, the model illustrates greater sensitivity to variations in recall rates, suggesting that we need to further optimize and adapt the model to promote its performance and stability. In the aspect of area calculation, OpenCV can quickly and accurately calculate the number of pixels in the contour, but the actual area represented by the pixels needs to be considered. On the basis of the experimental and assessment results, we verified the validity, accuracy and adaptability of this approach in marine aquaculture type identification and calculation. Finally, the total area of the aquaculture area is acquired by counting the actual area of each raft.

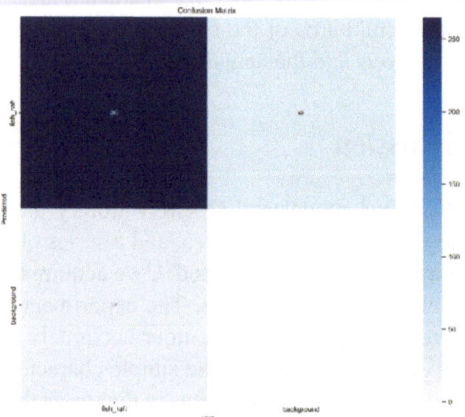

Fig. 5. The chaos matrix of the model

5 Conclusions

Through this research, we put stress on the importance of artificial intelligence technology in marine aquaculture type identification and calculation. The application of YOLOv8 model and OpenCV library not only promotes the efficiency of identification and calculation, but also reduces the burden of manual operation. Artificial intelligence technology can quickly and accurately monitor in large-scale and complex environment, and assist farm managers in finding problems and take corresponding measures in time. The future research directions include model optimization and adjustment, data enhancement and expansion, real-time performance and automation, so as to further promote the performance and stability of the model, improve the technical development on the scope of marine aquaculture, and make greater contribution to the sustainable development of the industry and the improvement of production efficiency. Combining the YOLOv8 model and the function expansion of OpenCV, the intelligent mariculture management system is developed to realize more accurate target detection and area calculation in mariculture area. The system can adopt YOLOv8 model to detect the target, identify the fish raft and other targets in the aquaculture area, and calculate the area and density of the fish raft quickly with the help of the function expansion of OpenCV, so as to realize the real-time monitoring and automatic management of the aquaculture area.

Acknowledgment. This study was supported by the Fujian Province Natural Science Foundation (No: 2023J011573, 2023J011402), the Fujian science and technology Major special project (No: 2022NZ033023), the Fujian Marine Economic Development Special Fund Project (No: FJHJF-L-2022-17), Fujian Province Key Science and Technology Innovation Project (2022G02009) and the MJU Scientific Research Foundation for Talents (No: MJY22033, MJY23014). This work was also Supported by the Opening Project (No. 2023QZJ01) of Sichuan Province University Key Laboratory of Bridge Nondestruction Detecting and Engineering Computing and the Scientific Research and Innovation Team Program of Sichuan University of Science and Technology (SUSE652B002).

References

1. Yu, W.: Discussion on the influence on mariculture on marine environment. Ocean Dev. Manag. (12), 113–117 (2008)
2. Yang, W., Gao, H., Zhang, Y.: Research progress on environmental influence on mariculture on coastal waters. Trans. Oceanol. Limnol. (01), 100–107 (2006)
3. Yan, J., et al.: Monitoring of harbor culture area on the basis of UAV remote sensing technology. J. Xiamen Univ. (Nat. Sci.) **55**(05), 742–748 (2016)
4. Mang, Q., Xu, G., Zhu, J., Xu, C.: Current situation and prospect of aquaculture development in China. Fishery Mod. **49**(02), 1–9 (2022)
5. Li, D., Liu, C.: Analysis and prospect of artificial intelligence in aquaculture. Smart Agric. (Chin.) **2**(3), 1–20 (2020)
6. Wang, X., Deng, Q., Wang, J., Fan, J.: Information extraction of marine raft culture with deep semantic segmentation MRF model. J. Shandong Univ. (Eng. Sci.) **52**(02), 89–98 (2022)
7. Cheng, T., Zhou, W., Fan, W.: Progress of remote sensing identification approaches for aquaculture areas. Remote Sens. Land Resour. **2012**(03), 1–5 (2012)

8. Shao, Y., Zhang, D., Chu, H., Zhang, X., Rao, Y.: Review of YOLO target detection on the basis of deep learning. J. Electron. Inf. Technol. **44**(10), 3697–3708 (2022)
9. Li, B.: Pattern classifier on the basis of convolutional neural network. J. Dalian Univ. (02), 19–23 (2003)
10. Ren, L.: Improved YOLOv8 neural network structure schematic diagram. Mendeley Data **V1** (2024). https://doi.org/10.17632/rbgkyhf6r9.1
11. Du, J.: Understanding of object detection based on CNN family and YOLO. J. Phys. Conf. Ser. **1004**(1), 012029–012029 (2018)

Robust Multi-agent Federated Reinforcement Learning for Task Offloading

Dibao Yan[1,2,3], Yongfeng Wang[1,2,3], Wenjing Hou[1,2,3], Huanhuan Song[1,2,3], Hong Wen[1,2,3(✉)], Wendi Ma[1,2,3], and Fan Sun[1,2,3]

[1] School of Aeronautics and Astronautics, University of Electronic Science and Technology of China, Chengdu 611731, People's Republic of China
{wyf_kt,hwj,hhs_communi,sunlike}@uestc.edu.cn
[2] Aircraft Swarm Intelligent Sensing and Cooperative Control Key Laboratory of Sichuan Province, Chengdu 611731, China
[3] Sichuan Intelligent IoT Communication Technology Engineering Research Center, UESTC, Chengdu, China

Abstract. With the proliferation of various mobile smart devices, edge computing and computational offloading technologies have emerged as pivotal support mechanisms, enhancing the service quality of these devices. To facilitate the learning of task offloading strategies across diverse and complex scenarios, this study introduces a federated learning strategy algorithm based on multi-agent deep reinforcement learning. This algorithm aims to aggregate the training strategies of multiple edge computing devices, thereby enabling the synthesis of superior task offloading strategies tailored to a wide array of environments while also providing safeguards against malicious nodes. This paper specifically examines the reward inversion attack, illustrating the algorithm's capability to identify and counteract such malicious threats effectively. Experimental results validate that the proposed algorithm not only robustly defends against these attacks but also adeptly learns the task offloading strategies pertinent to each node.

Keywords: task offloading · malicious node detection · federated learning · multi-agent deep reinforcement learning · insert

1 Introduction

With the advancement of 5G mobile communication technologies and the widespread adoption of smart terminals, Internet of Things (IoT) technology is increasingly being deployed in diverse applications, including smart robots, various cameras, sensors, and other devices. Mobile terminals have an increasingly high demand for computing power [1], and the current mobile terminal devices have a trend of miniaturization and lightweight development, so that they can be applied to more scenarios.

The lightweight of the Internet of Things terminal leads to its poor computing power. It is difficult to deal with a variety of complex computing tasks in a timely manner, and it is also facing the rapid loss of electricity. These problems seriously affect the operating

© The Author(s) 2025
P. Siarry et al. (Eds.): WCNA 2023, LNEE 1361, pp. 211–218, 2025.
https://doi.org/10.1007/978-981-96-2409-6_21

efficiency of the device and the user experience. In order to solve the above problems, edge computing and computational offloading technologies came into being and play a key role in the intelligent interconnection of the Internet of Things and the Internet of vehicles [2].

Edge computing positions resources at the edge of the network closer to users to provide IT service environment and cloud computing capabilities for mobile networks, thus providing users with ultra-low latency and high-efficiency network service solutions [3, 4]. As one of the key technologies in edge computing, computing offloading refers to the technology in which terminal devices transfer part or all computing tasks to the cloud computing environment to solve the shortcomings of mobile devices in resource storage, computing performance and energy efficiency.

The task offloading process always affected by many aspects, such as the external environment and the offloading system. Therefore, in order to enable edge computing devices to learn task offloading policies in a variety of complex task offloading scenarios, the training strategies of multiple edge computing devices are aggregated in combination with the characteristics of distributed security data sharing based on federated learning [5]. Learn better task offloading strategies to adapt to more diverse scenarios. At the same time, there are often multiple data nodes in federated learning, and whether each data node is safe and reliable is another key issue.

In this paper, a federated learning strategy algorithm based on multi-agent deep reinforcement learning is proposed. The main objective of this strategy is to aggregate the task offloading policies of each target agent, and detect the security of each node participating in federated aggregation to avoid damage to the system caused by malicious nodes.

Specifically, the contributions of this paper include:

(1) Federated Aggregation: By performing FedAvg (Federated Averaging) [6] on the data of participating aggregation nodes, each node can learn the strategies of other agents and improve the universality of the agent.
(2) Malicious Node Detection: The federated aggregation approach outlined in this paper extends beyond mere FedAvg; it also entails the detection of data involved in the aggregation. This is achieved through the use of Euclidean distance and Modified Z-score methodologies to identify and exclude data from abnormal nodes.
(3) Experimental Verification: The robustness of the algorithm is tested through reward flip attacks on individual agents participating in federated reinforcement learning. The comparative performance of the algorithm, both with and without the presence of malicious nodes, is empirically validated. The experimental data substantiate the method's resilience in the face of malicious nodes within the context of federated reinforcement learning.

2 System Model

2.1 Task Offloading Model

For each edge computing device, the optimization objective of task offloading is to minimize the computational delay and energy consumption of device terminal task processing within its respective region.

Regarding local computing, the computational delay for tasks completed locally is defined as:

$$T_n^{local} = \frac{Task_n(t)\sigma}{y_n^{local}} \tag{1}$$

The variable $Task_n(t)$ T_n^{local} represents the task load, y_n^{local} denotes the number of CPU cycles for device n, and σ signifies the number of CPU cycles required to complete a unit task.

The energy consumption for local computation is given by:

$$E_n^{local}(t) = \alpha(y_n^{local})^3 T_n^{local} \tag{2}$$

where α is the energy consumption coefficient.

For edge computing, the total latency is composed of transmission latency and computation latency. The transmission latency is related to the communication transmission rate. The maximum uplink rate, calculated based on the Shannon formula, is given by:

$$C_n(t) = W_n(t) \log_2(1 + \frac{p_n(t)h_n(t)}{N_0}) \tag{3}$$

where $p_n(t)$ is the transmission power, $h_n(t)$ is the channel gain, $W_n(t)$ represents the bandwidth, and N_0 is the Gaussian noise power spectral density. Consequently, the transmission latency is:

$$T_n^{edge\_trans} = \frac{Task_n(t)}{C_n(t)} \tag{4}$$

The computation latency is:

$$T_n^{edge\_comp} = \frac{Task_n(t)\sigma}{y_n^{edge}} \tag{5}$$

Thus, the total latency is

$$T_n^{edge} = T_n^{edge\_trans} + T_n^{edge\_comp} \tag{6}$$

The energy consumption for edge computing consists of transmission energy consumption and edge server computation energy consumption:

$$E_n^{edge} = p_n(t)T_n^{edge\_trans} + \alpha(y_n^{edge})^3 T_n^{edge\_comp} \tag{7}$$

Therefore, for the task offloading strategy:

$$x_n(t) = \begin{cases} 0 \ local \\ 1 \ edge \end{cases} \tag{8}$$

When the task needs to be offloaded to the edge server, $x_n(t)$ is set to 1, otherwise it is set to 0. Consequently, the total energy efficiency function is:

$$A(x) = \sum_{n=1}^{N} ((1 - x_n(t))(C_e E_n^{local} + C_t T_n^{local})$$
$$+ x_n(t)(C_e E_n^{edge} + C_t T_n^{edge})) \tag{9}$$

where C_e is the energy consumption coefficient, and C_t is the latency coefficient.

The task offloading optimization problem is to find the strategy $x_n(t)$ that minimizes $A(x)$. The energy efficiency function given by (9) cannot be directly solved. For such complex problems, deep reinforcement learning is more suitable for finding the solution.

2.2 Multi-agent Deep Reinforcement Learning Algorithm

The core concept of Twin Delayed Deep Deterministic Policy Gradient (TD3) [7] is to utilize two independent Q-networks and a delayed policy update mechanism to mitigate estimation bias and prevent excessive policy updates. Specifically, TD3 maintains two value function networks (Critic) and a policy function network (Actor). During the value update process, TD3 employs the minimum estimates from the two critic networks for updates, which helps to reduce bias in the estimates. Additionally, TD3 adopts a delayed update strategy, where the policy function is updated only after several value function updates, enhancing the algorithm's stability. The TD3 algorithm also introduces target policy smoothing, which involves adding limited-amplitude noise to the target policy during updates, further improving the stability of the learning process.

2.3 Malicious Node Detection

After each round of federated aggregation, the neural network parameters of each node are identical. As the nodes continue distributed training, it can be considered that all nodes are iteratively training in the same direction. Consequently, the Euclidean distances of the neural network parameters between nodes will diverge.

Specifically, during federated aggregation, the Euclidean distance of the neural network parameters of each node is first calculated. Then, the Modified Z-score algorithm is used for outlier detection. This algorithm is an improvement of the Z-score algorithm. The Z-score normalization method is as follows:

$$z_i = \frac{x_i - \mu}{\delta} \tag{10}$$

where μ is the mean, x_i represents the observed value of the sample, δ represents the standard deviation of all observed values, and z_i indicates how many standard deviations the sample point is from the sample mean. This is used to represent the relative position of each raw data point within the dataset.

Since the mean and standard deviation are highly sensitive to outliers, the conventional Z-score method can lead to biased results. Therefore, the Modified Z-score method is introduced:

$$z_i = \frac{x_i - median(x_i)}{k * MAD} \tag{11}$$

where x_i is the observed sample value, $median(x_i)$ is the median of all observed sample values, k is a constant factor, and MAD (Median Absolute Deviation) is defined as:

$$MAD = median|x_i - median(x_i)| \tag{12}$$

When the condition:

$$|z_i| > threshold \tag{13}$$

is met, the data can be considered an outlier, potentially originating from a malicious node.

3 Algorithm Design

This paper proposes a federated learning strategy algorithm based on a multi-agent deep reinforcement learning framework. The algorithm first identifies and eliminates potential malicious nodes by analyzing the model Euclidean distance. Following this, it aggregates the neural network parameters of each node, thereby effectively obtaining the learning parameters while mitigating the risk of attacks from malicious nodes.

The procedure of our algorithm is as follows:

(1) Set up multiple tasks offloading scenarios and establish a basic federated learning environment. Initialize the parameters of each edge computing device and the relevant parameters for federated aggregation.

(2) Set up the reinforcement learning environment, initializing two critic networks $Q_{\theta 1}$ and $Q_{\theta 2}$, as well as their corresponding target networks $Q_{\theta 1'}$ and $Q_{\theta 2'}$. Initialize an actor network μ_ϕ and its corresponding target actor network $\mu_{\phi'}$. Initialize the experience replay buffer, used to store experience tuples (s, a, r, s', d), where s is the state, a is the action, r is the reward, s' is the next state, and d indicates whether it is a terminal state. Finally, copy the current reinforcement learning network parameters to the remaining agents.

(3) The agent executes the task offloading decision, obtaining a set of samples (s, a, r, s', d) and storing them in the experience replay buffer. Samples are then drawn from the replay buffer to update the Actor and Critic networks. For each drawn sample (s, a, r, s', d), the next action a' is generated using the target actor network $\mu_{\phi'}(s')$ and action noise. The target values are calculated using the target critic networks $Q_{\theta 1'}$ and $Q_{\theta 2'}$. The critic networks $Q_{\theta 1}$ and $Q_{\theta 2}$ are updated based on the target values and minimizing the evaluation error. The actor network μ_ϕ is updated by optimizing ϕ through the maximization of $Q_{\theta 1}(s, \mu_\phi(s))$. The target networks' parameters are updated using a soft update strategy.

(4) Each node uploads its model data to the aggregation node. The aggregation node screens out outliers based on the Euclidean distance using the Modified Z-score method. The remaining reliable model data are then aggregated using the FedAvg algorithm, and the updated model data are distributed back to each node.

(5) Repeat steps (3)–(4).

4 Experimental Results Analysis

4.1 Basic Experimental Setup

The experiment set up five task offloading scenarios and five task offloading decision-making agents, with the total number of aggregation rounds set to 10,000. After each aggregation, each node performs 10 model updates, with a batch size of 128 for each update. Since the goal of reinforcement learning is to maximize the reward through learning strategies, it is necessary to invert the energy efficiency function in (9), transforming the minimization problem of the energy efficiency value into a maximization problem.

This paper addresses the participation of malicious nodes in federated aggregation by modifying the training set during the local training of the model. The following section presents the simulation experiment results and analysis in the presence of malicious nodes participating in the aggregation algorithm.

4.2 Malicious Node Attacks

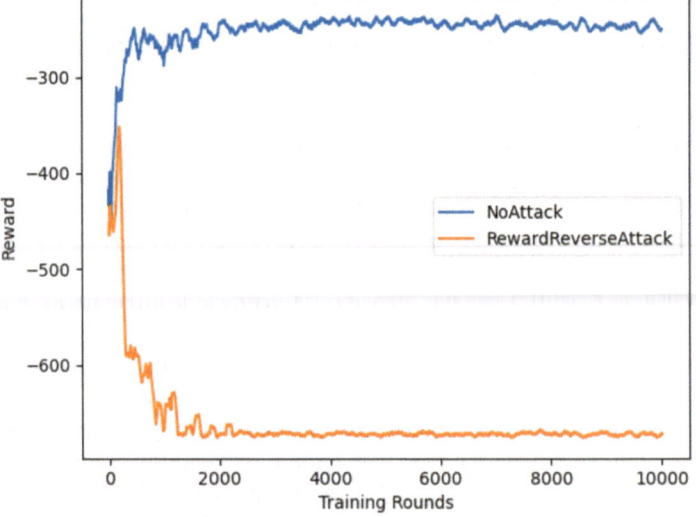

Fig. 1. Reward curves of normal nodes and malicious nodes

As shown by the blue line in Fig. 1, the reward value of the agent gradually converges to the optimal value with iterative training. The orange line demonstrates that the presence of malicious nodes gradually reduces the reward value as training progresses, eventually converging to the worst value. This indicates that malicious nodes consistently attempt to enforce the worst task offloading strategy.

Figure 2 illustrates the reward curve in the presence of malicious nodes among the nodes participating in federated aggregation. When there is no attack and the FedAvg

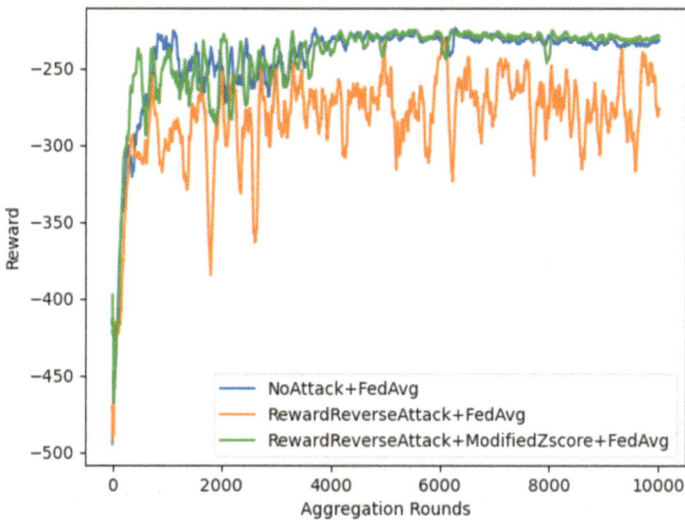

Fig. 2. Performance of the proposed algorithm in the presence of malicious node (Color figure online)

algorithm is used for aggregation, as shown by the blue line in Fig. 2, the reward curve converges to around -230 after 3700 aggregation rounds. In the presence of a malicious node participating in federated aggregation, the malicious node learns the worst strategy by inverting the reward in the experience replay pool. When this node participates in FedAvg, the aggregated reward curve, depicted by the orange line, fluctuates within the range of $[-350, 350]$ and shows no convergence trend. However, when the Modified Z-score method is applied in the scenario with an attack, as shown by the green line, the reward curve gradually converges at 3800 rounds, and the overall curve closely resembles the blue line in the no-attack scenario. This demonstrates that the federated learning strategy algorithm based on the TD3 deep reinforcement learning algorithm proposed in this paper can achieve similar performance to that in a no-attack scenario while resisting certain attacks.

5 Conclusion

This paper proposes a federated learning strategy based on a multi-agent deep reinforcement learning algorithm. The algorithm can effectively identify malicious attacks and maintain performance similar to that in scenarios without attacks. Through experimental validation in simulated attack scenarios, this paper demonstrates that the algorithm exhibits excellent defense performance in the presence of malicious nodes utilizing reward inversion attacks.

Acknowledgement. This work is supported by NSFC under Grant U23B2021 and 62201132.

References

1. Apostolopoulos, P.A., Fragkos, G., Tsiropoulou, E.E., Papavassiliou, S.: Data offloading in UAV-assisted multi-access edge computing systems under resource uncertainty. IEEE Trans. Mob. Comput. **22**(1), 175–190 (2023)
2. Yang, C., Liu, Y., Chen, X., Zhong, W., Xie, S.: Efficient mobility-aware task offloading for vehicular edge computing networks. IEEE Access **7**, 26652–26664 (2019)
3. Chen, X., et al.: Information freshness-aware task offloading in air-ground integrated edge computing systems. IEEE J. Sel. Areas Commun. **40**(1), 243–258 (2022)
4. Lai, X., Fan, L., Lei, X., Deng, Y., Karagiannidis, G.K., Nallanathan, A.: Secure mobile edge computing networks in the presence of multiple eavesdroppers. IEEE Trans. Commun. **70**(1), 500–513 (2022)
5. Qiao, D., Liu, G., Guo, S., He, J.: Adaptive federated learning for non-convex optimization problems in edge computing environment. IEEE Trans. Netw. Sci. Eng. **9**(5), 3478–3491 (2022)
6. Mahan, B., Moore, E., Ramage, D., Hampson, S., Arcas, B.A.: Communication-efficient learning of deep networks from decentralized data. Artif. Intell. Stat., 1273–1282 (2017)
7. Kang, C., et al.: TD3 algorithm based on dynamic delay policy update. J. Jilin Univ. (Inf. Sci. Ed.) **38**(4), 474–481 (2020)

Optimization of Real-Time Power Grid Measurement Data Scheduling Based on Genetic Algorithms

Bin Fang[1]([✉]), Shi Zhu[1], Hongyu Zhu[1], Ziqian Zhang[2], Ye Tao[1], and Yubin Sheng[3]

[1] Information and Communication Branch of State Grid Hunan Electric Power Company Limited, Changsha 410007, China
403183314@qq.com
[2] Nanjing NARI Information and Communication Technology, Nanjing 210000, China
zhangziqian@sgepri.sgcc.com.cn
[3] School of Software, Xinjiang University, Urumqi 830091, China
ybsheng@csu.edu.cn

Abstract. An increasing demand of real-time monitoring and data analysis for the power grid, is facing challenges from the processing rate of massive real-time measurement data such as voltage, current and power from billions of sensors. To guarantee process rate, many measurement data scheduling method had been studied, while the problem for the balance of realtime requirement and load balance has not been solved well, which leads to uneven load distribution and data congestion. Thus, this paper proposed a measurement data scheduling method based on genetic algorithm. The proposed method considers the characteristics of historical measurement data, task allocation matrix, and creates an adaptive parameter iterative solution based on using genetic algorithm. The results show that the proposed method can achieve appreciate load balance with time latency guarantee.

Keywords: Power grid · Measurement data · Data scheduling · Genetic algorithm

1 Introduction

The power grid system is a complex and massive entity, including many devices, whose normal functions are crucial for industrial production and daily life. In order to ensure the normal operation and safe stability of the power grid, real-time monitoring and data analysis of numerous equipment in the power grid system are crucial [1]. These data are known as power grid measurement data. Once a large amount of measurement data is gathered from the devices, it is imperative to transmit this data promptly to the data center for analysis and assessment of the grid's operational state, thus making the real-time transmission of grid measurement data of utmost importance. However, due to limitations in network resources such as channel capacity, existing methods for the transmission of power grid measurement data encounter problems like data congestion,

P. Siarry et al. (Eds.): WCNA 2023, LNEE 1361, pp. 219–229, 2025.
https://doi.org/10.1007/978-981-96-2409-6_22

load imbalance, and low network utilization rates. This scenario necessitates the design of efficient data scheduling and processing strategies.

Current grid measurement data scheduling methods have not fully considered the streaming nature and the disorder of equipment data reporting, nor have they fully analyzed and utilized the characteristics and patterns of measurement data. This has led to an imbalance in data processing load among different processors, thereby affecting the timeliness and accuracy of processing.

To address these issues, this paper proposes a measurement data scheduling optimization method based on genetic algorithms. Specifically, we divide a day into several equal-length time windows and analyze the resource consumption characteristics of historical measurement data for each window. Then, based on the resource consumption characteristics of the historical measurement data, we initialize the population of the genetic algorithm and construct a fitness function to evaluate the merits of the data scheduling scheme. Finally, through selection, crossover, and mutation operations, we iteratively generate an optimized scheduling scheme.

The main contributions of this paper are:

Proposing a data scheduling model tailored to the real-time measurement data characteristics of the power grid.

Designing a fitness function to accurately assess the performance of measurement data processing schemes.

Using genetic algorithms to achieve load balancing and real-time optimization of measurement data processing.

2   Related Works

2.1   Measurement Data

The power grid requires real-time monitoring and data analysis to ensure its normal operation and safety [?]. Grid measurement data, including signals such as voltage, current, and power, is collected through various sensors in the power system. This data, reflecting the status of the power system and the operation of individual devices in real time, are critically important for the safety and optimized management of the grid [3].

The main characteristics of measurement data include its real-time nature, high frequency, and large volume [4]. These characteristics present two main challenges: first, how to process a large volume of real-time flowing data quickly and effectively to ensure timely updates of information; second, how to accurately associate data with corresponding equipment records for further analysis and use.

Traditional methods for measurement data scheduling mainly involve distributing tasks among multiple servers, with each server processing data in parallel. However, due to the streaming nature and disorderliness of measurement data [5], these methods have not effectively coped with the uncertainty of data arrival, leading to an imbalance in processing load among different servers and affecting the timeliness and accuracy of data processing. In addition, traditional methods cannot fully utilize computing resources, and face problems like resources slack and processing bottlenecks [6].

In contrast, measurement data scheduling optimization methods based on genetic algorithms can better handle these challenges. Genetic algorithms optimize task scheduling strategies by simulating the genetic and selection mechanisms of biological evolution, achieving more balanced and efficient data processing. Our method leverages the global search capability of genetic algorithms to quickly adjust task distribution in dynamically changing environments. This ensures full utilization of computational resources, avoids idleness and bottlenecks, and provides a new solution for grid measurement data processing.

2.2 Genetic Algorithm

The Genetic Algorithm (GA) is a metaheuristic algorithm used in computer science and operations research. It is inspired by the process of natural selection and belongs to the broader category of evolutionary algorithm [7]. Genetic Algorithm is commonly used to generate high-quality solutions for optimization and search problems, relying on biologically inspired operators such as mutation, crossover, and selection. In a genetic algorithm, a population composed of candidate solutions evolves towards better solutions. Each candidate solution has a set of properties (chromosomes or genotypes) that can be mutated and altered, typically represented as binary strings, although other encoding methods are also possible.

Genetic algorithms improve the solution set through an iterative process, with each iteration known as a generation. In each generation, the fitness of each individual in the population is assessed, usually being the value of the objective function of the optimization problem being solved. More fit individuals are randomly selected from the current population and their genomes are modified through crossover and possible random mutation to form a new generation [8]. The advantage of genetic algorithms lies in their adaptability to solve large-scale and complex problems that traditional measurement data methods struggle with.

2.3 Genetic Algorithm in Electrical Grid Domain

In the field of power grid, Bharathi used genetic algorithm (GA-DSM) to optimize industrial load redistribution in demand side management [9]. LiF used an integer encoded multi-objective genetic algorithm to enhance reactive power compensation planning in power grids and demonstrated its superiority over traditional linear programming methods [10]. Askarzadeh introduced a memory based genetic algorithm (MGA) for effectively allocating energy in smart grids, with a focus on reducing energy production costs [11]. Korotunov et al. studied the application of genetic algorithms to optimize the charging infrastructure of electric vehicles in the context of smart grids. It aims to identify and propose an optimal method for enhancing the digital twin of the electric vehicle charging system [12]. Özdemir introduces an online genetic algorithm for optimizing electricity consumption tariff parameters, which was tested and validated in a real-world power trading agent competition, demonstrating its effectiveness in optimizing tariff parameters and reducing peak-demand charges [13]. Jeyaranjani introduces an enhanced Genetic Algorithm for the optimal scheduling of household appliances in Smart Grid systems, effectively reducing electricity costs and decreasing both the number of iterations and the execution time of the algorithm [14].

To our knowledge, there is currently no research on the application of genetic algorithms in the field of electrical grid measurement data. Compared to traditional methods, our proposed measurement data scheduling optimization approach based on genetic algorithms is better equipped to handle the dynamic changes in data processing, ensuring the full utilization of computational resources, and avoiding resource idleness and processing bottlenecks.

3 Overall Architecture

Our research adopts a time-window-based approach aimed at optimizing the processing and real-time nature of equipment measurement data. Initially, we collect historical measurement data reported by multiple devices over several days. By applying a time-window analysis, we ascertain each device's computational power requirements during different time windows, laying the groundwork for subsequent real-time optimization. In addition, we transform the task of processing parallel measurement data archives into a task allocation problem. Our optimization goal, achieved by using genetic algorithms, is to minimize the range of processing times for all time-window tasks. This method achieves task distribution balance within fine-grained time windows and overall load balancing, effectively enhancing the timeliness of measurement data processing. The overall architecture of our approach is shown as Fig. 1.

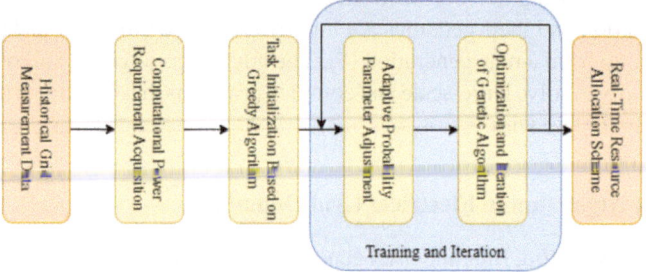

Fig. 1. Optimization process for real-time measurement data scheduling based on Genetic Algorithms

4 Methods

4.1 Computational Power Requirement Acquisition

In the context of real-time data processing across multiple devices, acquiring computational power requirements is a critical step to ensure the timeliness and efficiency of data processing. We propose a time-window-based method to obtain these requirements, enabling efficient processing of real-time measurement data through fine-grained analysis of historical data.

First, the system collects historical measurement data reported by each device over several days and evenly divides a day into K time windows, where K is the number of time intervals defined based on the system's operational requirements. For each device, we calculate the average computational power required in each time window, defined by H_n^k, representing the average computational power needed by device n in the k-th time window.

$$H_n^k = \frac{1}{I} \sum_{i=1}^{I} \sum_{t=1}^{x_n^{(i,k)}} c_n^t \tag{1}$$

where $x_n^{(i,k)}$ denotes the quantity of historical measurement data for device n in the k-th time window of the i-th day, and c_n^t represents the computational power required for the t-th historical measurement data in the k-th time window of the i-th day. Through this process, we determine the computational power requirements of each device in different time windows.

Considering the large number of devices and the relatively small computational power requirement of each, we further group the devices to enhance the subsequent algorithm's speed and accuracy. Devices are evenly divided into groups, with each group treated as a single task. We sum the average computational power requirements for multiple devices in the same time window within each group, determining each task's computational power requirements in different time windows. Consequently, in subsequent sections, H_n^k represents the total computational power for all devices assigned to the n-th task during the k-th time window.

4.2 Optimization Problem Formulation

To facilitate task allocation, we define an allocation matrix σ as follows to describe the distribution relationship between M processes and N grouped tasks.

$$\sum_{m=1}^{M} \sigma_n^m = 1,$$
$$\sigma_n^m \in \{0, 1\}, \forall n \in \{1, \ldots, N\}, \forall m \in \{1, \ldots, M\} \tag{2}$$

When an element σ_n^m in the matrix values 1, it indicates that the m-th process has been allocated the n-th group of tasks. Based on this allocation matrix and the computational power requirements of each task in different time windows determined from historical data, we can calculate the time D_m^k required by process m to handle tasks in the k-th time window as follows:

$$D_m^k = \sum_{n=1}^{N} \frac{H_n^k \times \sigma_n^m}{\omega_m} \tag{3}$$

where ω_m represents the computing power of process m. To make resource allocation and scheduling more efficient, the goal is to minimize the difference in time required to process tasks in the same time window. Based on the allocation matrix and the computational power requirements of tasks, we define the range d_k to represent the maximum difference in time required by multiple processes to handle tasks in the same time window k:

$$d_k = \max D_m^k - \min D_m^k \tag{4}$$

where $\max D_m^k$ and $\min D_m^k$ respectively represent the longest and shortest times taken by all processes to complete task processing in the k-th time window. We set the constraint $d_k \leq \alpha, \forall k \in \{1, 2, 3, \ldots, K\}$, where α denotes the maximum tolerable processing time difference for each time interval, d_k indicates the uniformity of task distribution within the k-th time window.

For all time windows, the smaller the sum of differences in processing times for different processes, the more balanced the task allocation and load balancing. Therefore, we define minimizing the sum of time range differences d across all time windows as the optimization objective as follows:

$$d = \sum_{k=1}^{K} d_k \tag{5}$$

Where K represents the total number of time windows, and d_k is defined as Eq. (4).

4.3 Solution Strategy

We adopt a strategy based on Genetic Algorithms, with Eq. (5) serving as the fitness function. Through design and iterative search, we obtain the optimal task allocation matrix, achieving rational allocation of computational resources and improving the efficiency and quality of data processing. Equation (2) indicates that each task can only be allocated to one process. Hence, we use integer encoding to represent chromosomes. A chromosome is represented as a sequence $\{o_1, o_2, \ldots, o_N\}$ where $o_n \in \{1, 2, \ldots, M\}$ indicates that task n is allocated to process o_n.

Population Initialization
Existing methods use completely random initialization to increase the diversity of solutions for the optimization problem. However, this often fails to meet constraint conditions and leads to slow convergence, making it difficult to find optimal solutions. We propose a random initialization method based on greedy algorithms to generate the initial population, improving diversity and accelerating convergence.

We calculate the average time avg required to process measurement data:

$$\text{avg} = \frac{\sum_{n=1}^{N} \theta_n}{\sum_{m=1}^{M} \omega_m} \tag{6}$$

where θ_n represents the average computational power requirement of each task across all time windows as follows.

$$\theta_n = \sum_{k=1}^{K} \frac{H_n^k}{K} \tag{7}$$

Firstly, a randomization strategy is used to shuffle the task index set to eliminate the impact of initial order, resulting in a randomized index set B. During initialization, we introduce a cumulative sum variable and a counter to indicate the current task index for use in subsequent task allocations.

In the allocation process, values from the randomized index set are obtained one by one. The chromosome position corresponding to the current task index is assigned the current task allocation marker, and the cumulative sum is updated based on the ratio of task requirements to available computational resources. If the cumulative sum has not reached the predetermined average value avg defined as Eq. (6), the current allocation sequence continues; otherwise, the task allocation marker is updated, and the cumulative sum is reset to start a new allocation sequence.

To ensure that the generated chromosome encoding meets the condition $d_k \leq \alpha, \forall k \in \{1, 2, 3, \ldots, K\}$, the algorithm employs a fine-tuning strategy. Based on the current encoding, a range vector is calculated. If the range d_k exceeds the predetermined threshold α, the algorithm randomly reallocates tasks from the process corresponding to $\max D_m^k$ to the one corresponding to $\min D_m^k$ until all range values do not exceed the threshold.

Furthermore, to avoid infinite loops, if the fine-tuning does not produce a viable solution after several attempts, the algorithm restarts. Considering that computational power supply usually exceeds task demand in practical applications, the algorithm is designed to ensure a high success rate and efficiency. Ultimately, this algorithm outputs a feasible chromosome encoding that satisfies all constraint conditions, facilitating the effective allocation of tasks among computational resources.

Optimization and Iteration

In our proposed genetic algorithm-based scheduling method for measurement data, the update process of the population is achieved through selection, crossover, and mutation operations. Specifically, the algorithm starts by identifying a set of winners with lower fitness values from the parent population P_t^p using a competitive selection mechanism, forming a winners' set. In this process, two individuals are randomly selected each time, and the one with a lower fitness value calculated using Eq. (5) is chosen as the winner. To ensure diversity, we make sure that each pair of parental individuals is distinct. This step is repeated until the length of the winners' set reaches $N_p/2$, where N_p represents the number of individuals of population P.

During the crossover process, we decide whether to perform a single-point crossover operation based on a preset crossover probability P_c. Two parent individuals have a probability P_c to crossover and produce new offspring; otherwise, the offspring directly inherit the genes of the parents.

Additionally, we introduce a mutation operation to increase the diversity of the population. Each gene position has a certain probability of mutation, as shown in the following formula:

$$o_n = (o_n + \lceil \text{rand}(P_m \times \beta, (1 - P_m) \times \beta) \rceil \% M) \tag{8}$$

where β represents the mutation scale and $\beta \in [1, M]$, % denotes the modulo operation, rand(a, b) is a random number between a and b, and $\lceil \ \rceil$ is the ceiling function. The values of the mutation probability P_m determines the range of mutation values. As shown in Eq. (8), a larger mutation probability leads to a smaller mutation range, while a smaller mutation probability may result in a larger range of mutation. This mechanism allows the algorithm to maintain search diversity while avoiding premature convergence. By repeating the aforementioned crossover and mutation operations for each pair of parent individuals, we generate a new population P_t^o consisting of N_p new offspring individuals, and then merge the offspring and parent populations to form $P_t = P_t^p \cup P_t^o$.

We select the $N_p/2$ individuals with the smallest fitness values from the population P_t and then choose another $N_p/2$ individuals based on the roulette wheel selection method from the remaining individuals. This method ensures that the optimal solution does not get lost during the evolutionary process, and it balances the quality and diversity of the retained individuals.

The above process is iterated repeatedly until the maximum number of iterations is reached. After iteration completion, the individual with the smallest fitness value that satisfies the constraints is selected as the optimal solution P_{best}, and the chromosome encoding scheme $\{o_1, o_2, \ldots, o_N\}$ of P_{best} represents the corresponding task allocation scheme. Here, the set of archives assigned to process m can be represented as $S_m = \{i | o_i = m \vee o_i \in P_{best}\}$.

5 Adaptive Adjustment of Parameters

To enhance the convergence speed and quality of solutions of the algorithm, our research adopts an adaptive parameter adjustment strategy to meet the specific needs of genetic algorithms at different stages.

5.1 Crossover Probability

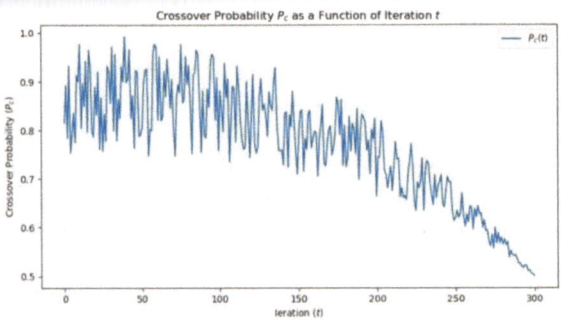

Fig. 2. Trend of crossover probability P_C with respect to iteration number t in Genetic Algorithms

The crossover operation plays a key role in genetic algorithms, as it helps generate new gene combinations and enhances the diversity of the population. Instead of a fixed crossover probability, we define the crossover probability P_c as a function that adapts with the iteration number t (Fig. 2):

$$P_c = \underline{P_c} + (1 - P_c) \times e^{\frac{t}{\max t}} \times \left(1 - \frac{t}{\max t}\right) \times \text{rand}(0.5, 1) \qquad (9)$$

where $\underline{P_c}$ is the lower bound of the crossover probability, t represents the current evolution generation, max t represents the total number of iterations, and rand(a, b) is a random number between a and b.

Equation (9) shows that the crossover probability decreases as the number of iterations increases. This design ensures that in the early stages of the algorithm, a higher crossover probability is maintained to rapidly increase the diversity of the population, thereby swiftly identifying superior solutions and accelerating the convergence speed of the population. As the population continues to evolve, the differences between individual members gradually diminish. At this point, gradually reducing the crossover probability helps in more quickly converging to optimal solutions.

5.2 Mutation Probability

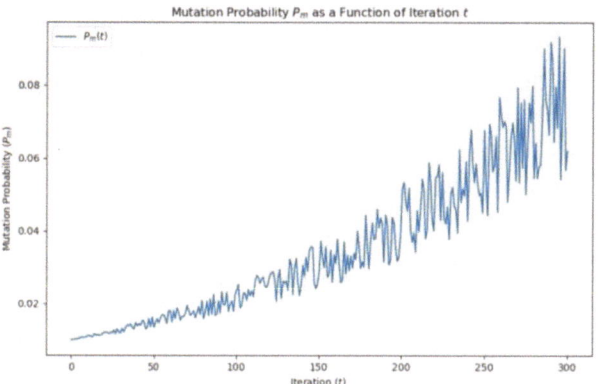

Fig. 3. Trend of mutation probability P_m with respect to iteration number t in Genetic Algorithm

The mutation operation introduces new gene variations, aiding the algorithm in escaping local optima and exploring a broader solution space (Fig. 3).

We define the mutation probability P_m as a function that adapts with the iteration number t:

$$P_m = \underline{P_m} + (\overline{P_m} - \underline{P_m}) \times e^{\frac{t - \max t}{\max t}} \times \frac{t}{\max t} \times \text{rand}(0.5, 1) \qquad (10)$$

$\underline{P_m}$ is the lower bound of the mutation probability, while $\overline{P_m}$ is the upper bound, t represents the current evolution generation, max t represents the total number of iterations, and rand(a, b) is a random number between a and b.

This design causes the mutation probability to increase in a trend that is initially slow and then accelerates. In the early stages of the algorithm, a lower mutation probability maintains the stability of the population; in the later stages, as the population converges, increasing the mutation probability effectively avoids the algorithm getting trapped in local optima.

6 Conclusion

This study successfully applied genetic algorithms to the real-time scheduling of power grid measurement data, significantly enhancing the efficiency and quality of data processing through an adaptive parameter adjustment strategy. The computational power requirement acquisition method and task allocation model proposed in this research fully consider the streaming nature and disorder of power grid data, ensuring balanced task processing. The genetic algorithm-based strategy for solving the optimization allocation matrix further improved the solution speed and accuracy of the algorithm. The proposed optimization method can effectively balance the load between different processors, reduce the disparity in data processing time, and enhance the overall performance of electric power system data processing. This achievement is significant for improving the monitoring and response capabilities of power systems and offers a new solution for real-time data processing in power grids.

Acknowledgment. This work is supported by State Grid Hunan Electric Power Co., Ltd research project No. 5216A8220004 and Hunan Key Laboratory for Internet of Things in Electricity, P.R. China No. 2019TP1016.

References

1. Amin, M., Ouinger, J.: The electric power grid: today and tomorrow. MRS Bull. **33**(4), 399–407 (2008)
2. Li, H., Dong, Y., Yin, C., et al.: A real-time monitoring and warning system for power grids based on edge computing. Math. Probl. Eng. **2022** (2022)
3. Albarakati, A.J., Boujoudar, Y., Azeroual, M., et al.: Microgrid energy management and monitoring systems: a comprehensive review. Front. Energy Res. **10**, 1097858 (2022)
4. Daki, H., El Hannani, A., Aqqal, A., et al.: Big data management in smart grid: concepts, requirements and implementation. J. Big Data **4**(1), 1–19 (2017)
5. Lu, P., Yue, Y., Yuan, L., Zhang, Y.: AutoFlow: hotspot-aware, dynamic load balancing for distributed stream processing. In: Lai, Y., Wang, T., Jiang, M., Xu, G., Liang, W., Castiglione, A. (eds.) ICA3PP 2021. LNCS, vol. 13157, pp. 133–151. Springer, Cham (2022). https://doi.org/10.1007/978-3-030-95391-1_9
6. Kolajo, T., Daramola, O., Adebiyi, A.: Big data stream analysis: a systematic literature review. J. Big Data **6**(1), 47 (2019)
7. Holland, J.H.: Genetic algorithms. Sci. Am. **267**(1), 66–73 (1992)
8. Faycal, T., Zito, C.: Direct mutation and crossover in genetic algorithms applied to reinforcement learning tasks. arXiv preprint arXiv:2201.04815 (2022)
9. Bharathi, C., Rekha, D., Vijayakumar, V.: Genetic algorithm based demand side management for smart grid. Wirel. Pers. Commun. **93**, 481–502 (2017)

10. Li, F., Pilgrim, J.D., Dabeedin, C., et al.: Genetic algorithms for optimal reactive power compensation on the national grid system. IEEE Trans. Power Syst. **20**(1), 493–500 (2005)
11. Askarzadeh, A.: A memory-based genetic algorithm for optimization of power generation in a microgrid. IEEE Trans. Sustain. Energy **9**(3), 1081–1089 (2017)
12. Korotunov, S., Tabunshchyk, G., Okhmak, V.: Genetic algorithms as an optimization approach for managing electric vehicles charging in the smart grid. In: CMIS, pp. 184–198 (2020)
13. Özdemir, S., Unland, R.: AgentUDE17: a genetic algorithm to optimize the parameters of an electricity tariff in a smart grid environment. In: Demazeau, Y., An, B., Bajo, J., Fernández-Caballero, A. (eds.) PAAMS 2018. LNCS, vol. 10978, pp. 224–236. Springer, Cham (2018). https://doi.org/10.1007/978-3-319-94580-4_18
14. Jeyaranjani, J., Devaraj, D.: Improved genetic algorithm for optimal demand response in smart grid. Sustain. Comput. Inform. Syst. **35**, 100710 (2022)

Research on Federated Radial Basis Function Neural Network Based on Genetic Algorithm

Yandong Ma[1,2], Gaifang Tan[1,2(✉)], Song Tang[1,2], Suli Ge[1,2], and Zhiqiang Wang[1,2,3]

[1] Institute of Applied Mathematics, Hebei Academy of Sciences, Shijiazhuang, China
1959280504@qq.com
[2] Information Security Authentication Technology Innovation Center of Hebei Province,
Shijiazhuang, China
[3] Julu County Institute of Applied Technology, Xingtai, China

Abstract. The radial basis function neural network model is optimized for application scenarios such as uneven data distribution, privacy sensitivity, or lack of direct access to the original data. Firstly, federated learning is applied to the radial basis function neural network, facilitating cross-multicenter distributed collaborative training. Then, genetic algorithms are used to optimize the hyperparameters of the federated radial basis function neural network model. Comparative experiments were conducted using the publicly available National health and nutrition examination survey 2013–2014 (NHANES) age prediction subset dataset and the proprietary HEART dataset. Compared with traditional centralized learning methods, the proposed model demonstrates superior performance, providing new ideas and methods for solving distributed learning and data privacy problems.

Keywords: genetic algorithm · federated learning · radial basis function neural network · distributed collaborative training · hyperparameter optimization

1 Introduction

Radial basis function based neural networks are widely used in multivariate problems due to their optimal approximation ability to capture complex nonlinear relationships [1]. However, problems such as the complex network topology and the difficulty of choosing a center point also impose some constraints on its development [2]. Especially in cases such as uneven data distribution, privacy sensitivity or lack of direct access to raw data, the computational efficiency, memory management, and communication overhead in distributed systems are under great pressure, and need to be combined with other optimization algorithms for more effective improvement.

Federated learning, as a distributed machine learning method capable of handling large-scale datasets, is expected to endow neural networks with low-cost and highly effective learning capabilities [3]. The potential of radial basis function neural networks in federated learning is mainly reflected in the following aspects: firstly, it has better global properties, which can effectively capture the intrinsic structure of the data and thus improve the generalization performance of the model; secondly, it has strong scalability,

P. Siarry et al. (Eds.): WCNA 2023, LNEE 1361, pp. 230–246, 2025.
https://doi.org/10.1007/978-981-96-2409-6_23

which can be easily applied to large-scale datasets; and thirdly, it can achieve parallel processing of data, which means that multiple devices or organizations can train the model at the same time.

Genetic algorithm is an optimization algorithm that can obtain the optimal solution through continuous iteration and evolution [4, 5]. In the training process of radial basis function neural network model, the global optimization search strategy with the help of genetic algorithm can find the global optimal solution of the parameters such as the center of the hidden layer activation function (e.g., Gaussian function), the width of the radial basis function, and the weights of the connection from the hidden layer to the output layer in a faster way. Finally, a federated radial basis function neural network model based on genetic algorithm (GA-FL-RBFNN) is proposed.

We applied the model to a disease classification scenario in the medical device regulatory process. Firstly, some heart disease diagnosis data from a medical institution were systematically combed to form a proprietary medical dataset HEART. Then the publicly National health and nutrition examination survey 2013–2014 (NHANES) age prediction subset dataset was called. Finally, the model was experimentally validated using these two datasets. The experimental results show that the developed GA-FL-RBFNN model is able to exhibit superior performance compared to centrally trained models as well as traditional federated learning models.

2 Model Basics

2.1 Radial Basis Function Neural Network

Radial basis function neural network (RBFNN) is a three-layer feed-forward neural network with a single hidden layer, and its basic idea is to map the sample data into a high-dimensional space, and ultimately make the data linearly distributed in the space [6]. RBFNN has a biological background and is compatible with function approximation theory, which can deal with complex laws that are difficult to explain in practical problems. It has good generalization ability and fast learning convergence speed, and has been successfully applied to complex nonlinear function approximation, data classification and prediction, and system modelling.

RBFNN aims to model the relationship between input and output parameters through the training process, and its great advantage lies in its powerful self-learning and nonlinear function approximation ability. Machine learning models constructed based on RBFNN are not only simple in structure and fast in convergence, but also quite accurate, thus enabling the development of a black-box model to capture and identify nonlinear relationships in complex physical systems. At this point, there is no need to consider the specific nature of the research problem, and all the influencing factors can be taken into account in order to obtain a complex joint interaction between the inputs and outputs. The advantage of RBFNN is that it can represent the real-world problem in a much simpler structure, at a lower cost, and in a shorter time. However, in order to obtain such a model that fits the real problem "perfectly", its structural design and related parameter configurations become a crucial part of the modelling process, which is also an important research content in this paper.

RBFNN is a typical feed-forward neural network, which consists of input layer, hidden layer and output layer, assuming that the number of nodes in each layer of RBFNN is I, H, O, and its structure is shown in Fig. 1.

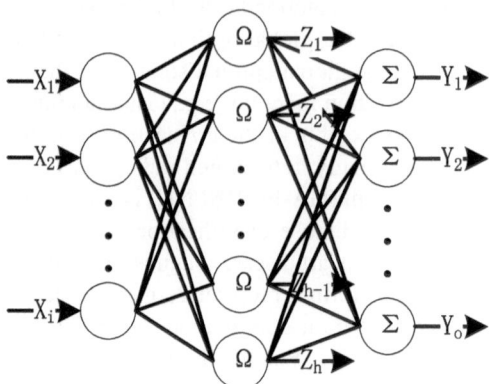

input layer   hidden layer   output layer

Fig. 1. Diagram of RBFNN structure

The first layer is the input layer, which contains I perceptual units that represent the input of the signal source node, and serves only as input data. The connection weights between the input layer to the hidden layer are constant 1, i.e., each component of the input vector is transmitted unchanged to each hidden layer node, as shown in Eq. (1):

$$w_{ij} = 1(i = 1,2, \ldots, I, j = 1,2, \ldots, H) \tag{1}$$

The second layer is the hidden layer containing H radial basis neurons (with RBF activation function), which maps the low-dimensional non-linearly divisible inputs to the high-dimensional linearly divisible space. The weights from the hidden layer to the output layer are trained using a supervised learning method as shown in Eq. (2):

$$w_{jk}(j = 1,2, \ldots, H, k = 1,2, \ldots, O) \tag{2}$$

The activation function of the hidden layer node responds locally to the input, when it is close to the center of RBF, the hidden layer node will produce a larger output; when it is far from the center, the output will decay exponentially. RBF is a non-negative nonlinear function that is radially symmetric to the center and attenuated, and a Gaussian function is used as the radial basis function of the model, at which time the output of the hidden layer node is as shown in Eq. (3):

$$Z_j = exp\left(-\frac{\|X_I - X_j^c\|^2}{2\sigma^2}\right)(j = 1,2, \ldots, H) \tag{3}$$

where $\| \cdot \|$ denotes the Euclidean paradigm, X_I is the Ith input sample, X_j^c is the centre vector of the radial basis function of the node, and σ is the width of the node, which is used to adjust the width of the Gaussian radial basis function.

The third layer is the output layer, containing O linear neurons (the activation function is linear), and the final output is a linear weighted sum of the outputs of the neurons in the hidden layer, as shown in Eq. (4):

$$Y_k = \sum_{j=1}^{H} w_{jk} Z_k (k = 1, 2, \ldots, O) \tag{4}$$

where w_{jk} is the weight of the hidden layer on the output layer.

Since the output of RBFNN exhibits a linear relationship with the network weights w_{jk}, the learning problem of RBFNN can be analysed with the help of the unified linear system theory, which allows the learning convergence performance of RBFNN to be fully guaranteed.

It can be seen that the RBFNN constructs an approximate functional relationship starting from the input layer and passing through the hidden layer until the end of the output layer. In this case, the number of input and output variables determines the number of neurons in the input and output layers of the model, respectively. However, there is still no general rule for the setting of the number of neurons in the hidden layer, which means that the optimal combination of the hyperparameters of the model is hard to find.

In summary, we introduce the data-driven model for RBFNN. As a result, the training process of RBFNN can be regarded as the learning of two sets of network parameters, one of which is the number of nodes in the hidden layer, the center of the radial basis function in the hidden layer and the radial basis width, and the other is the connection weights from the hidden layer to the output layer [7]. In general, the algorithm used to confirm the model network parameters is divided into two steps: the first step uses unsupervised machine learning algorithms such as clustering algorithms to train the radial basis function centers and radial basis widths; the second step uses supervised algorithms to train the connection weights from the hidden layer to the output layer. Among them, the global optimality-seeking property of genetic algorithms provides a new idea for RBFNN model parameter determination, i.e., genetic algorithms can adaptively determine the number of hidden layer nodes while calculating the connection weights from the hidden layer to the output layer [8, 9].

2.2 Federated Learning

Federated learning (FL) [10, 11] is a distributed machine learning approach that aims to construct machine learning models based on datasets distributed in multiple centers, and to effectively reduce the risk of data leakage and communication costs due to data transmission by designing encrypted parameter transfer instead of the original remote data transmission during the training process. The commonly used federated learning architecture adopts the client-server (CS) model, i.e., Federated averaging (FedAvg) [12], and the model structure is shown in Fig. 2.

A typical federated learning framework consists of multiple clients $C = c_1, c_2, \ldots, c_n$ that hold data D_n and a central server S. Firstly, federated learning selects k active clients to participate in training in each communication round of training; secondly, S distributes randomly initialized model parameters W to C, which performs local iterations and uploads the updated model parameters; finally, S aggregates the parameters W_t^i from all parties, updates the global model and distributes it again. The training

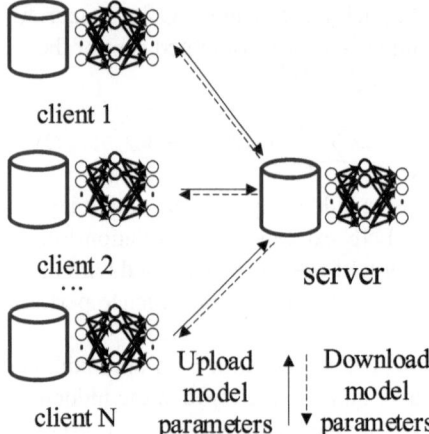

Fig. 2. Diagram of federation learning

process of the model from round t to round t + 1 is shown in Eq. (5):

$$W_{t+1}^i = W_t^i - \sum_{e=0}^i \eta_t \nabla F_i\left(W_{t,e}^i, X_{t,e}^i\right) \tag{5}$$

where η_t, W_t^i represent the learning efficiency and the received joint model in round t, respectively, e is the number of local training times, ∇F_i is the gradient update of the ith local client c_i in round t, and X_t^i is the training data.

The above process is repeated for several iterations of "global-local" model updating between "central server-participating clients" until the model converges or reaches a predefined condition and terminates. Finally, a global model is trained that can be used to perform the corresponding prediction tasks in external clients.

In order to solve the training and optimization of RBFNN parameters, the general idea of federated learning is: design the network structure of RBFNN on the client side according to the relevant situation of the objective to be optimized; use gradient descent method (e.g., fastest descent, stochastic gradient, etc.) to compute the network parameters of the RBFNN, which includes: uploading the gradient information of the parameters to the server; the server receives the gradient information and fuses it to form the global gradient, which is then sent to the client side. The training of the global model of RBFNN is finally achieved by this way. This training method can achieve network hyper-parameter optimization, but the need to transfer gradient information back and forth between the client and the server increases the risk of original data leakage.

In summary, by building the RBFNN model on a federated learning architecture, the central server can unite multiple clients to train the model together. At this time, the central server only aggregates parameters and does not participate in model training, and the clients participating in training are invisible to each other, which protects the data security of all parties. However, compared with centralized learning, federated learning causes a certain degree of performance loss and time delay [13]. Therefore, effective improvement of federated learning is needed. Genetic algorithms have the characteristics of parallelizability and global search. By introducing genetic algorithms on the basis of federated learning, the overall framework of federated learning can be improved [14]. In addition, the central server only aggregates the model parameters and still has enough arithmetic to support the operation of the genetic algorithm. Better network models can be obtained by the central server performing genetic operations among the clients involved in training.

2.3 Genetic Algorithm

Genetic algorithm (GA) is a stochastic global search optimization algorithm that simulates the natural evolutionary process, and its core idea is to regard the evolutionary process as a kind of natural selection, and search for the optimal solution through continuous iteration and mutation [15, 16].

Genetic algorithm takes the process of chromosome gene crossing and mutation in the process of biological evolution as a blueprint, abstracts the mathematical model, and achieves the solution of optimization problem through the operations of selection, recombination and mutation. During the algorithm, the chromosome represents a solution to the problem, and the initial population is operated in the order of chromosome crossover, mutation, fitness assessment, and selection to obtain a new generation of population. Since the selection operator always keeps the better chromosomes to the next generation, the good chromosomes will gradually occupy the main position of the population, and the population will move towards the direction of the optimal solution, thus obtaining a high-quality solution. The genetic operation is shown in Fig. 3.

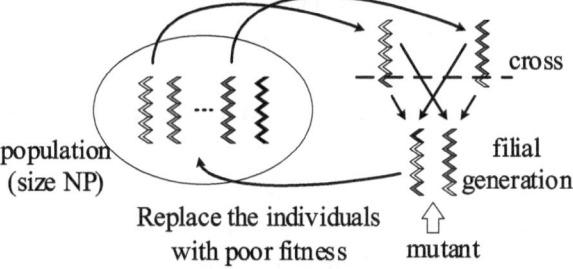

Fig. 3. Diagram of genetic operations

Like neural networks, genetic algorithms are adaptive global optimization search algorithms that seek to find the values of the decision state variables that optimize the objective function. Genetic algorithms can solve the problem of neural networks that tend to fall into local optima. The problem of solving the function minimum optimization is shown in Eq. (6):

$$\begin{cases} minf(X) \\ s.t.\begin{cases} X \in C \\ C \subseteq R \end{cases} \end{cases} \tag{6}$$

where $X = (x_1, x_2, \ldots, x_n)$ is the decision variable, f(x) is the objective function, R is the search space, C is the feasible domain, and X is the feasible solution, as shown in Fig. 4.

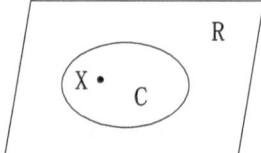

Fig. 4. Relationship between X, C and R

To sum up, genetic algorithm abstracts the actual problem by simulating the biological evolution process, and represents a feasible solution to the problem as an individual, and represents a collection of multiple feasible solutions as a population, and then evolves the population through genetic operation, thus obtaining the approximate optimal solution to the problem. Similarly, the training process of the neural network model can also be regarded as the process of updating the network parameters, which can be updated continuously to complete the optimal mapping of the input-output relationship. By applying the optimization feature of the genetic algorithm to the RBFNN model training process, the limitation of the difficulty in determining the parameters of the radial basis function can be broken. Further, because the genetic algorithm has excellent global search capability, it can also filter individuals in the population in parallel, thus speeding up the search. Applying genetic algorithms to federated learning can also alleviate the performance loss and communication pressure of federated learning relative to centralized learning to a certain extent.

2.4 Improved Genetic Algorithm for Federated Radial Basis Function Neural Networks

In order to construct a RBFNN model with better performance, the model parameters, i.e., the decision state variables of the objective function, need to be optimized, including the number of nodes in the hidden layer, the center of the radial basis function in the hidden layer, the width of the radial basis, and the connection weights from the hidden layer to the output layer. The discrete nature of the decision state variables makes its optimization a non-convex problem. Heuristic methods such as genetic algorithms can be used to solve such problems. Genetic algorithms search for feasible solutions in the domain, seeking to find the value of the decision state variable that optimizes the objective function. Starting from randomly generated candidate solutions, a new generation of solutions with modified objective values is extracted, and so on iteratively until a good enough solution is derived. As a result, the following improvements are made to the genetic algorithm for the federated RBFNN model.

1) *Chromosome coding scheme*

If the number of input nodes of RBFNN is I and the number of hidden layer nodes is H, the center vector of RBFNN can be encoded in real numbers, i.e., a chromosome is an array of real numbers of length I*H. The first I*H elements of the array store the components of the H center vectors with data dimension I. The weights of the hidden layer to the output layer of the RBFNN are obtained by using the pseudo-inverse method.

2) *Selection operator*

To ensure the diversity and representativeness of chromosomes in the population, the following chromosome selection strategy is designed:

a) Arrange the chromosomes in descending order according to their fitness function values;

b) Select the n chromosomes with the highest fitness function values to join the new population;

c) Divide the remaining chromosomes of the population into three parts according to the size of the fitness function value: high segment (first 30%), middle segment (middle 40%) and low segment (last 30%);

d) For the chromosomes in the high, medium and low segments, the roulette wheel selection method is used, where the probability of an individual is calculated based on the value of the fitness function of the individual, and then the individual is randomly selected based on this probability, so as to select the excellent chromosomes to be added to the new population.

3) *Crossover operator*

Randomly selects 2 chromosomes and performs a two-point crossover operation, thus generating 2 new chromosomes to join the new population.

4) *Variation operator*

One chromosome is randomly selected and a mutation operation is performed, resulting in a new chromosome that is added to the new population.

5) *Local optimization operator*
   a) Selects K chromosomes at random;
   b) Resolve the individual parameters of the RBFNN corresponding to it;
   c) Take the objective function of RBFNN as the optimization objective, use the above parameters as the initial point of search, and use Adm algorithm to search for the optimal parameter values;
   d) Replace the original chromosome using the optimal values of the parameters and update the value of the fitness function.

6) *Definition of the fitness function applicable to federated learning*

In order to measure the superiority of chromosomes, the central server of federated learning transforms the values of the fitness function uploaded by each client into the values of the global fitness function. The fitness function applicable to federated learning is defined as shown in Eq. (7):

$$fitness = \sum_{i=0}^{N-1} [(1 - \lambda)Acc_{train} \times Num_{train} + \lambda \times Acc_{test} \times Num_{test}] \qquad (7)$$

where, fitness is the value of the fitness function used by the central server to measure the chromosome's strengths and weaknesses, and N is the number of clients owned by the system. Acc_{train} and Acc_{test} are the classification accuracy scores of the RBFNN on the training and test sets, respectively. λ is the weight of the test set set to equalize the performance of the RBFNN on the training and test sets. Num_{train} and Num_{test} are the number of samples in the training and test sets, respectively.

In summary, genetic learning starts with the creation of an initial population consisting of randomly generated rules, each of which can be represented by a string of codes. The random population is used to define the state variables of the model architecture. Then a new population consisting of the most appropriate rules is formed. During each iteration, the genetic algorithm creates the next generation from the current population using three main types of genetic rules: crossover, mutation and selection. In crossover, genetic information from both parents of the selection operator is combined to explore the design space, while genetic information introduces diversity with mutation probability to prevent locally optimal solutions. At the same time, the selection operator is used to choose suitable individuals for the breeding offspring. In this way, the derivatives of the objective function are not required and a favorable choice of genetic algorithm is made for optimization problems with nonlinear and discontinuous functions. The fitness function (i.e., the response value of the objective) given above is used to measure the performance of the selected individuals compared to the whole. The process of generating a new population continues as long as each rule in the population satisfies a predetermined fitness threshold (unsatisfied termination check).

3 Federal Radial Basis Function Neural Network Model Based on Genetic Algorithm

3.1 Model Definition

For data privacy protection and multi-center client "distributed" collaborative training scenarios, this paper designs a RBFNN model using genetic algorithm and federated learning optimization, i.e., federated radial basis function neural network based on genetic algorithm (GA-FL-RBFNN) and makes the following definitions.

Assume that the multi-center collaborative research network consists of N clients, where the ith ($i = 1, 2, \ldots, N$) client exists a training sample $C_i = \{(X_i, y_i)\}_{i=1}^{N}$, whose feature vectors have a dimension of M, i.e. $X_i = \{x_i^1, x_i^2, \ldots, x_i^M, \} \in \mathbb{R}^M$. Among the clients used for external validation, there exist samples $D = \{X_{test}\} \in \mathbb{R}^M$ with the same feature dimensions. In addition, in the process of collaborative training of the multicenter model, each client can upload or download parameters from the central server S. The training objective of the model in this paper is: each participating client in the multicenter collaborative research network performs model training on the local dataset, and then the trained local model RBF_i^{local} is uploaded and downloaded from the central server S, and finally outputs the global model RBF^{global}. The global model will be used to execute the model on the external client test set $D = \{X_{test}\} \in \mathbb{R}^M$ to perform the corresponding classification prediction task.

The model training process is as follows:

Firstly, the multi-center collaborative research network includes N participating training clients $c_i(i = 1,2,\ldots,N)$; the original dataset $C_i(i = 1,2,\ldots,N)$ of each client.

Second, each participating client locally trains the corresponding local model RBF_i^{local} according to the update rules of the traditional RBFNN model.

Finally, after the local model training is completed, each client uploads the model RBF_i^{local} parameter Θ_i^{local} to the central server S. The central server S integrates the RBFNN model parameters computed by each client to construct the global model RBF^{global}, and finally outputs the global RBFNN model as shown in Eq. (8):

$$RBF^{global} = \left\{ RBF_1^{local}, RBF_2^{local}, \ldots, RBF_N^{local} \right\} \tag{8}$$

The model can be used to perform the prediction task on the external test set $D = \{X_{test}\} \in \mathbb{R}^M$ as shown in Eq. (9):

$$\hat{y}_{RF} = RBF^{global}\left(X_{test}; \Theta^{global}\right) \tag{9}$$

In order to construct the best performing machine learning model, suitable optimization algorithms are needed to optimize the model structure and network hyperparameters. Genetic algorithms can transform the parameter solution of the model into an optimization problem [17, 18]. The hyperparameter optimization problem of the RBFNN model can be defined as follows: for n hyperparameters $\theta \in R^n$ in the model, find a set of optimal hyperparameter configurations $\theta\prime$, and the network model based on the hyperparameters $\theta\prime$ has the best performance evaluation metrics λ. Generally speaking, it is assumed that there exists a function between the parameters θ and the performance evaluation metrics λ relationship, as shown in Eq. (10):

$$\lambda = f(\theta) \tag{10}$$

In the optimization scheme, various arrangements of neurons in the hidden layer are sought through the predefined RBFNN, and the number of neurons and activation function in each iteration are optimized to evaluate the fitness function. Similarly, the center and width of the radial basis function in the RBFNN are obtained.

3.2 Model Structure

Based on the above model definition, the GA-FL-RBFNN model is designed and the model structure is shown in Fig. 5. Among them, the central server is the algorithm initiator and the client is the main arithmetic execution end. Firstly, the central server generates the initial population of the genetic algorithm according to the relevant situation of the problem and sends the population to each client; after receiving the population, the client trains the model using local data, calculates the local adaptation values and uploads the list composed of the adaptation values to the central server. Then, after receiving all the fitness values, the central server fuses them into global fitness values, and then performs genetic operations such as selection, crossover, mutation, etc. on the populations, so as to update the information of the populations, and then sends the updated information of the populations to the clients; after receiving the updated information of the populations, the clients execute the local optimization operator of the genetic algorithm on the basis of the local dataset, so as to increase the progress of the optimization search of the genetic algorithm. With this cycle, the learning of each parameter of the model can be completed to train the best model.

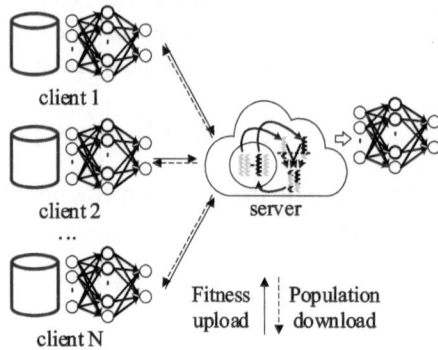

Fig. 5. Diagram of GA-FL-RBFNN

3.3 Model Algorithm

The detailed implementation of the GA-FL-RBFNN model is shown in Algorithm 1.
Algorithm 1 GA-FL-RBFNN
Input: number of input nodes I, number of hidden nodes H, number of output nodes O of RBFNN; population size F of GA, maximum number of iterations P.
Output: center and weight of RBFNN.

a) Based on the input parameters of the algorithm, an initial population code is generated at the central server.
b) The population codes are distributed to each client separately.
c) The client receives the population code, traverses each chromosome in the population, and calculates the fitness function values to form a list as follows:
   (a) Parsing the population code and synthesizing the center vector of the RBFNN.
   (b) Train the model based on the local dataset and use the pseudo-inverse algorithm to calculate the RBFNN model weights.
   (c) Calculate the fitness function values for each chromosome and compose a list of fitness function values.
d) Select the optimal M chromosomes in the client and execute the local optimization search operator with a certain probability.
e) Update the list of chromosomes and fitness function values according to the optimization result.
f) Clients encode and upload the fitness function values of the population to the central server.
g) The central server receives the fitness function values from all clients, decodes and generates the global fitness function values.

h) Determine whether the stop condition is satisfied:

① Yes, end.

② No, continue execution.

i) Execute the selection operator, crossover operator and variation operator of GA to update the population.

j) Jump to step b) to continue execution.

4   Experiments and Analysis of Results

4.1   Experimental Dataset

In this paper, some desensitized data related to heart disease in the process of medical device regulation in a medical institution was systematically sorted out to form the proprietary medical dataset HEART. The national health and nutrition examination survey 2013–2014 (NHANES) age prediction subset, a public dataset widely used in classification tasks was also used as a control dataset, as is shown in Table 1.

Table 1. Experimental datasets

Dataset	Charac-teristics	Number of samples	Number of train samples	Number of test samples
NHANES	8	2278	1822	456
HEART	13	303	243	60

When training the RBFNN model, the dataset is divided into two groups: the training dataset and the test dataset, where the training dataset is mainly used to train the network structure and approximate the connection weights, and the test dataset is used for cross validation of the model against unseen data to prevent possible model overfitting.

4.2   Experimental Environment

In order to verify the effectiveness of the model proposed, multiple training missions were held in the same environment. The hardware and software configurations are shown in Table 2. The RBFNN parameter configuration is shown in Table 3. Among them, the number of nodes in the input layer is determined by the feature dimensions of the dataset, the number of nodes in the output layer is determined by the output category of the model, and the number of nodes in the hidden layer and the center of the radial basis function are determined by the complexity of the data. The GA parameters are configured as shown in Table 4.

Table 2. Experimental configuration

Language/Environment	Version/Model
CPU	12th Intel(R) Core(TM) i9-12900 K
RAM	32.0 GB
Operating System	Windows 10 Professional
Python	3.7
Pytorch	1.13.1

Table 3. RBFNN parameter settings

Parameter type	Parameter value/parameter range	
	NHANES	HEART
Number of input layer nodes	8	13
Number of hidden layer nodes	150	90
Number of output layer nodes	2	2
Radial basis function centre	[−3,24]	[−3,6]

Table 4. GA parameter settings

Parameter type	Parameter value
Population size	50
Maximum number of iterations	20
Coding strategy	BG (Binary/Gray coding)
Selection operator	Random sampling selection operator
Crossover operator	Two-point crossover operator
Variation operator	Variation operator for binary chromosomes
Crossover probability	0.9
Mutation probability	1/Lind, Lind is chromosome length

4.3 Experimental Evaluation Metrics

The classification accuracy score (accuracy_score) is a function in the sklearn.metrics module used to calculate the accuracy of a classification model. Measuring the performance of a model usually depends on the difference between the predicted and target values, while accuracy_score accepts two parameters: true label and predicted label, and returns the value of accuracy. Meanwhile, in order to comprehensively measure the performance of the model on the training and test sets, this paper performs a weighted average of the training and test set accuracies, as shown in Eq. (11):

$$Acc_{weighted} = (1 - \lambda) \times Acc_{train} + \lambda \times Acc_{test} \qquad (11)$$

where $Acc_{weighted}$ is the weighted accuracy, Acc_{train} and Acc_{test} are the training set and test set accuracy respectively, and λ is the test set weight.

4.4 Analysis of Experimental Results

In order to verify the effectiveness of GA-FL-RBFNN adopted in this paper, RBFNN, GA-RBFNN, GA-FL-RBFNN are trained and tested on two datasets, NHANES and HEART, respectively, and the experimental results are analyzed in terms of accuracy scores and convergence speed.

1) *Model accuracy analysis*

As shown in Table 5, it can be found that RBFNN, GA-RBFNN and GA-FL-RBFNN can achieve high and similar accuracy rates on both datasets, which are analyzed as follows:

a) All models training set accuracy and test set accuracy on the public dataset NHANES can reach 96.27% and above. It shows that the inclusion of genetic algorithm and federated learning do not lead to the reduction of the model accuracy. Therefore, the model in this paper can be deployed in a distributed manner with data security;

b) On the proprietary dataset HEART, the equilibrium accuracy of GA-RBFNN has about 10.03% improvement compared to RBFNN. It shows that the strategy of using genetic algorithm to optimize the structure and parameters of RBFNN model in this paper is effective;

c) GA-FL-RBFNN and GA-RBFNN have similar accuracy rates, indicating that federated learning can achieve an accuracy rate similar to that of centralized learning. Although federated learning slightly reduces the accuracy of the model (in terms of the balanced accuracy, the public dataset NHANES is reduced by about 1.13%, and the proprietary dataset HEART is reduced by about 3.86%), the optimization strategy of federated learning in this paper is feasible in consideration of the greater communication pressure brought by the large-scale data transmission and the higher risk of data leakage that may be caused in the centralized learning;

d) GA-FL-RBFNN performs slightly differently on different datasets as shown in Fig. 6. It performs well on the NHANES dataset, reaching 98.19% and 96.27% accuracy in the training and test sets, respectively; however, although the classification accuracy of the training data on the HEART dataset reaches 93.80%, the classification accuracy of its test data is only 80.33%, which may be due to the phenomenon of overfitting.

2) *Analysis of model convergence speed*

The convergence curves of the fitness function values of GA-FL-RBFNN during model training based on the NHANES and HEART datasets are shown in Fig. 7. The model has levelled off after the 18th iteration on the NHANES dataset and after the 13th iteration on the HEART dataset, and the fitness function value does not increase any

Table 5. Comparison of model accuracy

Model	NHANES			HEART		
	Training set accuracy (%)	Test set accuracy (%)	Balanced accuracy (%)	Training set accuracy (%)	Test set accuracy (%)	Balanced accuracy (%)
RBFNN	98.89	96.86	97.27	94.19	72.52	76.85
GA-RBFNN	98.52	97.59	97.78	93.39	85.25	86.88
GA-FL-RBFNN	98.19	96.27	96.65	93.80	80.33	83.02

more with the increase of offspring. It shows that GA-FL-RBFNN can maintain a good level in terms of model convergence speed even under limited training time and iteration number.

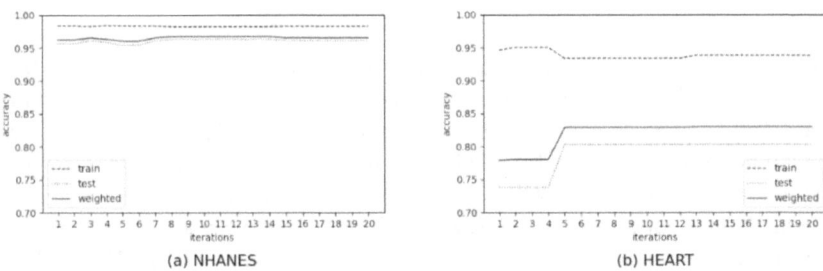

(a) NHANES   (b) HEART

Fig. 6. Accuracy curve

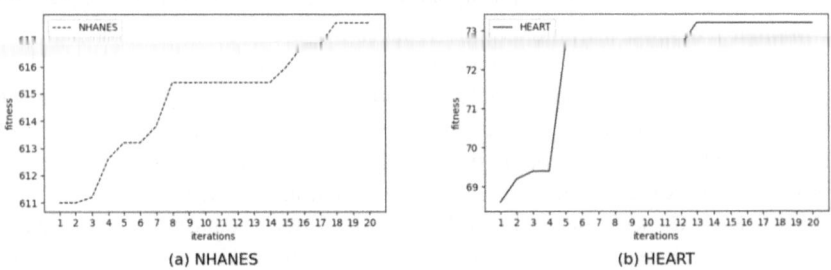

(a) NHANES   (b) HEART

Fig. 7. Fitness convergence curve

The above experimental results exemplify the superiority of the GA-FL-RBFNN model in this paper in terms of accuracy and distributed training. The GA-FL-RBFNN not only has higher prediction accuracy than the RBFNN, but also achieves similar prediction results to the centrally learnt GA-RBFNN, as expected.

Acknowledgment. Fund Project: Science and Technology Project of Hebei Academy of Sciences (23A04).

References

1. Qiu, L.F., Zhou, L.H., Peng, Z.G.: Radial basic function neural net. J. Hunan Indust. Polytechn. **2**(1), 8–9+40 (2002)
2. Li, Y.: Radial basis functions and their applications. Dalian University of Technology, Dalian (2009)
3. Wang, J.Z., Kong, L.W., Huang, Z.C., et al.: Research review of federated learning algorithms. Big Data Res. **6**(6), 64–82 (2020)
4. Wang, X.F.: Genetic algorithm and its application. J. Chin. Comput. Syst. **16**(2), 59–64 (1995)
5. Li, Y., Yuan, H.Y., Yu, J.Q., et al.: A review of the application of genetic algorithm to optimization problems. J. Shandong Indust. Technol. **2019**(12), 242–243+180 (2019)
6. Broomhead, D.S., Lowe, D.: Multivariable functional interpolation and adaptative networks. Complex Syst. **2**, 321–355 (1988)
7. Chai, J., Jiang, Q.Y., Cao, Z.K.: Function approximation capability and algorithms of RBF neural networks. Pattern Recogn. Artific. Intell. **15**(3), 310–316 (2002)
8. Zheng, B.G.L.: Radial basis function network configuration using genetic algorithms. Neural Netw. **8**(6), 877–890 (1995)
9. Hamdia, K.M., Zhuang, X., Rabczuk, T.: An efficient optimization approach for designing machine learning models based on genetic algorithm. Neural Comput. Appl. **33**(6), 1923–1933 (2020)
10. Konecný, J., Mcmahan H.B., Yu, F.X., et al.: Federated learning: strategies for improving communication efficiency (2016). https://doi.org/10.48550/arXiv.1610.05492
11. Zhou, C.X., Sun, Y.,Wang, D.G., et al.: Survey of federated learning research. Chin. J. Network Inform. Secur. **7**(5), 77–92 (2021)
12. Mcmahan, H.B., Moore, E., Ramage, D., et al.: Communication-efficient learning of deep networks from decentralized data. In: International Conference on Artificial Intelligence and Statistics (AISTATS), vol. 54, pp. 1273–1282 (2017)
13. Ryffel, T., Trask, A., Dahl, M., et al.: A generic framework for privacy preserving deep learning (2018). https://doi.org/10.48550/arXiv.1811.04017
14. Bian, W.D.: Research on federated learning optimization strategy based on genetic algorithm. Guangzhou University, Guangzhou (2022)
15. Holland, J.H.: Adaptation in natural and artificial systems: an introductory analysis with applications to biology, control, and artificial intelligence. The MIT Press, Massachusetts (1992)
16. Goldberg, D.E.: Genetic algorithm in search, optimization, and machine learning. Addison-Wesley Pub. Co, Massachusetts (1989)
17. Xu, X.Y., Chai, J.X., Yao, L.: A Model of gaussian radial basis function networks based on evolutionary computation. J. Xinyang Normal Univ. Natl. Sci. Edn. **27**(3), 425–428 (2014)
18. Nikbakht, S., Anitescu, C., Rabczuk, T.: Optimizing the neural network hyperparameters utilizing genetic algorithm. J. Zhejiang Univ.-Sci. A **22**(6), 407–426 (2021)

Improving Water Hyacinth Extraction from UAV Images Using Enhanced U-Net

Zhiru Niu[1], Wei Cai[1], Daowen Xu[2], and Shaofei Jin[2(✉)]

[1] Department of Digital China Research Institute, Fuzhou University, Fuzhou, China
[2] Department of Geography, Geography and Oceanography College Minjiang University, Fuzhou, China

jinsf@tea.ac.cn

Abstract. This paper addresses the issues of accuracy and efficiency in extracting the distribution areas of water hyacinth. A small-scale water hyacinth dataset was established based on drone-collected photos of water hyacinth in river channels, which was utilized for semantic segmentation tasks. A method based on deep learning was proposed to extract water hyacinth distribution areas from high-resolution drone remote sensing images. An efficient, accurate, and automated convolutional neural network called AttUNet was designed for this purpose, which eliminates the need for manually designed rules and can automatically learn remote sensing features of water hyacinth and extract distribution areas from images, thereby improving the accuracy and efficiency of acquiring relevant data. The research demonstrates that the proposed method can automatically extract features from massive high-resolution drone images, fully exploring complex nonlinear features, spectral features, and texture features in high-resolution drone images. The overall accuracy of extracting water hyacinth distribution areas in the study area reached 98.78%, with MIOU coefficient and mRecall coefficient of 95.86% and 98.01% respectively, both of which surpass the accuracy indicators of Deeplabv3+ and U-Net. The deep learning method for water hyacinth classification can fully exploit spectral, texture, and latent feature information in the data, and make it more suitable for extracting water hyacinth distribution information than traditional remote sensing classification methods.

Keywords: U-Net · Water Hyacinth Extraction · Semantic Segmentation · Attention Mechanism · UAV Images

1 Introduction

Water hyacinth is one of the invasive alien species and is recognized as one of the ten most pernicious weeds internationally [1]. The devastation caused by water hyacinth disasters disrupts aquatic ecosystems, obstructs river channels posing flood threats, and significantly impacts water resources, aquatic ecology, navigation, fisheries, etc., resulting in severe economic losses to China annually [2]. Previous studies primarily relied on traditional remote sensing methods for monitoring alien aquatic plants [3]. Remote sensing monitoring, characterized by large scale, long cycle, and timeliness, serves as

© The Author(s) 2025
P. Siarry et al. (Eds.): WCNA 2023, LNEE 1361, pp. 247–256, 2025.
https://doi.org/10.1007/978-981-96-2409-6_24

an effective means for monitoring aquatic invasive plants [4]. Previous study analyzed reservoir water hyacinth and water bodies using multisource satellite remote sensing data and water color remote sensing technology, obtaining chlorophyll concentration in the surface layer of water bodies [5]. Thamaga and Dube [6] accurately detected and mapped the spatial distribution and arrangement of water hyacinths in two freshwater lakes in Zimbabwe based on Landsat 8 remote sensing data using discriminant analysis and partial least squares analysis. Using Worldview-3 high-resolution remote sensing images and field survey data as sources, they evaluated the classification accuracy using a combination of visual interpretation and deep learning methods. They concluded that the growth area of water hyacinths in the river channel within 20m exceeded that of the river channel with a width greater than 20m. Furthermore, they found that traditional classification algorithms have limited feature extraction capabilities and insufficient representation of spatial information and complex regularities in high-resolution remote sensing images, leading to low classification accuracy. A remote sensing method [7] for monitoring water bodies, aquatic vegetation, and invasive water hyacinths nationwide, producing South Africa's first national-scale water hyacinth distribution map using medium-resolution satellite data was also proposed. Traditional remote sensing image-based water hyacinth extraction methods typically rely on manual thresholds or machine algorithms based on implicit and morphological features, resulting in low-dimensional feature acquisition and insufficient exploration of high-dimensional features [8], failing to achieve pixel-level extraction and having certain limitations, with a lower limit to accuracy.

In recent years, with the maturation of machine learning technology, remote sensing image classification tasks have also made certain research achievements at the semantic level. Mukarugwiro [9] used multi-temporal Landsat 8 images and a random forest classifier to assess the outbreak area of water hyacinths in Rwanda, providing a basis for governance for African countries plagued by invasive plant species. They constructed a deep learning remote sensing model for vegetation identification in karst wetlands based on SegNet, PSPNet, RAUNet, and DeepLabV3plus network architectures. Among these, texture features were found to significantly improve the accuracy of identifying cultivated land and water hyacinths, with an increase of approximately 0.05 in F1 scores based on both pixel and sample points.

In fact, the extraction of water hyacinths falls within the domain of image segmentation and can be categorized as a binary classification task [10]. By combining unmanned aerial vehicle (UAV) data with deep learning algorithms, the difficulty of obtaining high-resolution data samples and the high cost of training within the study area can be mitigated [11]. The complex terrain and fragmented land parcels in the distribution area of water hyacinths in Fujian Province pose additional challenges to the extraction of water hyacinth distribution areas, limiting the automation of extraction accuracy and efficiency. Deep learning technology provides a new theoretical approach for extracting water hyacinth distribution areas in the era of big data [12]. It addresses issues such as the intelligent extraction of image features and interpretation rules, offering novel solutions and insights for automated extraction processes.

2 Materials and Methods

2.1 Data Introduction

The study area is located along the Linwang Creek in Huian County, Quanzhou City, Fujian Province. Huian County is situated on the southeastern coast of Fujian Province, ranging between 118.783° to 118.785°E and 24.963° to 24.985°N. The county spans approximately 42 km east to west and 37 km north to south. The terrain inclines from northwest to southeast, characterized by hills, plateaus, with relatively less plains and basins. Rivers in Huian County are short in length and shallow in depth, mostly flowing independently into the sea. Due to the small watershed area and low flow volume of Linwang Creek, coupled with extensive agricultural water intake in the vicinity, flow interruption occurs during dry seasons. Linwang Creek features a flat terrain with slow flow velocity. Industrial and domestic effluents from both banks contribute to severe eutrophication in the water body, resulting in the perennial growth of water hyacinths in stagnant water areas or slow-flowing channels.

2.2 Data Acquisition

On October 21, 2021, aerial photography of Linwang Creek was conducted using a DJI Phantom 4 Pro UAV. The weather conditions during the shoot were favorable, with no clouds or haze affecting visibility. The UAV flew at an altitude of 200 m to ensure complete coverage of the river channel, following two flight paths with a 40% overlap rate. The sensor had an effective pixel count of 12.4 million and a shutter speed of 1/8000 s. The imagery comprised three bands: red, green, and blue, with a spatial resolution of 0.15 m. The data quality meets the requirements for research purposes.

2.3 Dataset Construction

The objective of this study is to achieve effective segmentation of UAV aerial images of small watershed rivers. However, existing public datasets are insufficient to meet this requirement. Therefore, a semantic segmentation approach is adopted to identify the target water hyacinths, necessitating the construction of a water hyacinth dataset tailored to the needs of this study.

From the captured data, 52 images were selected, each with an original resolution of 4836 pixels × 3648 pixels. However, these images cannot be directly used and need to be cropped into a standard dataset format. In this study, the PASCAL VOC format was chosen for dataset creation. The original images were segmented into 512 × 512 sizes and underwent filtering. Ultimately, 158 images that met the experimental requirements were selected. Labelme was employed for annotation to generate files in JSON format. However, these files cannot be directly used. Through Python scripts, the information in the JSON files was converted into corresponding PNG label images. After data augmentation, a total of 1422 images were obtained. Image augmentation techniques enhance the robustness of the network model and include methods such as rotation, flipping, random noise addition, stretching, filtering, etc. The visualization of the water hyacinth dataset creation with labeled images is depicted in Fig. 1.

Fig. 1. Samples in the datasets.

2.4 Construction of the AttUNet Model

U-Net was initially proposed for semantic segmentation of medical images in the field of medical imaging [13]. Its network structure is a U-shaped architecture, which gained significant attention in the field of image segmentation in 2015. U-Net was widely applied in segmentation tasks. Built upon the foundation of FCN, U-Net addresses the shortcomings of FCN in capturing contextual and positional information. One significant difference between U-Net and other common segmentation networks lies in its feature fusion mechanism: concatenation. U-Net concatenates features along the channel dimension to form thicker features, while FCN typically uses element-wise addition during feature fusion without thickening the features. One of the most prominent contributions of U-Net is its skip connection operation, which effectively addresses the loss of detailed information caused by downsampling operations, such as boundary information. This is crucial for dense prediction tasks like semantic segmentation, aiding the network in precise localization. Similarly, upsampling using deconvolution is generally more effective than direct bilinear interpolation. However, applying deconvolution in an overfitting model can have counterproductive effects. The network fully utilizes abstract features obtained from deep layers and image context information contained in shallow layers, employing a copy-and-add strategy for feature fusion, thereby achieving effective and accurate image segmentation.

Convolutional neural networks (CNNs) sometimes encounter issues such as distortion or blurring when processing image boundaries [14]. However, attention mechanism modules can effectively address these problems. The objective of this study is to achieve recognition of water hyacinths in small river channels. The collected images exhibit severe adhesion between water hyacinths, aquatic vegetation, and trees along the banks, with irregular distribution areas of water hyacinths. Various objects are stacked together, making semantic information complex and requiring more detailed image features for each segmentation object.

Therefore, it is necessary to introduce attention modules to improve model accuracy. This paper proposes to deepen the network depth based on the original U-Net network model. Additionally, to avoid training overfitting issues caused by deep network structures, the attention mechanism draws inspiration from human visual attention mechanisms. Its essence lies in allocating attention weights through a series of attention weight coefficients in neural networks to more accurately extract specific features.

Mnih et al. [15] first introduced the attention mechanism into the RNN model for image classification tasks, achieving good performance. Woo et al. [16] improved the Unet model by incorporating residual modules into the encoding structure and introducing attention mechanism modules into the skip connection part, significantly enhancing the accuracy of building extraction in high-resolution remote sensing images.

3 Model Training and Prediction

The experiment was conducted in a 64-bit Windows 10 environment, using the PyTorch 1.2.0 deep learning framework and Python 3.7 programming language. For hardware support, the CPU utilized an Intel(R) Core(TM) i7-8700 3.20GHz, and the GPU was an NVIDIA GeForce RTX 3080 with 8GB of memory. Model construction, training, and testing were performed using the PyTorch deep learning framework.

To facilitate training and testing, the dataset was first divided into training and testing sets. In accordance with practical considerations, the experimental sample data were divided into training and testing sets at a ratio of 8:2, comprising 1123 images for training and 281 images for testing. Care was taken to ensure that there was no overlap between the training and testing sets.

During model training, a combination of frozen training and unfrozen training was utilized. In the frozen stage, the feature extraction network remained unchanged to prevent weight destruction, with minimal memory consumption, and only minor adjustments were made to the network. In the unfrozen stage, the backbone network of the model was not frozen, and all parameters in the network were adjusted. The initial learning rate for the first 50 epochs of frozen training was set to 0.001, with a batch size of 2, a learning decay rate of 0.92, and the Adam optimizer was selected. Subsequently, the next 50 epochs were dedicated to unfrozen training, with an initial learning rate of 0.0001, a batch size of 2, a learning decay rate of 0.92, and the Adam optimizer was also chosen.

All code debugging and execution were implemented in PyCharm. Figure 2 depicts the training and validation loss curves for U-Net and AttUNet. The improved training and validation sets exhibited good fitting effects. Around 80 epochs, the curves began to stabilize, with loss values fluctuating around 0.05. This indicates that the learning rate was reasonably set, and the loss function achieved rapid convergence. It is evident that the AttUNet network effectively learned the water hyacinth dataset and demonstrated a certain level of stability in this network model.

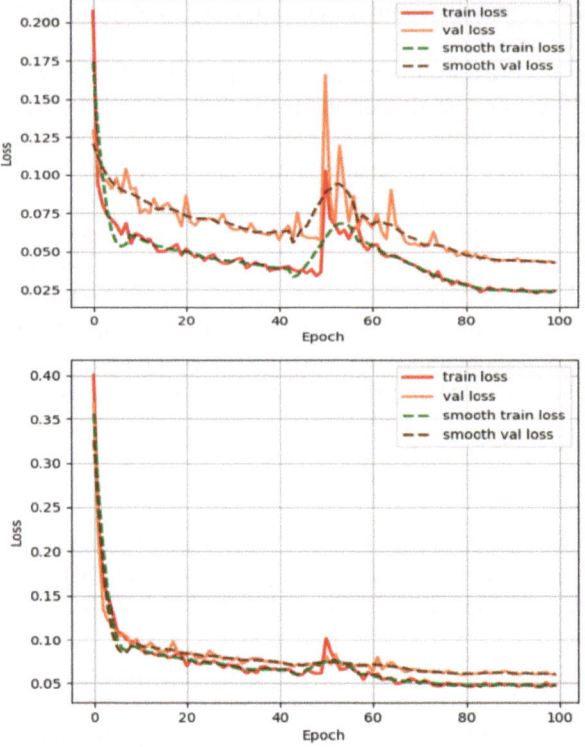

Fig. 2. AttUNet training set and validation set loss curves

4 Results

In many deep learning experiments and studies, after training and testing semantic segmentation networks, performance evaluation often requires comparison with other network methods. The evaluation metrics commonly used for comparison include Mean Pixel Accuracy (mPA), Mean Intersection over Union (MIoU), Intersection over Union (IOU), Precision, and Recall. Larger values of these evaluation metrics indicate better segmentation performance. Therefore, this paper utilizes the following three metrics for comprehensive evaluation and comparative analysis of different networks' performance in water hyacinth image segmentation: Accuracy, Recall, and MIoU. The formulas are as follows:

$$\text{Accuracy} = \frac{\text{TP} + \text{FN}}{\text{TP} + \text{TN} + \text{FP} + \text{F}} \tag{1}$$

$$\text{Recall} = \frac{\text{TP}}{\text{TP} + \text{FN}} \tag{2}$$

$$\text{Miou} = \frac{\sum_0^n \text{IOU}}{n} \tag{3}$$

Accuracy represents the percentage of correctly classified pixels.

Precision represents the percentage of correctly classified water hyacinth pixels out of the total water hyacinth pixels. MIoU evaluates the segmentation effect of water hyacinths and background classification, taking into account the misclassification of water hyacinth pixels as background pixels. TP (true positive) is the total number of pixels correctly classified as water hyacinth pixels. TN (true negative) is the total number of pixels correctly classified as background pixels. FP (false positive) is the total number of pixels incorrectly classified as water hyacinth pixels among background pixels. FN (false negative) is the total number of pixels incorrectly classified as background pixels among water hyacinth pixels.

To validate the effectiveness of the U-Net network model used in this paper, other commonly used classification models were employed for water hyacinth extraction and accuracy comparison. The trained models on the water hyacinth dataset were used to predict the test set, with input image resolution set to 512×512 and output results as water hyacinth extraction images. The three semantic segmentation network models compared in the experiment operate under the same environment and have identical network training optimization parameters. Some water hyacinth extraction accuracies are listed in Table 1, and extraction images are illustrated in Fig. 3.

Table 1. Model performance

Models	Miou%	Recall%	Accracy%
Deeplabv3	93.41	96.74	97.69
unet	95.24	97.53	98.36
AttUNet	95.86	98.01	98.78

From the statistical results in Table 1, it can be observed that all three semantic segmentation models exhibit good recognition capabilities for water hyacinths. However, there are differences among the specific methods of each model. The U-Net model performs the best among the three metrics, indicating its ability to accurately identify and segment targets in the image. The Deeplabv3+ model performs slightly lower than U-Net but still demonstrates high accuracy and stability. The PSPNet model shows relatively inferior segmentation effects, with instances of missegmentation, suggesting limitations in capturing detailed features of the target in image segmentation tasks, resulting in poorer accuracy and stability.

To further validate the segmentation effectiveness of the improved model, additional samples were predicted, and the prediction results are shown in Fig. 4. The segmentation effect of the proposed method for water hyacinth segmentation is significantly improved, without instances of missegmentation or undersegmentation, thereby further enhancing the accuracy of the segmentation results.

Fig. 3. Partial extraction diagram of water hyacinth. From the left to right is original figure, label figure, Deeplabv3 model, unet model, and the method in this study.

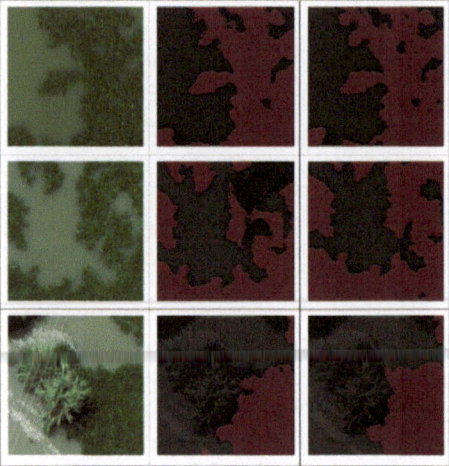

Fig. 4. Predication results of the method of U-Net and AttUNet. From the left to right is original figure, unet model, and the method used in this study.

5 Conclusion

This study utilized three semantic segmentation models to extract water hyacinths from small watershed rivers. In the absence of publicly available water hyacinth datasets, field surveys combined with visual interpretation were used to create deep learning training labels. By leveraging unmanned aerial vehicle (UAV) data and deep learning techniques, the study addressed challenges such as difficulty in obtaining high-resolution data samples and high training costs within the research area. It achieved the extraction of water hyacinths under the scenario of a small sample set, evaluated the accuracy of the extraction results, and improved the U-Net semantic segmentation recognition model. From the prediction results, it is evident that the proposed method is the optimal water hyacinth segmentation model, which holds significant reference value for subsequent remote sensing extraction studies of invasive plants.

Acknowledgment. Technology Innovation Center for Monitoring and Restoration Engineering of Ecological Fragile Zone in Southeast China, MNR (Grant NO. KY-030000-04-2023-021). N. Z. R. thanks the help of Professor Gaolong Zhu of the data collection.

References

1. He, C., et al.: Future global urban water scarcity and potential solutions. Nat. Commun. **12**(1), 4667 (2021)
2. Li, F., et al.: Water hyacinth for energy and environmental applications: a review. Biores. Technol. **327**, 124809 (2021)
3. Strittholt, J., Miles, L., Horning, N., Fosnight, E.: Sourcebook on remote sensing and biodiversity indicators. Techn. Series **32**, 203 (2007)
4. Bolch, E.A., et al.: Remote detection of invasive alien species. Remote Sensing of Plant biodiversity, pp. 267–307 (2020)
5. Fu, B., et al.: Multi-sensor and multi-platform retrieval of water chlorophyll a concentration in karst wetlands using transfer learning frameworks with ASD, UAV, and Planet CubeSate reflectance data. Sci. Total Environ. **901**, 165963 (2023)
6. Thamaga, K., Dube, T.: Remote sensing of invasive water hyacinth (*Eichhornia crassipes*): a review on applications and challenges. Remote Sens. Appl. Soc. Environ. **10**, 36–46 (2018)
7. Singh, G., Reynolds, C., Byrne, M., Reynolds, C., Byrne, M., Rosman, B.: A remote sensing method to monitor water, aquatic vegetation, and invasive eichhornia crassipes at national extents. Remote Sens. **12** (2020)
8. Datta, A., et al.: Monitoring the spread of water hyacinth (Pontederia crassipes): challenges and future developments. Front. Ecol. Evol. **9**, 631338 (2021)
9. Mukarugwiro, J., Newete, S., Adam, E., Nsanganwimana, F., Abutaleb, K., Byrne, M.: Mapping distribution of water hyacinth (*Eichhornia crassipes*) in Rwanda using multispectral remote sensing imagery. Afr. J. Aquat. Sci. **44**, 1–10 (2019)
10. Khanna, S., Santos, M.J., Ustin, S.L., Haverkamp, P.J.: An integrated approach to a biophysiologically based classification of floating aquatic macrophytes. Int. J. Remote Sens. **32**(4), 1067–1094 (2011)
11. Bouguettaya, A., Zarzour, H., Kechida, A., Taberkit, A.M.: Deep learning techniques to classify agricultural crops through UAV imagery: a review. Neural Comput. Appl. **34**(12), 9511–9536 (2022)
12. Zhang, W., Ching, J., Goh, A.T., Leung, A.Y.: Big data and machine learning in geoscience and geoengineering: introduction. Geosci. Front. (2020)
13. Alom, M.Z., Yakopcic, C., Hasan, M., Taha, T.M., Asari, V.K.: Recurrent residual U-Net for medical image segmentation. J. Med. Imag. **6**(1), 014006 (2019)
14. Borkar, T.S., Karam, L.J.: DeepCorrect: correcting DNN models against image distortions. IEEE Trans. Image Process. **28**(12), 6022–6034 (2019)
15. Mnih, V., Heess, N., Graves, A., Kavukcuoglu, K.: Recurrent models of visual attention. NIPS (2014)
16. Woo, S., Park, J., Lee, J., Kweon, I.: CBAM: Convolutional block attention module. In: Proceedings of the European Conference on Computer Vision (ECCV) (2018)

Video Codec Method for Compressed Sensing Theory

Haixia Yan[1] and Yanjun Liu[2(✉)]

[1] College of Electronic Science and Engineering, JiLin University, Changchun, China
`yanhx@jlu.edu.cn`
[2] Changchun Institute of Optics, Fine Mechanics and Physics, the Chinese Academy of
Sciences, Changchun, China
`liuyanjun@ciomp.ac.cn`

Abstract. In order to realize the rapid H.265 codec method, a new H.265 codec method based on Compressed Sensing Theory is proposed. The H.265 codec method are divided into three different types, intra-frame type, interframe type and mix frame type. Combined with the Compressed Sensing Theory, the input images linearly transform into the residual data, scale and quantify the transformed coefficients; after loop filtering and adaptive compensation, the H.265 coding data are got. When decoding, the process is reversed. Experimental results show that this method improves the image video codec quality and the image video codec speeds.

Keywords: compressed sensing · sparse reconstruction component · Video Codec · H.265

1 Introduction

Nowadays, the information technology is developing rapidly. The carrier of information transmission has also evolved from text and images to videos, and applications such as live streaming video and short video are popular [1–3]. With the development of the Internet, video information cause people's attention. Most of the traffic in network transmission is used for video information, and it will occupy more network bandwidth and memory space during data transmission [4, 5].

In general, surveillance video data needs to be stored for more than three months, and unprocessed massive video data is difficult to transmit and store. Especially with the development of high-definition video and the popularity of high-definition monitoring equipment, It is an urgent practical problem that achieving video encoding, real-time transmission and storage of video information [6, 7].

The effective ways in encoding videos, eliminating various redundant information in videos, and increasing efficiency in real-time video transmission, is that coding the video data and getting source videos through decoding [8].

Compressed Sensing (CS) theory is the technique that reduces the number of samples and the sampling time [9–11]. This theory is established within the theoretical framework of Shannon's sampling theorem. Compressed Sensing (CS) theory improves the

© The Author(s) 2025
P. Siarry et al. (Eds.): WCNA 2023, LNEE 1361, pp. 257–263, 2025.
https://doi.org/10.1007/978-981-96-2409-6_25

efficiency of sampling [12–14]. If the signal meets certain conditions, we can design an appropriate measurement matrix to collect samples with much lower original signal. The theory of Compressive Sensing has been widely applied since its inception [15–17].

2 Compressed Sensing Theory

2.1 Compressed Sensing Theory

The signal is not sparse in nature. In compressive sensing, the signal must be sparse, it is necessary to perform sparse representation before compressive sensing. There are many methods for sparse representation, if the signal is sparse and the matrix satisfies specific conditions, the image can be reconstructed [18, 19].

Compressed Sensing theory, if the signal $\alpha \in R^N$, where k $<<$ N, thus the signal α is sparse.

$$\|\alpha\|_0 = k \tag{1}$$

For a series discrete signal $x(n)(n = 0, 1 \cdots N - 1)$, which length is N, the signal X can be consider into a $N \times 1$ column vector. Signal X can be represented into linear combination of orthogonal base $\Phi = \{\varphi_1, \varphi_2, \cdots, \varphi_M\}$,

$$X = \Psi\Theta = \sum_{i=1}^{N} \theta_i \varphi_i \tag{2}$$

If the orthogonal basis is sparse, the original signals can be computed by the formula (2)

$$\Theta = \psi^T X \tag{3}$$

Then observed values Y of the original signal X can be represented as formula (3)

$$Y = \Phi X = \Phi X \Theta \tag{4}$$

In Compressed sensing theory, the sparse basis is necessary. In theory, there are many sparse basis. Generally, the signal is more sparse, the reconstruction error is more little. Thus the transformation is necessary to ensure that the signal is sparse and can be reverse transformation.

There are many types of sparse bases, and in addition to the wavelet transform and Fourier transform mentioned below, the total variation model is also an important sparse transform.

This type of model has an excellent effect on image noise reduction, which can effectively preserve the edge information of the image while filtering noise. It also has unique features in depicting image details. This type of model also has a wide range of applications. As long as the image has the property of local smoothness, the total variation can fully utilize its similarity, resulting in better reconstruction results.

2.2 H.265 Video Codec Method

High Efficiency Video Coding (HEVC) is the successor to H.264. HEVC improve the image quality, but also achieve a compression rate of twice that of H.264. It can support 4K resolution and even ultra-high definition television, with a maximum resolution of 8192×4320 (8K resolution).

There are three main coding methods, intra-frame coding, inter frame coding method, mix frame coding method.

The intra-frame coding method steps are follows.

(1) Firstly, we divide the input image sequence into coding unit;
(2) Then, we estimate the intra-frame data after segmentation;
(3) We predict the in frame image data;
(4) We get the residual data by subtracting block data from original image frame;
(5) We linearly transform the residual data and scale and quantify the transformed coefficients;
(6) We get the residual data by residual coefficients;
(7) We get the block data by adding residual data and inter frame predicted image frames;
(8) We get the image by loop filtering and adaptive compensation;
(9) We output the entropy encoding with inter frame encoded signal and residual signal coefficients.

The inter frame coding method steps are follows.

(1) Firstly, we divide the full image into small block image, then the motion estimation module computes code with the small block image and previous/next image;
(2) Then, we compensate the coding data by motion estimation;
(3) In order to obtain residual data, we subtract the inter frame prediction data from the original image block;
(4) The residual data in the frame is linearly transformed, and the transformed coefficients are quantified;
(5) We get the residual data by reverse process the residual data coefficients;
(6) We get the block prediction data by adding the predicted image frames and residual data;
(7) Finally, by using loop filtering and adaptive compensation, the "blocking effect" is removed, we obtain image frames that have not been segmented;
(8) The inter frame encoded signal and residual signal coefficients are encoded together using entropy and output.

The coding frame is shown in Fig. 1.

The decoding process, is the reverse process of encoding, mainly includes techniques such as transformation, inter frame prediction recovery, motion compensation, and inter loop filtering. The decoding process is as follows.

(1) We get the coding image data, then compute the entropy decoding with decoder and resort the image data;
(2) In order to get the processing data, we multiply the quantization result by the sorting data. We get the residual data by inverse transformation module

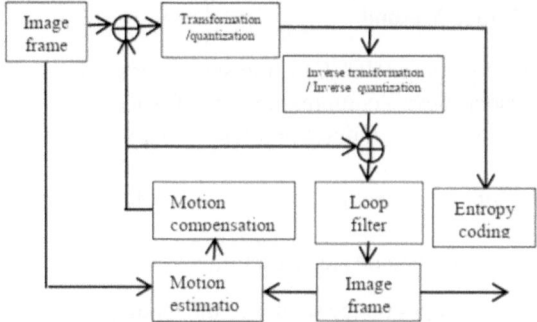

Fig. 1. H.265 encoding process

(3) We get the prediction block by step(1) and step(2), which is the same as the coding process.
(4) We send the prediction image frame into the loop filtering module, thus we eliminate the block effect, and obtain the reconstructed frame. This reconstructed frame is the next predicted frame.
(5) We repeat the above process until the stream ends.

The decoding process is shown in Fig. 2.

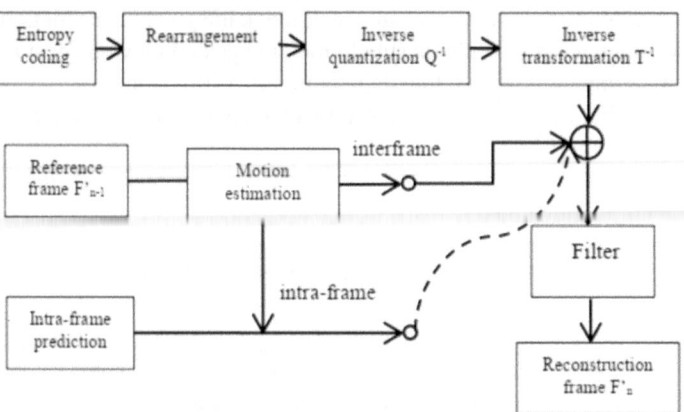

Fig. 2. H.265 decoding process

3 Rapid Video Codec Method for Compressed Sensing Reconstruction

3.1 EnCoding Process for Compressed Sensing

We design a codec method based the compress Compressed Sensing theory and traditional video codec technology. If the fame is the in frame, we code the frame by in frame method. If the frame is the inter frame, we code the frame by inter frame method. The coding process for compressed sensing is shown in Fig. 3.

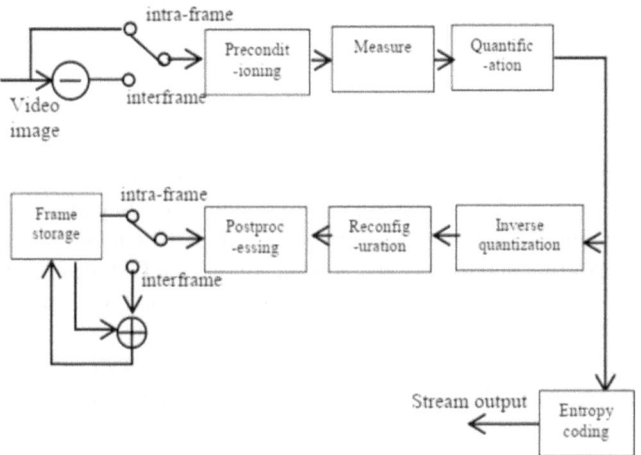

Fig. 3. Encoding process for compressed sensing

Due to the stronger sparse of residual images, the required number of measurement samples can be reduced. The encoded measurement values obtained in both modes can be quantified and entropy encoded to obtain the encoded stream. Like traditional video encoders, there is a local decoder inside the encoder to obtain a reference frame for inter frame encoding.

3.2 DeCoding Process for Compressed Sensing

When decoding, the received bit-stream data is entropy decoded firstly, and inverse quantized, then reconstruct image by underdetermined linear equations. If the obtained image is an in frame image, then the image is a reconstructed image; If it is a inter frame, then the image is a residual image. Here we add the residual image and reference image, thus we get the reconstructed image. The reconstructed image is needed by next frame as the reference image. The decoding process for compressed sensing is shown in Fig. 4.

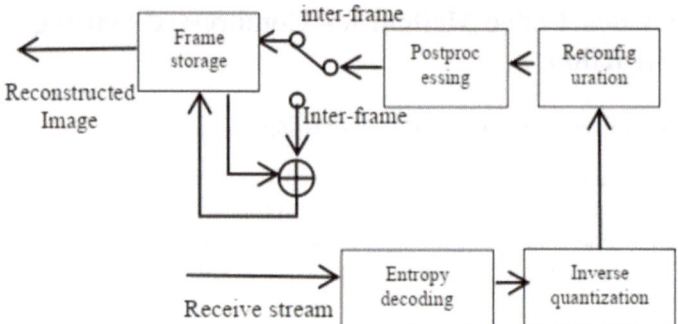

Fig. 4. Decoding process for compressed sensing

4 Experimet Results

To verify the effectiveness of the video encoder based on CS theory, two sets of video sequences were processed. The sparse basis used in the encoding measurement process is wavelet basis, and the measurement matrix is a 16×16 random perturbation block Hadamard matrix. The reconstruction algorithm adopts the GPSR algorithm.

In order to evaluate the image quality of decoding method, we introduce the PSNR and reconstruction errors. Experiment results show that the PSNR of decoding image1 and decoding image2 is 26.51dB and 27.29dB, the average PSNR of image is 27.35dB. Experiment results show that the Compressed Sensing theory based on H.265 Video Codec work effectively.

5 Summary

This article describes Video codec method based on compressed sensing theory. we introduce the compressed sensing theory and H.265 codec method, then we combine the H.265 codec with the compressed sensing method. Experimental results show that this method improves the image video codec quality and the image video codec speeds as possible.

Contact Author: LiuYanjun, E-mail: liuyanjun@ciomp. ac.cn.

References

1. Taha, M.A., Hamidouche, W., Sidaty, N., Viitanen, M., Deforges, O.: Privacy protection in real time HEVC standard using chaotic system. Cryptography **4**(2), 18 (2020)
2. Xu, H., Tong, X., Wang, Z., Zhang, M., Liu, Y., Ma, J.: Robust video encryption for H.264 compressed bitstream based on cross-coupled chaotic cipher. Multimedia Syst. **26**(4), 363–381 (2020)
3. Dolati, N., Beheshti, A., Azadegan, H.: 'A selective encryption for H.264/AVC videos based on scrambling.' Multimedia Tools Appl. **80**(2), 2319–2338 (2021)
4. Guan, B., Xu, D., Li, Q.: An efficient commutative encryption and data hiding scheme for HEVC video. IEEE Access **8**, 60232–60245 (2020)

5. Liu, B., Liu, J., Wang, S., Zhong, M., Li, B., Liu, Y.: HEVC video encryption algorithm based on integer dynamic coupling tent mapping. J. Adv. Comput. Intell. Intell. Informat. **24**(3), 335–345 (2020)
6. Ye, Q., Zhang, Q., Liu, S., Chen, K.: A novel chaotic system based on coupled map lattice and its application in HEVC encryption. Math. Biosci. Eng. **18**(6), 9410–9429 (2021)
7. Lee, M.K., Jang, E.S.: 'Cryptanalysis of start code-based encryption method for HEVC.' IEEE Access **9**, 92568–92577 (2021)
8. Piza, E.L., Welsh, B.C., Farrington, D.P., Thomas, A.L.: CCTV surveillance for crime prevention: a 40-year systematic review with meta analysis. Criminol. Public Policy **18**(1), 135–159 (2019)
9. Jones, M., Djahel, S., Welsh, K.: Path-planning for unmanned aerial vehicles with environment complexity considerations: a survey. ACM Comput. Surv. **55**(11), 1–39 (2023)
10. Notomista, G., Egerstedt, M.: Persistification of robotic tasks. IEEE Trans. Control Syst. Technol. **29**(2), 756–767 (2021)
11. Daryanavard, H., Harifi, A.: UAV path planning for data gathering of IoT nodes: ant colony or simulated annealing optimization. In: Proceedings of the 3rd International Conference Internet Things Applications, pp. 1–4 (2019)
12. Aggarwal, S., Kumar, N.: Path planning techniques for unmanned aerial vehicles: a review, solutions, and challenges. Comput. Commun. **149**, 270–299 (2020)
13. Donoho, D.L.: Compressed sensing. IEEE Trans. Inf. Theory **52**(4), 1289–1306 (2006)
14. Ding, X.: Compressed sensing image mapping spectrometer. IEEE Access **7**, 127765–127771 (2019)
15. You, D., Zhang, J., Xie, J., Chen, B., Ma, S.: COAST: controllable arbitrary-sampling network for compressive sensing. IEEE Trans. Image Process. **30**, 6066–6080 (2021)
16. Zhang, Z., Liu, Y., Liu, J., Wen, F., Zhu, C.: AMP-Net: denoisingbased deep unfolding for compressive image sensing. IEEE Trans. Image Process. **30**, 1487–1500 (2021)
17. Song, J., Chen, B., Zhang, J.: Dynamic path-controllable deep unfolding network for compressive sensing. IEEE Trans. Image Process. **32**, 2202–2214 (2023)
18. Shen, M., Gan, H., Ning, C., Hua, Y., Zhang, T.: TransCS: a transformerbased hybrid architecture for image compressed sensing. IEEE Trans. Image Process. **31**, 6991–7005 (2022)
19. Zhang, K., Gool, L.V., Timofte, R.: Deep unfolding network for image super-resolution. In: Proceedings of the IEEE/CVF Conference Computer Vision Pattern Recognition, pp. 3214–3223 (2020)

Research and Analysis of the Green Vision Rate of Street Space Based on Information Visualization Technology

Jun Wang, Hua Zhang$^{(\boxtimes)}$, and Zhulong Yan

Art School, Huzhou University, Huzhou, China
02526@zjhu.edu.cn

Abstract. Taking the research results of China National Knowledge Network (CNKI) from 1998 to 2023 as the data source, with the help of CiteSpace V and VOSviewer digital technology, the knowledge map on the research of China street space is drawn. The visualization technology shows the structure, evolution of street space, and the interrelationship of green vision rate research, and intuitively shows the research hotspot and frontier dynamics of green vision rate of street space in China. Research has learned that before 2017, the green vision rate of street space was in the bud, and then began to develop rapidly. At present, the street space green rate is mainly concentrated in the street roaming, street walkability, measurement method, evaluation index, healing environment, restorative environment, public space security, street image research, VR virtual reality research and green rate, green rate, green rate, etc. Through the visual map analysis of the green vision rate literature in the street space, it shows that the use of new technologies and cross-field integration will become the future research trend.

Keywords: street space · green visual index · date visualization · development trend

1 Introduction

With the rapid economic development and the continuous renewal of urban construction, the principal contradiction in Chinese society has become the contradiction between the people's ever-growing needs for a better life and unbalanced and inadequate development. At present, the street is one of the most frequently used public Spaces for residents' daily life and travel. The quality of its green landscape is the most easily perceived street element for residents, and also an important factor affecting the physical and mental health of residents. At present the main greening security means is through the formulation and implementation of related indicators provide management basis for greening construction, but the current greening index has certain limitations, more attention to two-dimensional plane and ecological benefits, the three-dimensional space and the subjective feelings of two attention, so the green rate as reflect the physical proportion of green plants, in recent years in the aspects of research and practice got more attention

© The Author(s) 2025
P. Siarry et al. (Eds.): WCNA 2023, LNEE 1361, pp. 264–272, 2025.
https://doi.org/10.1007/978-981-96-2409-6_26

and application. In order to realize the people-oriented street greening landscape environment shaping, this paper will analyze the research status of street space green vision rate in China, and discuss the research process of relevant scholars in this field, in order to provide reference for the subsequent research and design.

2 Date Sources and Study Methods

Please follow these instructions as carefully as possible so all articles within a conference have the same style to the title page. This paragraph follows a section title so it should not be indented.

2.1 Data Source

The title is set 17 point Times Bold, flush left, unjustified. The first letter of the title should be capitalized with the rest in lower case. It should not be indented. Leave 28 mm of space above the title and 10 mm after the title.

2.2 Study Methods

In recent years, information visualization has become a hot topic in information management research, and scientific knowledge graph has also become one of the hot methods of bibliometric research. At present, the mainstream knowledge graph software tools are CiteSpace V, Thomson Data Analyzer, VOSviewer, BibExcel, Pajek, etc. Relatively speaking, CiteSpace has better analytical functionality and compatibility. It has advantages in analyzing the structure, law and distribution of the tacit knowledge in the data, revealing the dynamic development of the discipline, and discovering the frontier of the discipline research. VOSviewer Can avoid the mutual coverage of important nodes and labels, and focus on the main information display of the data set. Therefore, VOSviewer and CiteSpace were selected for the visual analysis. Through the search of the research subjects, removing the newspapers and results, 283 papers were obtained as research objects, save all selected documents-export references-select Refworks format, and then import into VOSviewer and CiteSpace respectively to generate relevant data maps. At the same time, download the number of papers, subject distribution, journals, keywords, publishing institutions, fund projects, highly cited papers and other information.

3 Publication Volume and Circulation

The trend chart of the number of papers from 1998 to 2023 (Fig. 1) shows that the number of papers studied on the street space went through several stages. At first, from 1998 to 2016, the number of papers was relatively rare, and the annual number of published papers did not exceed 10, indicating that the attention to the green visual rate of street space in China was still relatively low during this period. In the next stage, from 2017 to 2023, post began to increase, especially in 2020 began to have rapid growth, and reached the peak in 2022, 64, from 2017 to 202064 shows that street space green rate research by the attention of more researchers, heat gradually increased and faster, although the number of papers published slightly decreased in 2023, the number of 42, but still can explain street space green rate research still remain in a relatively popular environment.

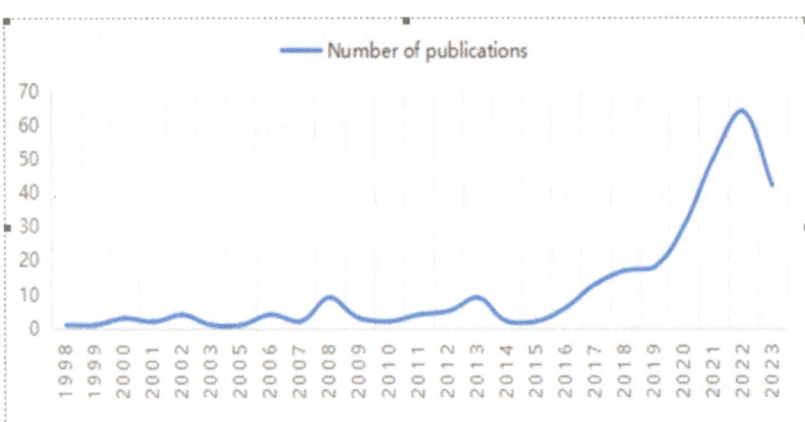

Fig. 1. Changes in the number of publications on the study of street spatial green visibility from 1998 to 2023. (Color figure online)

The above data show that in China, the research of green vision rate of street space gradually attracts attention. In recent years, China's urban development is transforming from incremental expansion to stock optimization. In the process of material environment construction, more attention is paid to the real feelings of people in the environment, and takes this as the starting point of the construction of material space environment, and the planning and construction concept of harmonious coexistence between people and environment has become a consensus. The 2015 Central Urban Work Conference clearly put forward the concept of "people's city" and required the path of urban development with Chinese characteristics, which is the concrete embodiment of the "people-centered" development thought in urban work. The party's 20th annual report for 2022 reaffirms this concept and requirement, and realizing China's urban modernization is an important part of Chinese-style modernization. At present the main green security means is through the formulation of relevant indicators and implementation to provide management basis for greening construction, the current greening index pay more attention to the ecological benefit of two-dimensional plane, so the green rate as reflect the field of physical quantity of green plants, can supplement the three-dimensional space and the subjective concern of two aspects. At present, the street is one of the most frequently used public Spaces for residents' daily life and travel. The quality of its green landscape is the most easily perceived street element for residents, and also an important factor affecting the physical and mental health of residents. Therefore, it will be important to strengthen the study of street space. As can be seen from the proportion analysis chart of street space in China from 1998 to 2023 (Fig. 2), compared with other journals, there are many articles in Urban Architecture and Landscape Architecture. Published 11 related articles (3.8%) and 9 articles (3.1%) respectively, and the other journals accounted for no more than 3.0%.

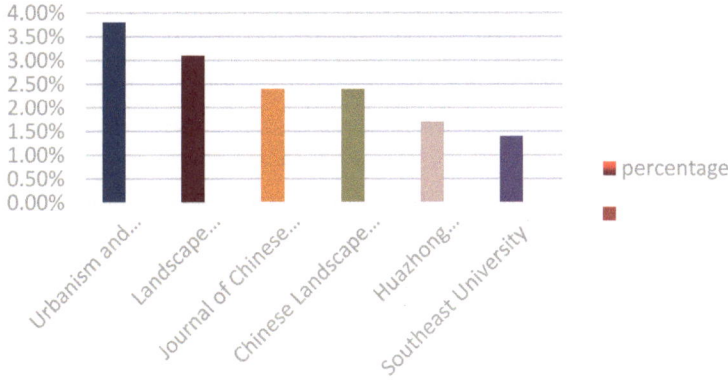

Fig. 2. Analysis of the proportion of research journals on street spatial landscape in China from 1998 to 2023

4   Main Content Analysis

4.1   High-Cited Paper

The top 10 papers in the number of quotes and downloads from 1998 to 2023 selected by the Knowledge Network of China (CNKI) were analyzed. Due to space constraints, only the top 5 papers will be listed. See Figs. 3 and 4 for more details. Among them, the highest number of citations was 207 times and the number of downloads reached 5791 times; the average citation was 18 times and the average download was 730 times.

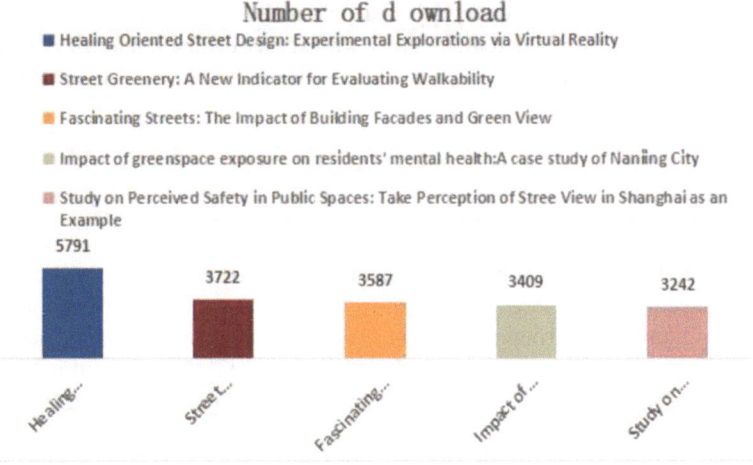

Fig. 3. From 1998 to 2023, it ranked among the top five cited papers in the field of street space green view rate research in China

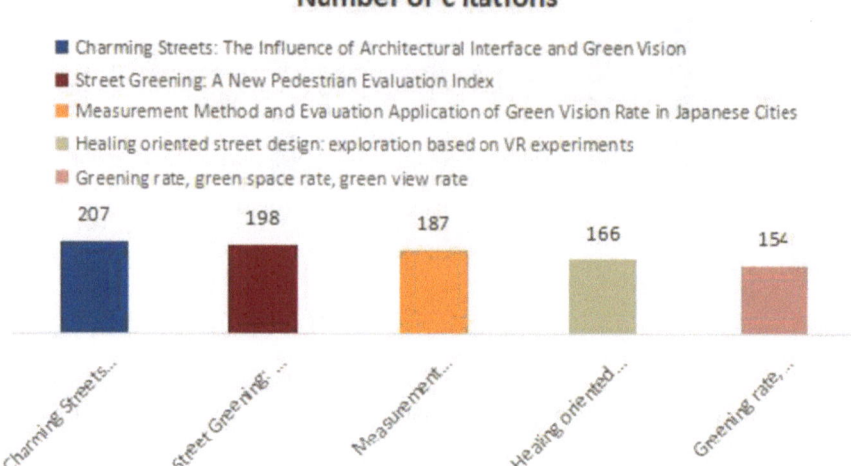

Fig. 4. From 1998 to 2023, it ranked among the top five downloaded papers in the field of street space green view rate research in China

These papers were mainly published in Beijing Forestry University, Shanghai Institute of Urban Planning and Design, China Institute of Urban Planning and Design, Huazhong University of Science and Technology and other major publishing institutions. These top 10 cited papers mainly focus on street walkability research, measurement methods, research, evaluation indicators, research, healing environment research, public space security research, street view research and VR virtual reality research. This shows that the concentration of the green vision rate of street space is not too high in the whole field, and it is more studied as an ornament in the large frame. Including Hao Xinhua wrote the street greening: a new walkability evaluation index of references and downloads are among the top three, the article by building an automatic method of large-scale, fine scale of street greening quantitative evaluation, the street walking system planning, street quality improvement work has certain guiding significance.

4.2   Identify the Headings

In a certain period of time, a group of literature research topics or perspectives have internal connection, and a large number of topics are known as the research hotspot of this group of literature. Through the keyword co-occurrence analysis and adjustment of relevant parameters, the high-frequency keywords were highlighted and improved for readability, and the keyword co-occurrence knowledge graph was formed (see Fig. 5).

In the figure, the colored circular nodes represent the different keywords. The size of nodes indicates the frequency of keywords and font size represents centrality. Connec-connection between nodes indicates the degree of correlation between keywords. The figure contains 71 keyword nodes and presents a relatively dispersed distribution. Most of the hot keywords are generic words, covering a wide range, and have not yet formed

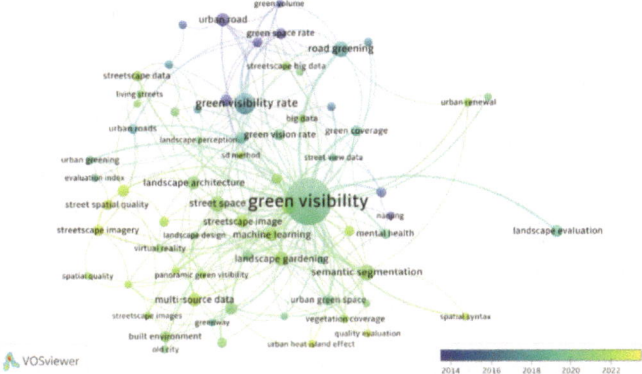

Fig. 5. Analysis of hotspots in the research of street spatial green visibility rate in China from 1998 to 2023. (Color figure online)

a relatively concentrated core hot spot area. As can be seen from the high-frequency keyword map, "green visual rate", "semantic segmentation", "street view image", "deep learning", "street view data", "landscape evaluation" and so on are all high-frequency hot keywords, which belong to the current research hotspot.

4.3 Analysis of Research Knowledge Evolution

Different research hotspots often arise in different time periods. In order to more clearly show the evolution trend of the research hotspots in this field, the street space green visual rate study drew the time zone knowledge map of the evolution of the hotspots (Fig. 6).

Fig. 6. Time zone knowledge graph of the evolution of hotspots in the study of street spatial green visibility in China from 1998 to 2023

Research is mainly divided into four stages, the first stage for 1998 to 2007, the street space green rate research basic stay in the concept of propaganda stage, the researchers combined with other green space index of green management, so the paper number is less, including "Nanjing" is the green rate research and one of the highest city, as early as 2003, Nanjing greening committee has said to improve the construction of urban green rate is an important link of building green Nanjing, so in Fig. 6 in 2003 highlights the word Nanjing. From 2008 to 2014, researchers gradually calculated and quantified the green vision rate, and the majority of the research subjects were urban streets. Meanwhile, landscape aesthetics was included in the research scope, and the evaluation and analysis of landscape preference were conducted. From 2015 to 2018, the domestic research on green view rate turned to the evaluation of street quality, the study of residents' psychological perception and the exploration of new measurement methods of green view rate. At this stage, Chinese scholars proposed that the green vision rate can be used as a new index to study the feasibility of the street, the index of urban street environment to study the psychological satisfaction of the population, and the automatic identification and calculation of the green vision rate based on the street view data, which expanded the depth and breadth of the green vision rate. At the same time, in other aspects of urban research, the green vision rate has also been included in the research system as an important indicator to measure the quality of cities, and has been widely used. Since 2019, with the machine learning method of convolution neural network and the popularization of street view images, green visual rate has achieved a breakthrough in automatic calculation method, and thus shows an obvious trend of data and automation, and tends to become one of the indicators of urban comprehensive evaluation model. At the same time, the research direction of mental health has gradually entered the research field. The gradual start began to associate with more fields, which also corresponds to the surge in the number of papers starting in 2019.

5   Conclusion

Based on the literature visualization software VOSviewer and CiteSpace, we analyzed 283 valid documents in the field of street space green visual rate research in the CNKI database from 1998 to 2023. First of all, it can be seen from the literature volume, keywords and timing development of the green vision rate that these three have a large consistency in time. In general, the green vision rate research has a short time span, which shows the characteristics of niche, but the research heat has gradually increased in recent years. Finally, in terms of scientific research institutions, the current literature research is scattered, all conducted by universities or individuals, and the interaction between teams is not strong. Multi-field and multi-type research modes are of great significance for increasing the breadth and depth of research results. Therefore, strengthening the cooperation between university-enterprise scientific research and various disciplines in this field will become a new driving force for the future development. With the machine learning method of convolutional neural networks and the popular use of street view images, the measurement of green vision rate will become more convenient, which makes our measurement range become larger and more feasible. With the emergence of the concept of people-oriented high-quality development in China, the research on the green vision rate of street space will gradually begin to intersect with more fields, especially some aspects that focus more on human perspective. For example, the research on human mental health and other fields will have more research investment and continue to heat up.

All of these will be an important direction of the future street space green vision rate research.

Acknowledgment. This research was supported by pre-research project of Institute of "Two Mountains" Theory (Huzhou University) under Grant No. LSY2302.

References

1. Xu, L., Meng, R., Chen, Z.: Charming streets: the construction interface and the influence of green rate. J. Landscape Architect. (10), 27–33 (2017)
2. Zhang, H., Xiong, M., Chen, B., et al.: Influence of tropical cyclones on outdoor wind environment in high-rise residential areas in Zhejiang Province, China. Sustainability **14**(7), 3932 (2022)
3. Deng, X., Wang, H.: Green rate Green land rate green sight rate. New Construct. **06**, 75–76 (2002)
4. Hao, X., Long, Y.: Street greening: a new walkability evaluation index. Urban Plan. Rev. Shanghai (01), 32–36+49 (2017)
5. Rui, L.: Review of domestic research status of green vision. Build. Sci. Technol. **5**(06), 64–67 (2021). (in Chinese)
6. Zhang, W., Zhou, Y., yang, M.: Research on automatic recognition and calculation of panoramic green viewing rate. J. Landscape Architect. **26**(10), 89–94 (2019)
7. Xu, L., Meng, R., Chen, Z.: Charming streets: the impact of architectural interface and greenness. J. Landscape Architect. **10**, 27–33 (2017)
8. Xiao, X., Wei, Y., Li, M.: Metrology and evaluation of urban greenness in Japan. Urban Plan. Int. **33**(02), 98–103 (2018). (in Chinese)
9. Zhao, Q., Tang, H., Wei, D., et al.: Spatial green quantity visibility characteristics of urban greenway based on green visibility. J. Zhejiang A F Univ. **33**(02), 288–294 (2016). (in Chinese)
10. Wu, L.: Research on Urban road green space Design based on green vision. Shanghai Jiao Tong University (2008)
11. Li, M.: Visual evaluation of green quantity of urban pedestrian space based on green visual rate. Chinese Academy of Forestry (2018)
12. Zheng, Y., Yang, J.: Research on refined urban repair method based on Artificial Intelligence analysis of large-scale street view images. Chin. Garden **4**(8), 73–77 (2020). https://doi.org/10.19775/j.carolcarrollla2020.08.0073
13. Li, Z., He, Z., Zhang, Y., et al.: Effects of green environment exposure on residents' mental health: a case study of Nanjing. Prog. Geogr. **39**(05), 779–791 (2019)
14. Sun, G.: Evaluation of urban street Walkability based on multi-source big data: a case study of the central city of Nanjing. Mod. Urban Res. (11), 34–41 (2020)
15. Li, M., Yang, Z., Xue, F.: Urban street greening quality measurement and planning design improvement strategy based on multi-source data: a case study of Fuzhou City. Landscape Architect. **28**(02), 62–68. https://doi.org/10.14085/j.fjyl.2021.02.0062.07
16. Li, Y., Huang, J.: Evaluation of pedestrian environment green vision perception in historical blocks based on green vision decay curve: a case study of Tongwen District, Zhongshan Road, Xiamen. Landscape Architect. **27**(11), 110–115 (2020). https://doi.org/10.14085/j.fjyl.2020.11.0110.06

Analysis of Spatial-Temporal Evolution of Water Surface Area in DaSuGan Lake Based on Multi-source Satellite Images and Computer Vision Technology

Liping Shang[1], Yu'e Du[1], and Baokang Liu[2(✉)]

[1] Gansu Natural Energy Research Institute, Lanzhou, China
[2] Tianshui Normal University, Tianshui, China
345337505@qq.com

Abstract. Lakes are sensitive indicators of climate change, and obtaining the information of lake water quickly and accurately is of great significance for regional climate change research, the protection and governance of ecological environment. This article is based on multi-source satellite images such as Landsat and GF-6, and uses the method of empirical threshold and single band threshold and water index and other methods to extract the water area of DaSuGan lake. Based on the accuracy evaluation results, the optimal water identification model is selected, and the spatial-temporal evolution law of the water area of DaSuGan lake is analyzed from 1991 to 2020. The results show that:

Keywords: Multi source satellite imagery · DaSuGan lake · Water Area · Temporal and Spatial Variation · computer vision technology

1 Introduction

The change has a significant impact on activities of ecology and human in lake area. It affects the strength of water storage capacity and local ecological balance and climate. Meanwhile it poses a flood risk during flood season. At present, researchers use remote sensing images to extract lake surface information, monitor and analyze the spatial-temporal dynamic changes of lakes. Scholars at home and abroad have conducted research on lakes. Mcfeeters (1996) created the Normalized Difference in Water Index (NDWI) and found that using the ratio of green light and near-infrared bands can greatly suppress vegetation information [1]; Frazier PS (2000) applied the Landsat TM 5 mid infrared band threshold method and maximum likelihood method to extract water bodies from the Marambi River in Australia, and the results showed that the extraction accuracy of the two methods was basically the same [2]. In 2021, Gu Zhenkui,et al. conducted large-scale surface water monitoring based on long sequence Landsat images. In long-term water monitoring, the combination of Landsat image data and Lem can effectively remove cloud cover images. [3] Gu Zhenkui et al. (2021) utilized long sequence Landsat images and LEM technology to effectively monitor surface water based on

P. Siarry et al. (Eds.): WCNA 2023, LNEE 1361, pp. 273–283, 2025.
https://doi.org/10.1007/978-981-96-2409-6_27

long sequence Landsat images. However, the water quality and geographical location of lakes vary in different regions, resulting in differences in the optimal water body models applicable to each region. At present, there is relatively little research on lakes based on GF-1 satellite images in arid areas of northwest China [4]. Studying the spatial temporal dynamic changes of lake water area is one of the important contents for predicting global environmental changes.

This article combines the Landsat series and GF-6 WFV satellite images, applies multiple methods to extract the water area of the DaSuGan Lake, and selects the best model through accuracy evaluation. This study estimated the changes over the past 30 years and analyzed the spatial temporal dynamic characteristics in water area. This study aims to provide reference for the changes of lakes in arid areas of northwest China, and to provide theoretical support and decision-making basis for local ecological civilization construction, environmental governance, and water resource management.

2   Data Sources and Methods

2.1   Overview of the Research Area

DaSuGan Lake is the largest lake in Gansu Province, belonging to the saltwater lake category. It is located at east longitude 93°46′-94°01′, North latitude 38°51′-38°55′.Akse Kazakh Autonomous County located in the western part of Haizi Grassland in Heping Township, as shown in Fig. 1. To the north is the Al Ayiger Daban Mountains, to the west is the Kunlun Gobi Mountains, and to the east is the Liuge Expressway, with an altitude of 2795–2808 m. The water source comes from the surface water storage in the Halteng Basin. The surface water first flows into XiaoSuGan Lake, and then into DaSuGan Lake from XiaoSuGan Lake. The climate in area belongs to the inland high-altitude and semi-arid climate.

Fig. 1. Geographical Location of DaSuGan Lake

Aksaiqin has a relatively high terrain and belongs to the high-altitude region. For example, in Kangxiwa, the average temperature is −0.6 °C in a year, −11.3 °C in January,9.8 °C in July, there are the frost free period of 10 days in a year, and it is 20 °C the accumulated temperature above 10 °C. The annual average temperature of the observatory is −9.8 °C, −21.0 °C in January, and 3 °C in July.

2.2   Data Sources and Preprocessing

The download website for this series of data is from the United States Geological Survey (http://EarthExplorer.usgs.gov), with a download period from 1991 to 2020. The satellites used include Landsat 5, Landsat 7, and Landsat 8. Most of the image data is from August to November (except for 2019 and 2020), and the cloud cover is less than 10%; GF-6 data comes from China Resources Satellite Application Center (36.112.130.153:7777/DSS Platform/index. ml), with image coverage from 2019 to 2020. The cloud cover is all below 10% (except for June 23, 2020). GF-6 is an optical remote sensing satellite used in low orbit, and it is also China's first high-precision agricultural observation satellite with a designed service life of 8 years [4]. The GF-6 satellite is equipped with 2-m panchromatic and 8-m multi-spectral high-resolution cameras, a 16 m wide multi spectral medium resolution camera, a 2-m panchromatic and over 8-m multi spectral camera with an observation width of 90 km, and a 16 m multi spectral camera with an observation width of 800 km.

2.3   Research Methods

• Empirical threshold method

The segmentation threshold of the empirical threshold method is determined based on existing research and relevant background knowledge. Therefore, this method is the most convenient, comprehensive, and easily achievable among all methods. Meanwhile, although this method does not require other auxiliary data, the results obtained vary depending on the researchers, and it has strong subjectivity.

$$\text{Nfloat(NIR)/float(Green)} < C_1 \text{ and NIR} < C_2 \tag{1}$$

Among them, NIR is in the near-infrared band, Green is in the green band, and C1 and C2 are the optimal thresholds.

• Single band threshold method

The single band threshold method is to use a single band as the identification parameter, which is often the band with the most significant water characteristics, while other land features are not obvious (such as near-infrared and mid infrared bands). Finally, by determining the threshold, the water in the image is judged. This method utilizes the strong absorption characteristics of water in the near-infrared and mid infrared bands, while also reflecting the high reflection characteristics of vegetation and soil in the near-infrared and mid infrared bands.

$$\text{NIR} < T_1 \text{ and NIR} > T_2 \tag{2}$$

Among them, NIR is in the near-infrared band, and T_1 and T_2 are the optimal thresholds.

• Multi Band Spectral Relationship Method

The multi band spectral relationship method is one of the methods for digital processing of multi band remote sensing images. The basic principle is to calculate the ratio of brightness values of each pixel in two different bands, and then use this ratio to construct a new image (ratio image). In theory, the ratio of any two bands can be calculated, as well as the ratio of a certain band to the average value of each band, or the ratio of band sum to band difference. The specific ratio to be used depends on the experimental purpose and effectiveness.

$$(NIR)/float(Green) < C_1 \text{ and } NIR < C_2 \tag{3}$$

Among them, NIR is in the near-infrared band, Green is in the green band, and C_1 and C_2 are the optimal thresholds.

3 Extraction Methods and Accuracy Evaluation of Lake Water Area

3.1 Extraction Process for the Area of DaSuGan Lake

After preprocessing the Landsat 5, Landsat 7, and Landsat 8 downloaded from the United States Geological Survey and the GF-6 data downloaded from the China Resources Satellite Application Center, the data will be cropped. Finally, the multi band spectral relationship method combined with ArcMap will be used for visual interpretation to extract the water area of the DaSuGan Lake. The specific process is as follows:

In the classic mode of ENVI5.3, preprocessed images are first loaded, and the selected bands are near-infrared, red, and green. The bands for Landsat 5, Landsat 7, and GF-6 are 4, 3, and 2, and the bands for Landsat 8 are 5, 4, and 3. In Band math, enter the formula: (b4)/float (b2) lt 1 and b4 lt 900, as shown in Fig. 2. This paper uses near-infrared and green light bands to obtain gray scale images.

The extracted vector data contains other non lake features, such as rivers, snow capped mountains, clouds, noise, etc., which require manual visual interpretation to correct. The extracted vector data is overlaid on the original image, and non lake features are corrected through visual inspection and some complex boundaries are modified. Use the geometric calculation tool of Arcgis 10.8 to calculate the final lake area. Figure 2 shows the original grayscale image of DaSuGan Lake in September 2019. It can be seen from Fig. 2 that the extracted lake in the area is mixed with the shadows of rivers and some mountains. Figure 2 are vector images after visual interpretation and regional filtering. The vectors in non lake water areas are basically eliminated, which can ensure the accuracy of water extraction to varying degrees.

3.2 Evaluation of Accuracy in Extracting the Area of the DaSuGa Lake

- **Visual verification**

This article combines software such as ENVI 5.3, Arcgis 10.8, and Shangcheng Map to verify the accuracy of extraction by visually interpreting and overlaying the original images. Through visual interpretation, it was found that the boundaries of larger water bodies are closer to reality, and there are basically no fractures on the water surface. The main reason for the poor accuracy of partially extracted images is that: (1) the resolution of the image is 30 m, and satellite remote sensing images with a resolution of 30 m are difficult to display very small water bodies; (2) There is eutrophication in the water bodies of some lakes, and the spectral characteristics of the water bodies undergo slight changes, resulting in similar spectral characteristics to soil and vegetation. Although there are some cases where the extraction accuracy is slightly poor, due to the large number of bands and rich spectral information, the extraction of most water bodies is relatively accurate.

- **Quantitative validation**

(1) The overall classification accuracy [5] refers to the proportion of classified samples correctly among all samples, which is an accuracy evaluation index obtained by dividing the correctly classified image pixels into molecules and the total number of pixels. The formula is as follows:

$$P_0 = \frac{\sum_{i=1}^{c} \alpha_{ii}}{N} \tag{4}$$

Among them, P_0 represents the overall classification accuracy, α The elements on the i-throw and i-th column of the i-mixed latent matrix are the correct number of i-th class samples, where N represents the total number of test samples and c represents the number of categories [6].

The Kappa coefficient was first proposed by Cohcn [7] in 1960 to analyze the consistency of remote sensing images before and after classification. The sampling conditions for Kappa coefficients are relatively relaxed, and the description of classification results is more objective and completion [8]. The formula is as follows:

$$K = \frac{P_0 - P_e}{1 - P_e} \tag{5}$$

Among them, K is the Kappa coefficient, P_0 is the overall classification accuracy, assuming that the true number of samples for each class is x, x, x. The predicted number of samples for each class is Y, Yz, Y. N is the total number of samples, $P_e = X_1 * Y_1 + X_2 * Y_2 + ... X_i * Y_i / N_2$ [9].

The Kappa calculation result is $-1-1$, but typically Kappa falls between $0-1$ and can be divided into five groups to represent consistency at the same level [10].

① 0.00–0.20: extremely low; ② 0.21–0.40: General; ③ 0.41–0.60: moderate; ④ 0.61–0.80: Consistency in height; ⑤ 0.81–1.00: Almost the same level;

Leakage [11] refers to pixels that originally belonged to the actual classification of the surface, but were not classified into the corresponding class, while leakage error occurs in the sliding array. The formula is as follows:

$$P_m = \frac{\sum_{i=1}^{c} \alpha_{+i} - \alpha_{ii}}{\sum_{i=1}^{c} \alpha_{i+}} \tag{6}$$

Among them, Pm is the missed fraction error, α_{ii} is the element on the i-th row and i-th column of the slip matrix, $\alpha i +$ is mixed Clear the elements in the i-th column of the matrix Calculate based on the overall accuracy, Kappa coefficient, leakage error, and other calculation formulas introduced earlier, The accuracy verification results of the extracted results are shown in Table 1.

Table 1. Extraction accuracy results of DaSuGan Lake in September 2019

Extraction model	Overall accuracy	Kappa coefficient	leakage error
Single-band threshold method	95.6914%	0.8893	5.33
Empirical threshold method	96.8586%	0.9260	0.03
Multiband interspectral relation method	99.0315%	0.9763	0.07

Through the analysis and verification in the previous text, it was found that the spectral relationship model has a relatively good extraction effect on the DaSu Gan Lake. The accuracy results of the DaSuGan Lake extraction in September 2019 are shown in Table 1. According to Table 1, the overall classification accuracy of the spectral relationship model extraction can reach over 99%, with a Kappa coefficient greater than 0.97 and a leakage error of less than 0.07%. Through comprehensive comparative analysis, the Kappa coefficients are all above 0.88, and coefficients between 0.81 and 1.00 are considered almost identical, which proves that the extraction accuracy of the spectral relationship model is optimal.

4 Analysis of Spatial-Temporal Dynamic Changes in Lake Area

4.1 Dynamic Changes in Lake Water Area

- **Inter annual changes**

This article uses a spectral relationship model to establish vector maps of water bodies in different time periods, and combines Arcgis 10.8 with visual interpretation to eliminate redundant interference and preserve the lake water boundaries of the DaSuGan Lake in different periods. The results of extracting water body information from the DaSuGan Lake are shown in Fig. 2.

The lake area obtained by reading the quantity data attribute table. The statistical results of the calculation of the lake area of DaSuGan Lake from 1991 to 2020 are shown in Fig. 4-1. From Fig. 2, it can be seen that from 1991 to 2020, the minimum area of DaSuGan Lake was 98.50 km^2, and the maximum area was 130.75 km^2 (in 2020). From 1991 to 2020, the lake area of DaSuGan Lake increased from 99.813 km^2 to 130.748 km^2, expanding by 30.935 km^2. However, by 2004, the lake area had shrunk to its lowest value of 98.5 km^2 in nearly 30 years. The years with an area below 100 km^2 included 1991, 2001, 2002, 2003, 2004, 2005, 2006, and 2007. From 1991 to 2009, the area of DaSuGan Lake entered a process of expansion, contraction, and cross expansion. After 2009, the area increased from 100.803 km^2 to 116.431 km^2 in 2018, showing an upward trend in these 8 years. From 2019 to 2020, the area increased from 122.0621 km^2 to 130.748 km^2, with an increase of 8.686 km^2, making it the fastest growing year in 30 years.

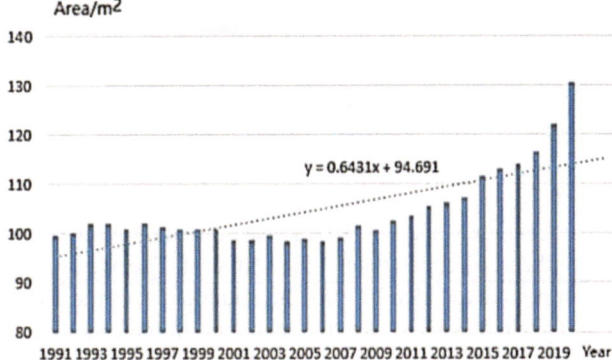

Fig. 2. Changes in Lake Area of DaSuGan over 30 Years

- **Inter monthly changes**

This article uses a multi band spectral relationship model to extract vector maps of water bodies in different time periods. Arcgis 10.8 is combined with visual interpretation to eliminate redundant interference and preserve the lake water boundaries of the DaSuGan Lake in different periods. The results of extracting water body information from the DaSuGan Lake are shown in Fig. 3.

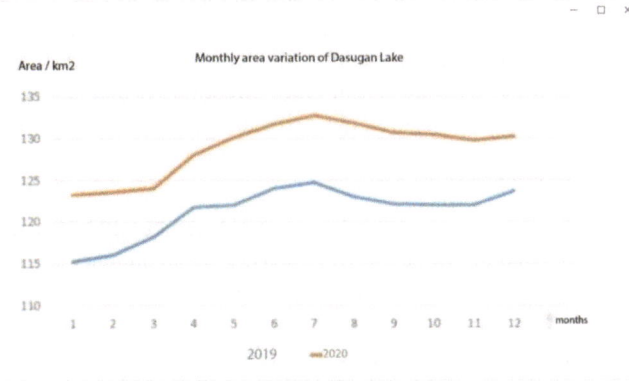

Fig. 3. Inter month Area Changes between 2019 and 2020

In December, the lake area will increase. The area of the lake is obtained by reading the area attribute of the extracted lake vector. By extracting the area of each month in 2019 and 2020, as shown in Fig. 3, it is found that the area of the DaSuGan Lake has expanded and also shrunk during the year. From two years of data, we can see that the maximum value of the year often occurs in July of summer, and the minimum value is in January. From January to July, the area shows an upward trend, and from August to November.

4.2 Spatial Changes in Lake Area

- **Inter annual changes**

From 1991 to 2020, the overall expansion of DaSuGan Lake to the east showed a trend of expansion, retreat, and expansion. DaSuGan Lake added new water bodies to the east, with the widest point reaching an astonishing 6796 m and the narrowest point reaching 1812 m. There was also slight expansion in other directions. From 1991 to 1995,the DaSuGan Lake expanded in the due east direction, with a more pronounced expansion in the southeast direction than in the northeast. The narrowest point expanded by 42 m, the widest point expanded by 1062 m, and the rest of the directions remained largely unchanged. From 1995 to 2000, the lake area in the east direction of the lake experienced a retreat of 100 m, with the maximum retreat reaching 407 m. There was no significant change in other directions. From 2000 to 2005, there was a significant retreat of 609 m at the edge of the lake in the southeast direction, and a retreat of 100 m in the northeast direction, with no significant changes in other directions. From 2005 to 2010, the lake began to expand eastward, with a much slower expansion speed in the northeast direction than in the southeast direction. The widest point reached 1120 m, and the narrowest point also reached 170 m. The shoreline of the lake can basically overlap with that of 1995, with no significant changes in other directions. From 2010 to 2015, the shoreline of the lake rapidly expanded eastward, with the expansion speed in the northeast direction being greater than that in the southeast direction. The maximum change point expanded by nearly 2000 m, the minimum also reached 500 m, and there was no significant change in the other directions. From 2015 to 2020, the east bank of the lake experienced a sharp expansion, with new water bodies emerging in the northeast direction and a large amount of water bodies also emerging in the due east and southeast directions. The expansion in the due east direction is greater than 1000 m, and the widest point is nearly 5000 m. The expansion in other directions is also about 30 m outward. The 30-year spatial-temporal variation of DaSuGan Lake is shown in Fig. 4.

- **Inter monthly changes**

By overlaying the vector areas from January to December 2020, it can be concluded that the lake experienced slight eastward expansion from January to March, with the largest expansion occurring at 523 m and minimal changes in other directions. From March to June, the lake showed significant changes in the due east direction, with the largest expansion reaching 1604 m. There have also been slight changes in the north and south directions, with changes of less than 100 m, while there has been no change in the west. Between June and July, new water bodies appeared in the northeast direction, and

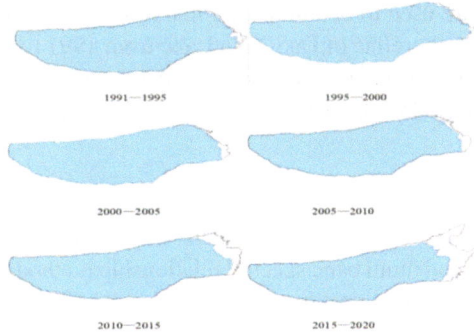

Fig. 4. Spatial and temporal variation of DaSuGan Lake in the past 30 years

the lake expanded 2236 m in the northeast direction. Expansion and retreat also occurred in other directions to the east, with the maximum retreat reaching 313 m. From July to October, the overall lake area showed a downward trend, with a relatively small retreat rate and a maximum retreat of less than 50 m. The remaining changes in direction are all within 100 m. From October to December, the newly added water in the northeast direction disappeared, while the changes in other directions were around 100 m. The spatial changes of DaSuGan Lake in 2020 are shown in Fig. 5.

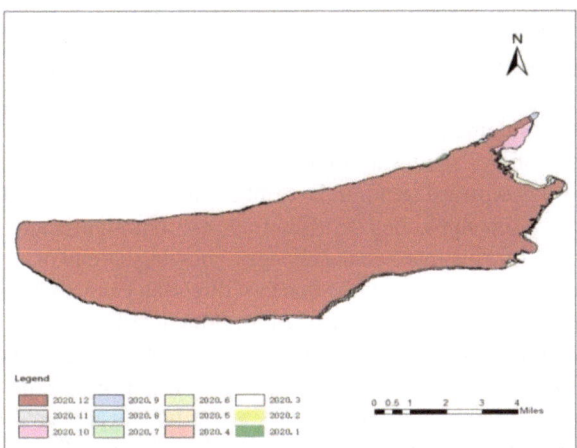

Fig. 5. Spatial Change Map of DaSuGan Lake in 2020

5 Conclusion

This article uses remote sensing data from the Landsat series and GF-6 satellites as data sources, and takes the DaSuGan Lake as the research object. Three models, namely empirical threshold method, single band threshold method, and multi band spectral relationship method, are used to obtain the optimal model through precision. Then, the changes in the area and shoreline of DaSuGan Lake from 1991 to 2020 and the changes in the area and shoreline of each month from 2019 to 2020 are estimated, and the spatial-temporal dynamic changes in the water area of DaSuGan Lake over the past 30 years are analyzed.

(1) Extraction models for three types of water bodies.

This article compares three water body models: single band threshold method, empirical threshold method, and multi band spectral relationship method. The overall accuracy of the multi band spectral relationship method reaches 99.0315%, which is better than 96.8586% and 95.6914% of the empirical threshold method and single band threshold method. The Kappa coefficient single band threshold method, empirical threshold method, and multi band spectral relationship method are 0.8893, 0.9260, and 0.9763, respectively. So overall, the multi band universal relationship method is the optimal water extraction model.

(2) The overall change of the lake in DaSuGan Lake shows an almost flat curve followed by a straight upward trend, with the lake area increasing year by year. From 1991 to 2020, the lake area in DaSuGan Lake increased by 30.935 km^2.Between 1991 and 2020, the minimum area of DaSuGan Lake was 98.50 km^2 and the maximum area was 130.75 km^2. The change in the area of DaSuGan Lake can be divided into three stages. From 1991 to 2009, the water area increased from 99.81 km^2 to 100.80 km^2, which was the basic stable period; From 2009 to 2018, the water area increased slowly from 100.80 km^2 to 116.43 km^2, marking a period of gradual increase; From 2018 to 2020, the water area expanded rapidly from 116.43 km^2 to 130.75 km^2. From the perspective of monthly changes, the maximum value within the year often occurs in July of summer, while the minimum value occurs in January. From January to July, the area shows an upward trend, and from August to November of the same year, it shows a downward trend. In December, the lake area will increase.

(3) In terms of spatial changes, the DaSuGan Lake exhibits significant expansion in the due east direction during inter annual changes, showing a trend of expansion, retreat, and expansion. The expansion in the due east direction exceeds 1000 m on average, with the largest expansion in the northeast reaching nearly 5000 m. The direction of the water outlet has added new water bodies, while other directions have expanded outward by 30 m. During the year's changes, there was a slight expansion of the lake in all directions in December. In July, new water bodies appeared in the northeast direction, and the lake expanded by 2236 m in the northeast direction. However, by November, the newly added water bodies will disappear, and there will be significant expansion and retreat in the due east direction in the remaining months.

Acknowledgment. Support: This paper is supported by the Natural Science Foundation project of Gansu Provincial Department of Science and Technology, titled "Research on the Mechanism of Ice phenological changes in DaSuGan Lake under the Background of Climate Change" (23JRRA1655).

References

1. McFEETERS, S.K.: The use of the Normalized Difference Water Index (NDWI) in the delineation of open water features. Int. J. Remote Sens. (7) (1996)
2. Frazier, P.S., Page, K.J.: Water body detection and delineation with Landsat TM data. Photogramm. Eng. Remote. Sens. **66**(12), 1461–1468 (2000)
3. Gu, Z., Zhang, Y., Fan, H.: Mapping inter- and intra-annual dynamics in water surface area of the Tonle Sap Lake with Landsat time-series and water level data. J. Hydrol. **601** (2021)
4. Chen, H., Wang, J.: Comparison of methods for extracting water body information from TM images in mountainous plateau regions. Summary of Papers at the 14th National Remote Sensing Technology Academic Exchange Conference [Publisher unknown], vol. 20 (2003)
5. Wang, L., Liu, J., Yang, F., Fu, C., Teng, F., Gao, J.: Early identification of winter wheat area based on GF-1 satellite remote sensing. J. Agric. Eng. **31**(11), 194–201 (2015)
6. Li, J., Jiang, Z., Yao, L., Jian, T.: A ship target fusion recognition algorithm based on deep learning. Ship Electron. Eng. **40**(09), 31–35+171 (2020)
7. Cohen, J.: A coefficacy of agreement for nonlinear scales. Educ. Psychol. Measure. **20**(1), 37–46 (1960)
8. Huang, J., Hou, Y., Su, W., et al.: A method for extracting the planting area of corn and soybean based on GF-1 WFV data. J. Agric. Eng. **33**(7), 164–1 (2017)
9. Nie, X.: Research on lake area extraction and dynamic changes in nanchang urban area based on landsat TM/OLI images. Donghua University of Technology (2018)
10. Huang, S., Liu, L., Dong, J., Fu, X.: Overview of ground filtering algorithms for vehicle mounted LiDAR point cloud data. Optoelectron. Eng. **47**(12), 3–14 (2020)
11. Zhuoma, L., et al.: Comparative validation study on cloud removal algorithm process of MODIS daily snow products on Xizang Plateau. Glacier Permafrost **38**(01), 159–169 (2016)

A Comparative Study of Inference Models Based on Image Recognition

Deling Zhao$^{(\boxtimes)}$

Internet Information Research Institute, Communication University of China, Beijing, China
zhaodeling@21cn.com

Abstract. Due to the limitation of web-side processing capabilities, the existing mainstream inference models adopt the client-to-server operation mode, and the emergence of WebAssembly has brought opportunities for the client to run the inference model alone for image recognition, and soon, the browser will fully support the standard. Therefore, on the basis of image recognition, this paper runs the mainstream inference model on the client and server side, and then compares the inference rate of different inference models at different batch sizes and the inference efficiency of the same inference model on the client and server side, conducts a large number of comparative experiments to record the experimental results, and then uses the method of averaging the experimental data to obtain a set of reliable experimental conclusions. Experimental results show that the performance of server-side TensorRT is better than other models under the same configuration conditions, ONNXRuntime is better than PyTorch in small batches, and PyTorch's performance improvement in accuracy has more room for optimization. The results of web-side experiments show that the inference model has a large difference in GPU usage, ONNXRuntime shows a greater performance advantage in the efficiency of GPU inference, while TensorRT will show better performance with the increase of batch. Compared with server-side inference, web-side inference shows stronger stability because it relies less on network transmission, so it can be preferred to direct inference when the batch is small, and server inference is preferred when the client configuration does not meet the conditions.

Keywords: Inference models · WebAssembly · Image recognition · TensorRT · PyTorch

1 Introduction

1.1 Research Background and Significance

At present, image recognition is a more mature branch of the field of artificial intelligence, image recognition can not be separated from the support of inference models, major companies launched inference models in the processing of a large amount of data performance difference is obvious, in order to ensure the effective use of resources, it is urgent to compare and experiment on commonly used inference models, based on real

© The Author(s) 2025
P. Siarry et al. (Eds.): WCNA 2023, LNEE 1361, pp. 284–293, 2025.
https://doi.org/10.1007/978-981-96-2409-6_28

data to find out the applicable scenarios of different inference models, so as to reasonably optimize the configuration and use of resources. At the same, time, in the process of studying the model, it was found that the model has high requirements for the device, so it is difficult to run independently on the client, most of the mainstream inference models are running on the server side or embedded devices, and then through the client-to-server interaction mode to achieve image recognition function, and the emergence of WebAssembly has brought opportunities for the client to run the inference model independently, therefore, It is also the focus of this research to explore the performance of the client-side independent operation of the inference model based on WebAssembly, and then point out the direction for the optimization of the subsequent inference model on the client.

1.2   Research Status at Home and Abroad

Current mainstream inference frameworks include TensorRT, ONNXRuntime, PyTorch, and others. Among them, TensorRT is a deep learning inference tool developed by NVIDIA, only supports inference does not support training, the underlying layer has been optimized for NVIDIA graphics cards in many aspects, and can be used in combination with the CUDA CODEC SDK to quickly and efficiently run the trained network on the GPU to generate results, usually in combination with mainstream training frameworks; The Open Neural Network Exchange (ONNX) format is a standard for representing deep learning models that enable models to be transferred between different frameworks. ONNXRuntime is an inference framework officially launched by Microsoft that can run on multiple platforms, with the best support for ONNX format input models, both CPU and GPU inference capabilities, the early ONNXRuntime only has inference functions, with the iteration of the latest version, it gradually has model training functions; PyTorch is a Python-first deep learning framework launched by Facebook's artificial intelligence research and development team in 2016, The advantages of Pytorch are that it has high flexibility, easy debugging, and fast speed, and the disadvantages of PyTorch use a lot of GPUs for training, and if the memory is not properly managed, there will be a memory overflow (OOM) problem, and CUDA unified memory allows oversubscribing tensor objects in the GPU, but the performance loss is serious [1]. In recent years, PyTorch has continued to iterate on versions and has now become one of the most popular deep learning frameworks in the world.

Image classification is one of the core tasks in computer vision and is the basis for other tasks in computer vision [2], and common vision converters often need to be pre-trained on large-scale natural image datasets (e.g., ImageNet) to achieve satisfactory performance. For natural images, the labels of the pre-trained dataset can be efficiently obtained through crowdsourcing, because even ordinary people have the ability to effectively identify and annotate objects in natural images [3]. The goal of image classification is to correctly classify images into predefined classes or categories. Categories can be anything from simple objects like "dogs", "cats", "cars" to complex scenes like "cities", "beaches", etc. With the development of embedded and the improvement of server performance, image recognition technology has made a great breakthrough in 2010, the accuracy of image recognition has reached unprecedented accuracy, and the workload of inference is mainly concentrated in the embedded system/server side, since then image recognition has been applied to mobile phones and embedded devices to bring convenience to people's lives. Image recognition is still evolving, and needs to serve more fields.

1.3   Research Content Arrangement

Based on the pre-trained model, this paper studies the performance of the inference model based on image recognition in five chapters, as shown in Fig. 1.

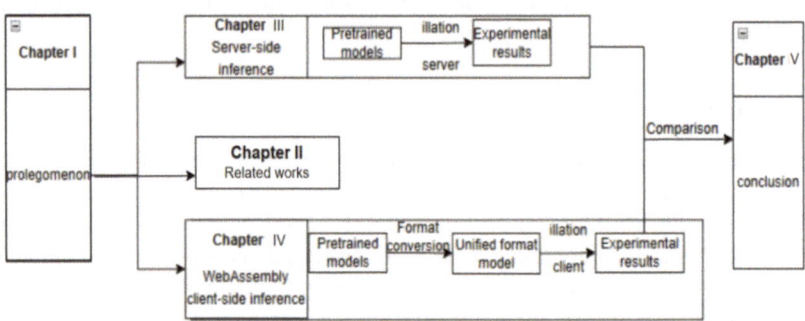

Fig. 1. Chapter organization chart

The first chapter is an introduction, first introducing the research background and significance of this paper, followed by introducing the development process and current status of the current mainstream reasoning framework, and then revealing the relationship between the development of image recognition and the inference model.

The second chapter Sect. 2 introduces the application of classical inference models in the field of image recognition and the commonly used methods of inference model comparison based on existing research cases. By evaluating the previous research, a more scientific and rigorous performance comparison method of the reasoning model was obtained for this experiment, and provided research and reference for future generations.

The third chapter is the reasoning research of the inference model on the server side, and different experimental results are obtained by selecting different inference models under the same experimental conditions for many times, and then the experimental results are averaged; Therefore, the performance differences of different inference models in the server environment are analyzed, and the applicable scenarios of different inference models on the server side are summarized.

The fourth chapter is the client-side inference mode, because the mainstream inference framework has extremely high requirements for computing power when inference, and image recognition needs to interact with the user on the client, so the inference model is deployed locally to allow users to operate locally. C/C++ and Python cannot run directly in the browser, so this experiment uses the latest standard WebAssembly, which the browser will soon support, to convert the language that the browser cannot run directly into the wasm binary format locally, so that the model can run at a near-native rate on the browser, and analyze the difference in the final experimental results.

The fifth chapter is an overall conclusion to the results of this experiment, summarizing the applicable scenarios of the mainstream inference framework used in the experiment, clarifying the performance differences, and explaining the areas worthy of improvement in the future, pointing out the direction of optimization for users.

2 Related Works

In the field of image recognition, the comparative study of inference models is a major direction of current research. This chapter will review past advancements in the comparison of inference models based on image recognition, focusing on the application of different methods, their performance comparisons, and key issues within the domain.

2.1 Introduction to the Environment

In past research, researchers have proposed various methods to assess the performance differences between different models. We will conduct a comprehensive evaluation of models based on their runtime memory and GPU memory usage. Image recognition is a fundamental task in the field of computer vision, involving the mapping of input images to predefined categories. In recent years, with the rise of deep learning, models such as convolutional neural networks (CNNs), recurrent neural networks (RNNs), and Transformers have achieved significant success in the field of image recognition. Based on the classical inference models ONNXRuntime, Pytorch, and TensorRT, we will explore their performance differences in the field of image recognition.

2.2 Comparison Methods for Inference Models

In the comparison study of inference models, researchers have proposed a variety of methods to evaluate the performance differences between different models, some common comparison methods, such as accuracy, recall, F1 score, etc., as well as applicability in different scenarios. We will perform a comprehensive evaluation of the model based on its runtime memory and video memory footprint.

2.3 Performance Comparison and Evaluation

In related work, researchers usually evaluate the advantages and disadvantages of different inference models by comparing their performance on specific tasks or datasets. This study will focus on the performance differences and limitations of each model in terms of figure-out and different application scenarios. This helps to provide readers with a reference when selecting and applying an inference model.

2.4 Performance Comparison and Evaluation

The key problem of the inference model based on image recognition is that the recognition accuracy of the lightweight model is not high, and the dependence of the heavyweight model on the video memory is high, which is difficult to run independently on the web, so it is very important to optimize the code of the model and improve the performance of the basic language.

3 Server-Side Inference Performance Research

3.1 Introduction to the Environment

In this experiment, mainstream inference frameworks were selected: ONNXRuntime, TensorRT, PyTorch;

In order to exclude the influence of other factors, the operating system used is unified as CentOS Linux 7, the server has an 8-core 32G processor, the GPU is "Nvidia 3090Ti" with ResNet, of which the cuda version is 11.6, and the CPU version used is "Intel(R) Core(TM) i7-10700 CPU @ 2.90 GHz 2.90 GHz"; The models used in the experiment are all pre-trained models with the same data set to ensure the uniformity of variables. In order to make a more fine-grained division, the experiment is divided into two different intensive reading methods: FP16 and FP32.

3.2 Experimental Results

The ONNXRuntime model was 1.13, PyTorch 2.0, and TensorRT8.2.1.8, and the comparison results are shown in Fig. 2.

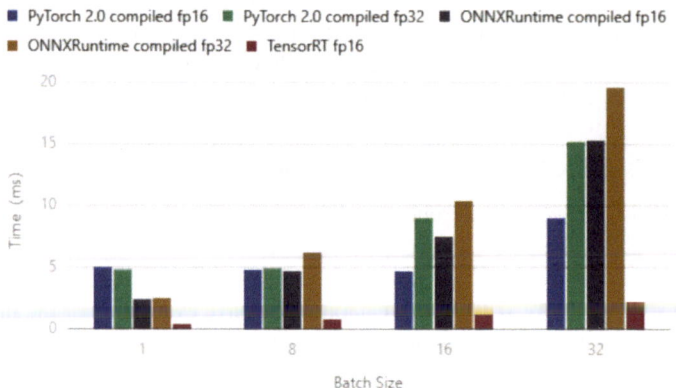

Fig. 2. Server-side inference comparison

From the comparison results, it can be seen that the larger the batch, the more obvious the speed improvement of PyTorch 2.0. The accuracy of fp16 is more efficient than the fp32 compilation in large batches, because Pytorch 2.0 compilation is mainly designed for training, and the batch size of training is generally higher than inference. Therefore, the optimization of this model can start from FP16, because large models generally use mixed precision.

ONNXRuntime performs better than PyTorch 2.0 at smaller batch sizes, while the opposite is true for larger batch sizes. This is also because ONNXRuntime is designed primarily for inference.

TensorRT performs far better than others in both small and large batches. As the batch size increases, the relative speed becomes faster. This shows that NVIDIA is able to make better use of hardware caching when inference, as the memory consumed by activation grows linearly with the batch size, and proper memory usage can greatly improve performance.

4   Research on Client-Side Inference Model Based on WebAssembly

4.1   Implementation Principle

Despite the tremendous and continuous improvement of JavaScript engines, JavaScript's performance is still insufficient for compute-intensive applications such as cryptography, graphics processing, or physics engines [4]. This is not due to inadequacies in runtime engine design but rather the inherent flaws of the JavaScript language itself [5]. Therefore, to enhance the visual experience when processing the model on the client side, this experiment integrates an advanced object detection and image segmentation model for image inference and recognition. This model is a new version introduced in early 2023, following the upgrades of its previous versions, adopting more efficient algorithms and frameworks. With the support of new features such as the Web Graphics Library (WebGL) and WebGpu, artificial intelligence neural network computing on web browsers has been widely applied in various fields, such as natural language processing [6], face recognition [7], target recognition, and chatbots [8].

Since browsers cannot run C/C++ and Python directly, they need to use model conversion techniques. At present, the model conversion technology is mainly divided into two ways, the first is the direct conversion technology [9], and the second is the model conversion technology using ONNX as the intermediate key [10]. This experiment uses the second method to convert the above inference model into a unified binary format file, and then call it through ONNX.js, which contains the WebAssembly standard, which is a new portable binary code form jointly introduced by multiple browser vendors, which can provide Web side applications with execution speed close to local layer code, and can be used as a compilation target for languages such as C/C++. Because of this, it provides a solution for applications that were previously difficult to port to the web and provides an efficient operating environment [11]. The conversion process is shown in Fig. 3.

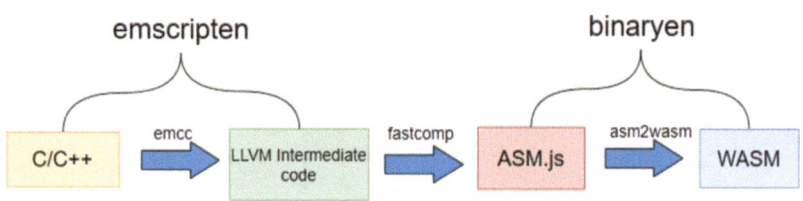

Fig. 3. How WebAssembly works

After the binary file is generated, the browser can use the ONNX js file locally to access the ONNX/WASM binary format file to run the inference model. Taking Python as an example, the process of the browser running the binary file is shown in Fig. 4.

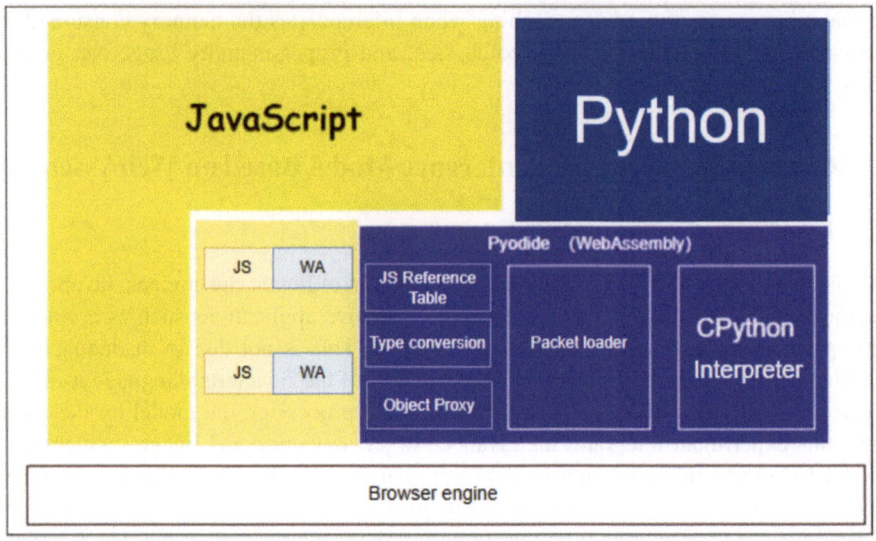

Fig. 4. How browsers run Python

4.2 Analysis of Results

Experiments are performed with the same batch volume. The experimental interface is shown in Fig. 5.

The experimental results show that when the browser has a small batch processing volume, whether it is on the client or the server, the execution efficiency of the same inference model is closer, the difference is that the direct call to CPU and GPU processing through the browser saves network overhead and reduces data transmission, so the local processing time from uploading the picture to getting the result is much smaller than that of the server's remote processing.

When comparing the processing efficiency of different inference models on the client side when they are converted to a unified time, it is found that there is little difference in CPU when the client configuration is the same. Table 1 is a comparison chart of differences on GPUs.

It can be seen that the overall performance of ONNXRuntime is better when the client memory is sufficient. TensorRT will improve performance significantly with the increase of batch processing. PyTorch's overall row performance is not high.

Fig. 5. Advanced object detection and image segmentation model application interface

Table 1. 'Web-based inference comparison

Batch		Pytorch	ONNXRuntime	TensorRT
1	Time	14.2 ms	1.5 ms	2.5 ms
	Memory	1290 M	1420 M	856 M
	GPU	14%	68%	62%
8	Time	1.8 ms	0.2 ms	0.3 ms
	Memory	1356 M	1426 M	856 M
	GPU	18%	85%	83%
16	Time	0.9 ms	0.13 ms	0.17 ms
	Memory	1446 M	1434	858 M
	GPU	25%	88%	94%

5   Conclusion

This experimental study finds that TensorRT is the best for inference when the memory is large, and when the memory is limited, PyTorch should be preferred when processing large batches of data, and ONNXRuntime should be preferred for low-batch data.

ONNXRuntime has the best performance when the client memory is sufficient. TensorRT can be considered when the video memory is small and the batch processing volume is large; PyTorch, on the other hand, is suitable for use with very low memory configurations.

In the case of small batch size and web-side support, web-side inference can be preferred to reduce network overhead and save user time.

Acknowledgements. At the completion of this paper, I would like to express my sincere gratitude to all those who have supported and assisted me throughout the research process.

First and foremost, I would like to thank my supervisor, Professor Hao Shen. Thank you for your meticulous guidance and patient explanations throughout the entire research process. Your expertise and academic insights have provided a solid foundation for my study, and I have greatly benefited from them.

I also want to express my gratitude to my fellow colleagues in the laboratory, thanking them for their assistance in academic exchanges and experimental design. Through our collective efforts, we have overcome various challenges and successfully completed this research.

Furthermore, I want to thank my family for their continuous support and encouragement. In times of difficulty and challenge, it is your unwavering support that has given me the strength to persevere until the end.

Lastly, I extend my thanks to all scholars who have contributed to the progress of humanity in the research field. It is because of your work that I have been able to stand on the shoulders of giants, delve into research, and achieve some results.

Thank you for your support and encouragement, which have allowed this paper to come to fruition. I sincerely appreciate everyone who has cared for and supported me.

This research was financially supported by "the Fundamental Research Funds for the Central Universities".

References

1. Choi, J., Yeom, H.Y., Kim, Y.: Improving oversubscribed GPU memory performance in the PyTorch framework. Clust. Comput. **26**(5), 2835–2850 (2023)
2. Chen, L., Li, Q., Dai, Q., et al.: Review of image classification algorithms based on convolutional neural networks. Remote Sens. **13**(22), 4712 (2021)
3. Li, Y., Huang, Y., He, N., et al.: Improving vision transformer for medical image classification via token-wise perturbation. J. Visual Commun. Image Represent. 104022 (2023)
4. Reiser, M., Bläser, L.: Accelerate JavaScript applications by cross-compiling to WebA ssembly. In: Proceedings of the 9th ACM SIGPLAN International Workshop on Virtual Machines and Intermediate Languages, pp. 10–17. ACM (2017)
5. Selakovic, M., Pradel, M.: Performance issues and optimizations in JavaScript: an empirical study. In: Proceedings of the 38th International Conference on Software Engineering, pp. 61–72. ACM (2016)
6. Otter, D.W., Medina, J., Kalita, J.K.: A survey of the usages of deep learning for natural language processing. IEEE Trans. Neural Networks Learn. Syst. (2020). https://doi.org/10.1109/TNNLS.2020.2979670
7. Madhavan, S., Kumar, N.: Incremental ethods in face recognition: a survey. Artif. Intell. Rev. **54**(1), 253–303 (2021)
8. Matsuura, S., Omokawa, R.: Being aware of one's selfin the auto-generated chat with a communication Robot. In: International Conference on Human-Computer Interaction, pp. 477–488 (2020)

9. Wang, Y., Yang, Z., Liu, T., et al.: Passivity and synchronization of multiple multi-delayed neural networks via impulsive control. Discret. Dyn. Nat. Soc. **8**, 1–11 (2020)
10. Lin, W.F., Tsai, D.Y., Tang, L., et al.: ONNC: a compilation framework connecting ONNX to proprietary deep learning accelerators. In: 2019 IEEE International Conference on Artificial Intelligence Circuits and Systems (AICAS), pp. 214–218. IEEE (2019)
11. Heil, S., Siegert, V., Gaedke, M.: ReWaMP: rapid web migration prototyping leveraging WebAssembly. In: Mikkonen, T., Klamma, R., Hernández, J. (eds.) Web Engineering. ICWE 2018. LNCS, vol. 10845. Springer, Cham (2018). https://doi.org/10.1007/978-3-319-916 62-0_6

Research on Intelligent Monitoring System for Double-Boom Floor-Standing Derrick Based on Distributed Wireless Monitoring Technology

Lijiang Sun[✉], Yongjun Xia, Fanhao Meng, and Wei Zhou

China Electric Power Research Institute Co., Ltd, Beijing, China
sljiang1990@163.com

Abstract. Employing double-boom floor-standing derrick in tower erection construction involves complexities in tension and challenges in achieving real-time perception, potentially complicating construction tasks or leading to accidents. In order to enhance the safety of tower erection construction with the use of double-boom floor-standing derrick, this paper proposes an intelligent, information-based wireless monitoring system based on distributed wireless monitoring technology. Through an analysis of the tower erection construction process with the use of double-boom floor-standing derrick, this study identifies safety factors and monitoring parameters that impact the working conditions of double-boom floor-standing derrick. Given the extensive and dispersed nature of monitoring points within the holding pole system, a wireless, distributed, and multi-level monitoring scheme is adopted, which can satisfy the construction requirements on tower erection construction sites while ensuring safety and stability throughout the construction process.

Keywords: Double-boom floor-standing derrick · Tower erection construction · Intelligent monitoring · Tension sensor

1 Introduction

Electricity plays a uniquely important role in safeguarding China's energy security, with the construction of power grids serving as a crucial component in power construction. The recent years have been a critical phase in the development of China's ultra-high voltage transmission and transformation projects. These projects are marked by their significant construction tasks, substantial construction challenges, and the extensive use of construction machines. Among the five primary factors that affect the construction quality and safety of these projects - "man, machine, material, method, and environment" - construction machines rank as a close second to construction personnel in terms of importance. As China increasingly prioritizes work safety, any safety accidents can lead to enormous economic losses and negative social impacts. The safety and reliability of construction machines, particularly the holding poles utilized in the construction of overhead transmission lines (hereinafter referred to as "holding poles"), are crucial for

© The Author(s) 2025
P. Siarry et al. (Eds.): WCNA 2023, LNEE 1361, pp. 294–302, 2025.
https://doi.org/10.1007/978-981-96-2409-6_29

maintaining the construction quality of projects and the safety of construction personnel. To meet the construction schedule, it is common practice to work on multiple power transmission towers concurrently. The safety monitoring of the construction process, which involves multiple holding poles working in parallel during the construction of overhead transmission lines, has garnered increasing attention [1, 2].

The double-boom floor-standing derrick serves as a crucial piece of construction equipment during tower erection construction for overhead transmission lines. Given the limitations imposed by terrain, factors such as the weight, position, and angle of the loads being hoisted on both sides of the holding pole's boom affect the lifting torque, which can negatively impact construction personnel, the process of construction, and overall safety. Currently, tower erection construction sites that use double-boom floor-standing derrick rely solely on the subjective judgment of construction personnel, based on their experience, to determine the working conditions of these holding poles, a practice fraught with subjectivity and significant safety risks [3]. The recent advancements in wireless monitoring technology for industrial applications have made the intelligent monitoring of holding poles' working conditions a reality. This paper introduces an intelligent, information-based wireless monitoring system based on distributed wireless monitoring technology, which enables real-time monitoring and early warning of parameters such as tension, inclination, height of the free section, and imbalance torque throughout the entire process of tower erection with the use of double-boom floor-standing derrick.

2 Tower Erection Construction Technology with Double-Boom Floor-Standing Derrick

The tower erection construction with the use of double-boom floor-standing derrick is primarily accomplished by components such as the amplitude luffing system, hoisting system, rotation system, guy line system, waist hoop system, and holding pole lifting system. The site layout is illustrated in Fig. 1, where the double-boom floor-standing derrick is situated at the center of the tower, with its top extending above the tower to be erected. The height of the holding pole is increased through methods such as inversion jacking or extension lifting, while its stability is ensured through the use of internal guy lines and waist hoops. Boom luffing is facilitated by the power system and luffing wire ropes, and the boom's rotation around the slewing mechanism is achieved through manual dragging or an electric slewing mechanism. Lifting is carried out through the pulley block, hook, and motorized winch located at the end of the boom [4].

3 Analysis of Safety Factors Affecting the Working Conditions of Holding Poles

3.1 Lifting Weight and Boom Inclination

Lifting weights exceeding the rated lifting capacity will result in insufficient safety reserves for the holding pole. The power industry standard DL5009.2-2013 *Code of Safety Operation in Power Engineering Construction* and various corporate safety regulations strictly mandate that the lifting capacity must be controlled within the rated limit.

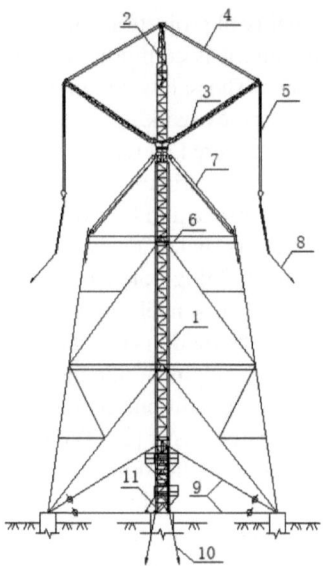

1 - Pole Body; 2 - Mast; 3 - Boom; 4 - Luffing Pulley Block; 5 - Lifting Pulley Block; 6 - Waist Hoop; 7 - Holding Pole Guy Line; 8 - Control Rope; 9 - Anchoring Rope; 10 - Lifting Traction Rope; 11 - Hydraulic Jacking Frame

Fig. 1. Layout Diagram of A Double-boom floor-standing derrick for Tower Erection

Additionally, the lifting weight and the boom inclination can be jointly used to determine the holding pole's imbalance torque. If the imbalance torque during the operation of the holding pole exceeds its rated imbalance torque, the holding pole could experience significant eccentric loading, which will affect its stability and potentially cause accidents. Therefore, it is essential to monitor both the lifting weight and the boom inclination of the holding pole.

3.2 Verticality of Pole Body

As the holding pole is a towering structure, both its body and boom are subject to deformation from lifting loads and their own weight. Insufficient verticality can cause the holding pole to experience eccentric compression under working conditions, creating an eccentric distance that impacts its normal operation.

This paper introduces an electrical lifting scheme for the double-boom floor-standing derrick. Following the elevation of the holding pole, this approach can be used to monitor its balance condition and aid in aligning the pole into a vertical stance. Currently, following the hoisting of the holding pole, its verticality is monitored in both the along-track and cross-track directions using a theodolite. Once the holding pole is aligned vertically, the top layer waist hook and the luffing pulley blocks on its booms are tightened and secured. However, employing a theodolite for measurements demands a high level of operator skill, entails an intricate procedure, and its accuracy largely depends on the subjective judgment of operators, leading to results that are not intuitively evident. Therefore, there arises a necessity to develop a new type of monitoring device specifically tailored for monitoring the verticality of the pole body.

3.3 Tension in Guy Line and Inclination of Internal Guy Line

During tower erection, the internal guy line serves to prevent bending and deformation of the holding pole at its standalone height. The guy line bears substantial stress during this process. Should the load on the guy line exceed the limit, it could result in a swift change in the condition of the holding pole, potentially leading to instability. The power industry standard DL/T 319-2018 *General Technical Conditions and Test Methods for Holding Pole of Overhead Transmission Line Construction* mandates that the safety factor of pole components and flexible attachments must not fall below 3.

The inclination of the internal guy line relative to the horizontal plane impacts the lateral displacement of the free section, which in turn affects the stability of the holding pole body and the tension in the internal guy line. Thus, monitoring both the tension and the inclination of the internal guy line is essential for double-boom floor-standing derrick.

3.4 Height of Free Section

The cross-sectional area of the holding pole remains constant. As the height of its free section increases, this leads to an increase in the slenderness ratio, consequently impacting the stability and safety factor of the holding pole. The power industry standard DL/T 319-2018 *General Technical Conditions and Test Methods for Holding Pole of Overhead Transmission Line Construction* mandates that the stability and safety factor for a dual-boom holding pole must not fall below 2.

3.5 Wind Speed

According to the power industry standard DL 5009.2-2013 *Code of Safety Operation in Power Engineering Construction - Part 2: Power Transmission Line*, operations such as outdoor hoisting, tower erection, and line erection are prohibited under conditions of force 6 winds or stronger. Therefore, monitoring wind speeds at construction sites is crucial to ensure that operators can adjust their operations promptly.

The abovementioned analysis underscores the critical importance of monitoring the lifting weight of the double-boom floor-standing derrick, its boom inclination, the verticality of the pole body, the tension in the guy line, the inclination of the internal guy line, and the height of the free section. Besides, the impact of wind speed on construction must also be taken into account.

4 Overall System Scheme Design

4.1 System Overview

The intelligent monitoring system for double-boom floor-standing derrick primarily consists of the edge computing module, wireless force sensor for lifting weight, wireless sensor for boom inclination, wireless sensor for pole body inclination, wireless force sensor for guy lines, integrated sensor for guy line's tension and inclination, wind speed sensor, data reception module, and display terminal (see Fig. 2). This system enables remote wireless monitoring of factors affecting the safety of construction employing double-boom floor-standing derrick, including imbalance torque, inclination of the pole body and the boom, guy line's tension and inclination, lifting weight, and wind speed. Moreover, it supports edge analysis of monitoring data and, utilizing built-in algorithms and safety thresholds, triggers audible and visual alarms for warnings. This system is applicable to various models of double-boom floor-standing derrick.

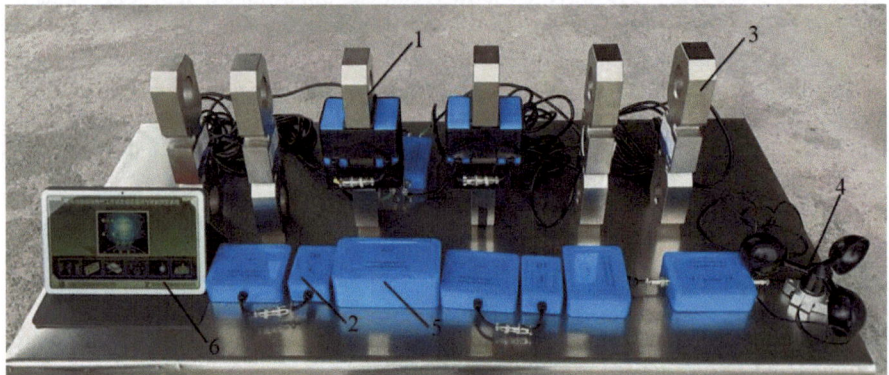

1- Wireless Force Sensor for Lifting Weight; 2-Wireless Sensor for Boom inclination; 3-Wireless Force Sensor for Guy Lines; 4. Wind Speed Sensor; 5-Gateway; 6-Display Terminal

Fig. 2. Composition of An Intelligent Monitoring System for Double-boom Floor-standing derrick

The system implements a layered architecture for real-time, online safety monitoring of tower erection construction using holding poles. Designed to meet engineering requirements, this system offers multiple categories of monitoring data, rapid data transmission, extensive communication range, low power consumption, multi-dimensional energy management, and plug-and-play among a diverse array of features. It not only meets the needs of on-site construction personnel for safe construction informed by monitoring data but also preserves the monitoring data, making it accessible for remote management personnel to query history. The system enables the optimization of construction plans and aids in accident prevention.

4.2 System Scheme Design

The intelligent monitoring system for double-boom floor-standing derrick adopts a five-layer architecture, comprising the data acquisition layer, data computation layer, data transmission layer, on-site man-machine interaction layer, and data storage and analysis layer. Figure 3 illustrates the configuration of the safety monitoring system employed during constructions involving double-boom floor-standing derrick. The data acquisition layer acquires critical data throughout the operation of holding poles (lifting weight, boom inclination, posture of the holding pole, tension in the guy line, inclination of the guy line, height of the free section, wind speed, etc.). The data computation layer processes acquired sensor data, which involves filtering and other computational analysis. It generates safety-related conditions of the holding pole for respective nodes, including lifting weight, boom inclination, posture of the holding pole, tension in the guy line, inclination of the guy line, height of the free section, and wind speed. Such data is then encoded and transmitted wirelessly. To support extended construction periods, energy management is implemented for monitoring nodes on the holding pole to achieve prolonged standby operation. The data transmission layer wirelessly transmits the results from the data computation layer to data receiving and processing units for subsequent analysis and display. The on-site man-machine interaction layer facilitates main-machine interaction. It allows on-site personnel to configure system parameters and provide on-site alarms. The data storage and analysis layer stores on-site data, enabling future remote access and analysis through internet, and providing remote alarms as well [5–7].

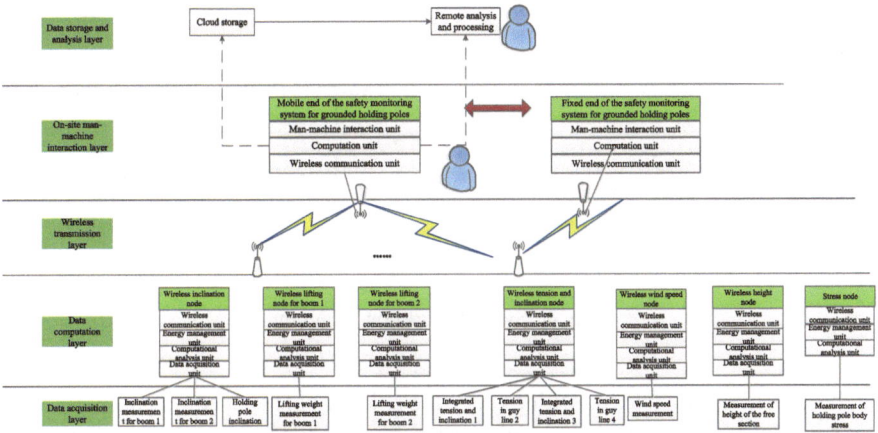

Fig. 3. General Scheme of the Intelligent Monitoring System

Figure 4 shows the workflow of the system, which utilizes a 433 MHz frequency band for wireless communication to receive data from wireless nodes. The data is then transmitted to the LCD screen via the 232 port for display. These nodes can be programmed through a USB serial port and feature energy management capabilities.

The system comprises two main components: wireless nodes and wireless man-machine interaction modules. Upon being mounted on the holding pole and powered on, the wireless nodes start by clearing the EEPROM's contents. The EEPROM is designed to retain its contents during power loss, making it ideal for storing parameters of the holding pole, benchmark information, calibration results, and other data. However, it is necessary to clear these contents before new measurements are taken to ensure that the monitoring of a previous holding pole construction does not affect subsequent monitoring data. Parameter settings, including calibration values, dimensions, thresholds, and the range of safe postures, are entered into the control unit through both offline and online methods. In offline mode, settings are uploaded to the controller using a programming interface as MCU firmware. This method allows for configuration only while offline

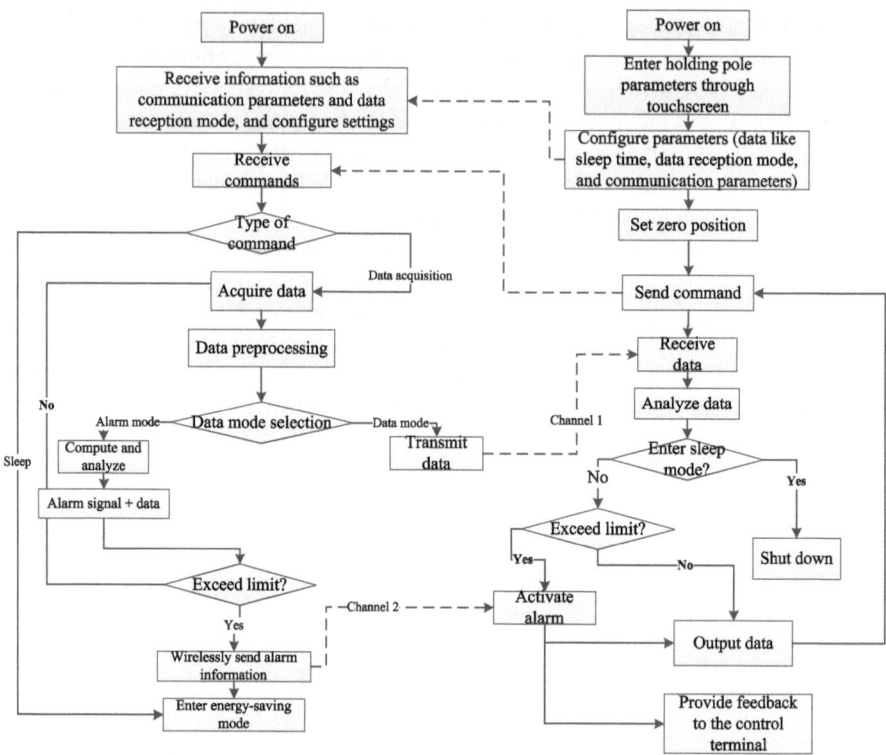

Fig. 4. System Workflow Diagram

and becomes inaccessible for settings once the holding pole is operational. In the online mode, settings are configured through a centralized controller that communicates with nodes wirelessly at a 433MHz frequency. This approach facilitates both online and offline settings. For offline calibration, the system acquires initial force data and the gesture of the holding pole as benchmarks, and stores such information in EEPROM. It receives online calibration signals, retrieves the calibration start signal wirelessly, and stores the stress and posture of the holding pole at that moment in the EEPROM as benchmarks. Given that the holding pole works mid-air, online calibration is activated by construction personnel and executed through commands sent to the nodes. Data acquisition is accomplished through the I/O port of the controller, and the data values are computed. These values are then compared against preset values to determine whether they exceed the safety threshold. Should they exceed the threshold, alarm information along with the relevant data is sent. Otherwise, only the data is sent. The energy-saving mode employs a two-tier power consumption management approach: the first tier utilizes MCU's power-down mode; simultaneously, the second tier turns off the power supply to the controller, thereby reducing minimum power consumption to microampere levels. Following a preset power-down time, the controller is scheduled to start. This approach ensures that all wireless nodes have an extended operating time [8, 9].

Upon powering on the wireless on-site interface module for holding poles, users can configure the pole and man-machine interaction parameters (display method, data accuracy, etc.) through the touchscreen. The set parameters of the holding poles are then wirelessly sent to stress sensor nodes at a frequency of 433 MHz. Upon reception, the stress nodes configure their measurement parameters accordingly. On-site personnel, according to the working conditions of the holding poles, can activate an online calibration signal through the touchscreen. This signal is wirelessly transmitted to the wireless nodes to start the calibration process. Following this, the system receives data from the nodes and activates the energy-saving mode. Meanwhile, the LCD screen displays various parameters such as the lifting weight, tension, boom inclination, posture of the holding pole, height of the free section, wind speed, and more. Should any of these values exceed the safe operating range, an alarm is triggered to alert on-site personnel. In the future, received data can be uploaded to clouds/computers/mobile phones through means like Wifi/Wan/3G/4G. Remote management personnel can access the data through the network, engaging in data management and analysis to offer big data support for the overall operation of the entire holding pole [10].

5   Conclusion

This paper investigates the process of tower erection construction with the use of double-boom floor-standing derrick, along with safety factors throughout the entire process. It identifies monitoring parameters that affect the working conditions of holding poles and develops an intelligent monitoring system for double-boom floor-standing derrick. This system enables real-time monitoring and early warning of critical information throughout the process of tower erection construction with the use of double-boom floor-standing derrick, thus enhancing the safety of tower erection construction using holding poles.

Acknowledgments. Authors gratefully acknowledge the financial support of Science and Technology Research Project of State Grid, grant number 5108-202218280A-2-327-XG.

References

1. Zhang, F., Wan, J., Chen, G., et al.: Optimization research on technical standard system for assembly and erection construction of the steel towers of overhead transmission line. Electric Power **50**(11), 59–64 (2017)
2. Zhao, S., Li, X., Zhang, C.: Construction technology and control method of derrick supported by inner suspension & outer backstay for tower erection of 750kV transmission line. Electric Power **42**(6), 89–91 (2009)
3. Sun, W., Duan, F., Peng, L., et al.: The Safety planning for erection of the tower for Zhoushan island-mainland large-crossing transmission project. Electric Power Construct. **32**(4), 113–116 (2011)
4. Zhao, Y.: Research and application of tower erection technology of UHV AC line. North China Electric Power University, Beijing (2014)
5. Yu, S.: Application of wireless video monitoring in transmission and distribution constructions. Electric Power Construct. **2**, 41–43 (2008)

6. Xu, G., Lv, C., Xiao, G., et al.: Development and application of monitoring system for assembling tower with inside suspension pole. Electric Power Construct. **33**(9), 106–108 (2012)
7. Marcus, H., Donald, H.: Underground wireless data transmission using 433-MHz LoRa for agriculture. Pubmed **19**(19) (2019)
8. Gambi, E., Montanini, L., Pigini, D., et al. A home automation architecture based on LoRa technology and message queue telemetry transfer protocol[J]. SAGE Publications, 2018,14(10)
9. He, C.: Research on the energy efficiency of MAC protocol in wireless sensor networks. Harbin Institute of Technology, Harbin (2011)
10. Huang, M., Tang, B., Li, F.: Design of wireless intelligent monitoring system for levitating braced transmission tower. Mach. Des. Manufac. **07**, 206–212 (2022)

Indoor UWB Localization Algorithm for Multi-robot and Its Application

Shuo Wang, Zhihao Xu, Kaiwen Liu, Yuanjiang Liao[✉], Xiaorong Li, and Yule Zhong

Institute Robotics, Ningbo University of Technology, Ningbo, China
liaoyuanjiang734@163.com

Abstract. Multi-robot are difficult to accurately locate in complex indoor space environments, with high real-time requirements and the inability to utilize satellite signals, thus limiting their application range. This article studies relevant algorithms based on the demand for indoor localization solutions in open intelligent manufacturing factories, existing UWB localization methods and principles. By measuring the distance value between the tag and the base station, and then using these distance values to calculate the tag coordinates, localization function can be achieved. Designed a localization scheme, built an indoor multi-robot UWB localization system, achieved precise positioning, and successfully verified the effectiveness of its algorithm. By combining multi-robot motion control technology with UWB positioning, it can be applied to indoor logistics transportation, achieve automatic data collection, and reduce costs and increase efficiency.

Keywords: uwb targeting scheme · localization algorithm · indoor localization · multi-robot

1 Introduction

With the wide application of multi-robot systems in smart manufacturing factories, a large number of application scenarios based on location awareness have emerged, and the related localization technology plays an increasingly important role. Further, in indoor environments, the continuous and reliable provision of location information can bring more application scenarios for various users.

This paper studies the high-precision indoor positioning technology of multi-robot system. The main application scenario is the spatial positioning of multi-robot in the intelligent manufacturing workshop, such as accurately transporting materials to the required position in the workshop.

Existing indoor positioning technologies include RFID positioning, ZigBee indoor positioning technology, LED positioning system and UWB technology. RFID has the advantages of high data transmission rate, good security and free from non-line-of-sight communication problems; The disadvantage is that a large number of readers and electronic tags need to be deployed in the location area. ZigBee Master' To be used in the field of low cost, low data transmission and low power consumption wireless

© The Author(s) 2025
P. Siarry et al. (Eds.): WCNA 2023, LNEE 1361, pp. 303–313, 2025.
https://doi.org/10.1007/978-981-96-2409-6_30

sensor networks, the vast majority of user intelligent devices in the general business field do not support this protocol. The advantage of the LED positioning system is that it can rely on the LED lights already deployed indoors, and is not affected by the interference of radio waves; The disadvantage is that LED lights are required to have flicker coding function, and the positioning accuracy will drop sharply due to non-line-of-sight communication problems [1]. UWB technology has the advantages of high multi-path resolution and strong anti-interference ability, so UWB is used as the core to research positioning algorithms and design positioning systems. The diversified application of UWB technology can deeply participate in core links such as production, transportation, supervision and safety, help customers operate efficiently and work safely, achieve cost reduction and efficiency increase, and provide more accurate, convenient and reliable services for indoor application scenarios.

2 UWB Positioning Method

2.1 Localization Algorithms

UWB technology uses TWR ranging algorithm and trilateral positioning algorithm to locate.

The trilateral location algorithm is based on the number of base stations in the range. If the number of base stations is less than 3 [2], it cannot be solved; If the number of base stations is greater than 6 [3], you need to exclude the base stations with the largest ranging value each time until the number of base stations is 6. After obtaining all the base stations involved in the solution, combine every three.

There are three non collinear base stations and a label X on the plane. The measured distances from the three base stations to the label X are r1, r2, r3, respectively. With the coordinates of the three base stations as the center of the circle and the distance from the three base stations to the unknown terminal as the radius, three intersecting circles can be drawn. As shown in Fig. 1, the coordinates of the label are the intersection points of the three circles [4].

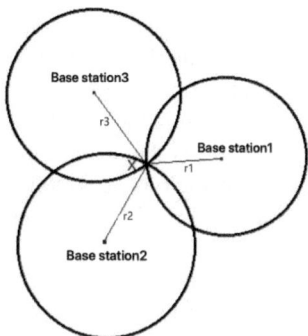

Fig. 1. Schematic diagram of the trilateral localization algorithm

Three base stations to first determine whether the three base stations can form a triangle, used to exclude the three base stations of the same straight line, and then to determine whether the maximum length of the triangle formed by the three base stations will be less than all base station ranging value, if so, the ranging does not meet the conditions, the label is not surrounded by the three triangles in the surface. Finally, determine whether the distance measurement circle between the two base stations intersects, if they intersect, the coordinates can be solved.

N coordinate points (xi, yi) can be calculated from the combination of n base stations that meet the filtering conditions, and then the filtering method is as follows:

1. The distance between the base station and the tag obtained from the ranging algorithm is ri (i = 1, 2, 3), then Eq. (1) can be listed:

$$\begin{cases} (x - x_1)^2 + (y - y_1)^2 = r_1^2 \\ (x - x_2)^2 + (y - y_2)^2 = r_2^2 \\ (x - x_3)^2 + (y - y_3)^2 = r_3^2 \end{cases} \tag{1}$$

2. Differentiating each of the first 2 equations from the third equation yields:

$$AX = B \tag{2}$$

Where A, X and B are respectively

$$A = 2 \begin{pmatrix} x_1 - x_3 y_1 - y_3 \\ x_1 - x_3 y_1 - y_3 \end{pmatrix} \tag{3}$$

$$X = \begin{pmatrix} x \\ y \end{pmatrix} \tag{4}$$

$$B = \begin{pmatrix} x_1^2 - x_3^2 + y_1^2 - y_3^2 + r_3^2 - r_1^2 \\ x_2^2 - x_3^2 + y_2^2 - y_3^2 + r_3^2 - r_2^2 \end{pmatrix} \tag{5}$$

3. Use the least squares method to calculate the label coordinate X:

$$X = (A^T A)^{-1} A^T B \tag{6}$$

Calculate the difference between the real coordinates of the localization target and the position coordinates estimated iteratively. If the value is larger than the preset threshold, the coordinate points are eliminated, and the above steps are repeated after elimination; otherwise, the first coordinate points are output as the initial value of the Taylor's solution [5]. After obtaining the initial value of the Taylor's solution, the solution is performed iteratively according to the minimum quadratic variation criterion. If the error of the

Taylor solved value is smaller than the threshold value or larger than the number of iterations, the iteration is stopped and the final coordinates are output.

2.2 UWB Communication Principle

UWB is different from the traditional communication technology that uses a high-frequency carrier to modulate a narrowband signal. It realizes wireless transmission by sending and receiving extremely narrow pulses with nanosecond or below microsecond. Therefore, it can realize ultra-wideband in the spectrum, and the bandwidth used is more than 50 MHz. Even if wireless communication is used, the data transmission rate can still reach hundreds of megabits per second.

UWB technology can be used to transmit signals over a very wide bandwidth.

In HDS-TWR mode:

The work flow of the master base station: After the ranging is enabled, select the tag currently measured as the ID, send an Inform packet to the peer, and enable the receiver to wait for the tag to respond to the Poll packet. After receiving the packet, send the Resp packet back according to the set delay, and open the Final packet sent by the receiver tag. After receiving the final packet, the primary base station sends the distance echo request packet to the secondary base station participating in the ranging in turn according to the current secondary base station situation, and receives the reply packet from the secondary base station to obtain the ranging value of the secondary base station. After all the secondary base station ranging is obtained, if the ranging is performed, it will be reported directly, and the coordinates will be uploaded after solving the coordinates.

The workflow of the secondary base station: it is always in the monitoring state and performs different actions according to the received packets. It will respond to the Inform packet of the primary base station. It will judge the involvement of the base station in ranging based on the byte corresponding to the packet. If yes, after receiving the Poll packet of the tag, it will send the Resp packet according to the set delay time, and wait for receiving the Final packet to calculate the ranging value. If not, it will not be sent; Secondly, it will respond to the request packet sent by the primary base station and send back the reply packet of the current ranging.

The tag's workflow: After receiving the correct Inform packet sent by the master base station, the tag will send a Poll packet to enable this ranging. After sending the Poll packet, the receiver will open for about 10ms to receive the Resp packet. After 10ms, the final packet sending time will be estimated and written into the final packet, and the final packet will be sent later, as shown in Fig. 2.

A. Ranging algorithms

UWB is effectively an electromagnetic wave signal. The specific process of unilateral ranging is to use two devices that support UWB, one is used to transmit UWB signals and the other is used to receive UWB signals. The two devices record the time of transmission or reception respectively.

The propagation time of UB signals can be obtained according to the difference between the two times. The distance between the two devices is the product of this time and the propagation rate of electromagnetic waves. This method requires high time synchronization of equipment transmission and reception, and the use of bidirectional

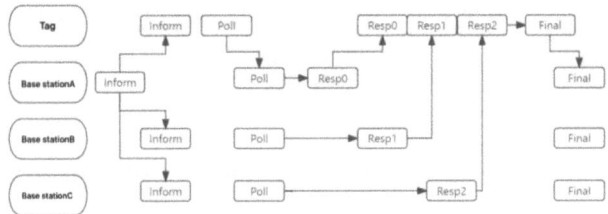

Fig. 2. Communication illustration

ranging algorithm can reduce the error caused by time synchronization. Bilateral bi-directional ranging on the basis of unilateral bi-directional ranging, but also increased the process of sending a signal to the receiving device, which can reduce the impact of clock drift on the delay accuracy, so this experiment selects the bilateral bi-directional ranging, as shown in Fig. 3.

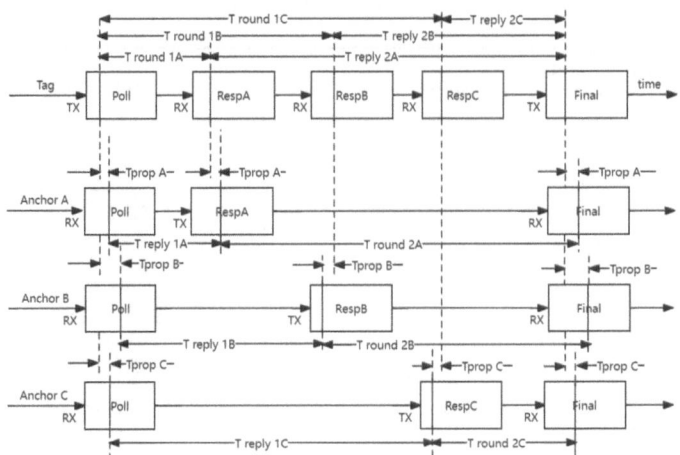

Fig. 3. Ranging schematic

The time of T round is the time from the start of sending P to receiving R to replying, and the time of T reply is the time from receiving P to sending R. According to the schematic diagram, A receives the response again to complete a ranging, when A and B send and receive data each time. The current timestamp must be recorded. The transmission time difference can be obtained by subtracting the time stamp:

$$\text{TpropA} = \frac{T_{\text{round1A}} * T_{\text{round2A}} - T_{\text{reply1A}} * T_{\text{reply2A}}}{T_{\text{round1A}} + T_{\text{round2A}} + T_{\text{reply1A}} + T_{\text{reply2A}}} \tag{7}$$

$$\text{TpropB} = \frac{T_{\text{round1B}} * T_{\text{round2B}} - T_{\text{reply1B}} * T_{\text{reply2B}}}{T_{\text{round1B}} + T_{\text{round2B}} + T_{\text{reply1B}} + T_{\text{reply2B}}} \tag{8}$$

$$T_{\text{propC}} = \frac{T_{\text{round1C}} * T_{\text{round2C}} - T_{\text{reply1C}} * T_{\text{reply2C}}}{T_{\text{round1C}} + T_{\text{round2C}} + T_{\text{reply1C}} + T_{\text{reply2C}}} \tag{9}$$

TpropA, TpropB, and TpropC are the flight times of UWB signals from labels to base stations A, B, and C. Multiplying the flight times and the electromagnetic wave transmission rate is the distance between nodes.

3 System for the Validation of Localization Algorithms

3.1 Hardware Systems

1. Self-balancing robots

The self-balancing robot uses STM32F405RGT6 as the controller, based on high-performance Arm ® Cortex B- M4 32-bit RISC core, operating frequency up to 168 MHz [6], The DSP instruction and memory protection unit enhance the security of the application [7].

Table 1. Hardware list

The name of the item	Quantity	Remarks	Appearance
The main base station	1	Base stationA	
Sub base station	4	Base station B、C、D	
Tags	2	Tag 0、1	
Self-balancing robots	2	Without	

2. Bluetooth module

Through the Bluetooth module, the mobile phone APP can be used to adjust PID parameters, monitor the self-balancing robot and control its movement.

3. Primary and secondary base stations

The positioning system takes the main base station as the control core, the main base station first sends out the ranging start information, and then the label and the base station start ranging.

After the completion of ranging, the main base station sends a request packet to the secondary base station, and all the secondary base stations return the ranging to the main base station for processing and output.

4. Labeling

The tag sends a Poll packet to all base stations and measure the distance between the base stations (Table 1).

3.2 Communication Between the Localization System and the Bluetooth Module

The AT instruction set is sent from the terminal equipment to the terminal adapter or data circuit terminal equipment [8]. The AT command function can set the parameters of the Bluetooth module, but it can only be used when the Bluetooth is not connected, and the parameters can be set through serial port transmission [9].

4   Positioning Validation Platform

The positioning principle is to locate the label within the range of the manually deployed base station. The user needs to measure the coordinate position of each base station after determining the location of the base station [10]. When building the system, as the physical properties of the UWB signal will be affected by the conductor (human body, ground wall), the equipment shall be installed in a 30cm environment as free as possible. In order to obtain good data results, open signal transmission space shall be kept between modules as far as possible.

Two dimensional positioning requires at least three base stations and one label [11]. It is recommended to build it in an open area of more than 5 * 5m. This ranging experiment uses four base stations for three-dimensional positioning, as shown in Fig. 4:

When building, the main base station (A) uses a USB cable to connect to the PC end. The USB port is also used for communication and charging functions. The secondary base station (B, C, D) and tags are directly connected to the mobile power supply. After the site is arranged, the upper computer is used to configure simple parameters and locate the main base station. The main base station is connected to the PC end using the equipped Android data cable.

The tag can obtain the last ranging information result sent by the master base station during ranging and positioning, and the serial port outputs the positioning and ranging information of this time. Two data output formats can be selected: ASCII code direct output and Modbus protocol output. Ranging and positioning data can be obtained through

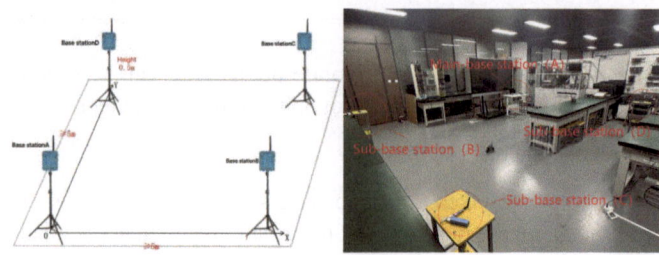

Fig. 4. Three-dimensional construction modeling diagram and physical schematic

manual access. In order to ensure the real-time positioning data, this experiment adds the mode of automatic data output on the original basis, which does not follow the Modbus question and answer mode, and saves communication overhead.

As a one-to-one protocol, Modbus the master station to send a message, and the slave station therefore sends back relevant information. Therefore, a Modbus class is designed, which contains the Modbus ID, register address, register number and other relevant information required by the relevant Modbus. In the main program of software, the instance of this class can be regarded as an interaction of information between fixed function codes [12].

The protocol for automatic data output also follows the Modbus protocol. After the positioning command is sent, the equipment will continuously carry out ranging/positioning, and will automatically output the measured data. The specific example is as follows:

Configure the master base station automatic output protocol to enable the output of positioning, ranging, and received signal information.

Send (Hex): 01 10 00 72 00 01 02 00 07 ED 40

Reception (Hex): 01 10 00 72 00 01 AI D2

In 2.2, the communication between PC upper computer and self-balancing robot HC-08

Bluetooth module has been established. Send the positioning data from the master base station (base station A) to the Bluetooth communication program through the automatic output data protocol of the Link PG system, and use the automatic output data protocol to achieve positioning data transmission, Write the interface program of the automatic output data protocol into the transmission data part of the Bluetooth communication program to realize the transmission of the tag positioning data calculated by Base Station A to the Bluetooth module of the self-balancing robot. The core idea of the program is as follows:

- 1. Assign values to interface variables.
- 2. Send positioning commands.
- 3. Receive command replies.
- 4. Receive automatic output of positioning data.

The software development environment is WIN10 + VS2019 + Framework 4.7.2 [13].

The main test platform used in the experiment is the upper computer software programmed by C # Winform. The main function of the upper computer is to control the positioning system and display the positioning results by connecting with the serial port of the main base station, and the data sent and received by the serial port. The design function of the upper computer is mainly serial port Modbus communication and data transmission drawing display. The device parameter interface of the upper computer is shown in Fig. 5. The location mode, ranging mode, the distance between the tag and each base station, the coordinates of the tag and other parameters are displayed.

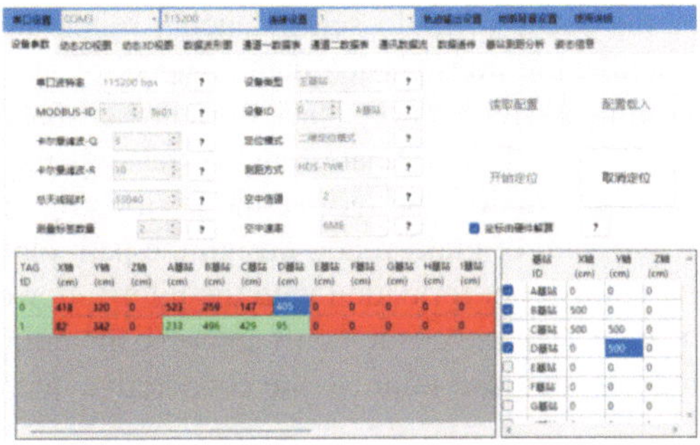

Fig. 5. Interface diagram of the parameters of the upper computer device

This experiment through the control of the self-balancing robot carrying tags to the specified location to simulate the transportation and delivery of materials in the factory scene and real-time localization feedback, the experimental process and localization feedback is shown in Fig. 6.

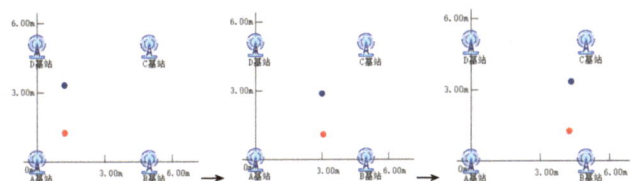

Fig. 6. Displacement diagram of self-balancing robots carrying tags for transportation

As shown in Fig. 7, the real-time localization data displayed in the test platform is accurate and the error is small, indicating that the localization algorithm validation

Fig. 7. Upper computer dynamic 2D view interface

platform effectively verifies the relevant localization algorithms, and if it is applied in open indoor environments it can timely and accurately reflect the positional information to the monitors, which can be applied in the self-balancing robots transported in the logistic information and also cooperate with the users to interact with each other.

5   Summarizing and Looking Forward

In this paper, the indoor positioning algorithm is verified by a positioning platform built by multi self-balancing robots. Based on the UWB positioning algorithm, the platform simulates the positioning scene of multi self-balancing robots in an open intelligent factory, and uses the automatic transmission data protocol of the LinkPG positioning system to achieve real-time positioning feedback of multi self-balancing robots [14].

Subsequent experiments will investigate how to transmit the positioning data of multi-robot to the main base station and receive the positioning data of multi-robot in real time through the program of multi-robot; design the path planning of multi-robot to form the intelligent multi-robot formation with the pilot-follower strategy and other functions. This application scenario can not only improve the efficiency of multi-robot transportation of materials, but also improve the cooperative operation of multi-robot [15].

References

1. Ten domain science and technology Comparison of mainstream indoor positioning technologies in 2020. https://zhuanlan.zhihu.com/p/150056792
2. Miao, X., Yang, M.: Chapter 165 Interference Analysis Between Satellite and 5G Network. Springer Science and Business Media LLC (2021)
3. Brito, J.M. C.: Technological trends for 5G networks influence of e-health and IoT applications. Int. J. E Health Med. Commun. (2018)
4. Jiang, Y., Hao, L., Han, X., Tian, H.: Indoor and outdoor joint location algorithm for power smart supply chain park. In: 2022 4th International Conference on Intelligent Information Processing (IIP) (2022)
5. Zuo, P., Li, Y.: A hybrid TOA/TDOA localization algorithm based on modified taylor weighting technique under NLOS conditions. In: 2023 4th Information Communication Technologies Conference (ICTC) (2023)
6. Skordilis, Z.I., Tsiami, A., Maragos, P., Potamianos, G., Spelgatti, L., Sannino, R.: Multichannel speech enhancement using MEMS microphones. In: 2015 IEEE International Conference on Acoustics, Speech and Signal Processing (ICASSP), 2015Hong Ai, Xuebin Cheng. "Research on embedded access control security system and face recognition system", Measurement (2018)
7. Shen, H.: Application of STM32 single chip microcomputer and full case practice. Electronic Industry Press, Beijing (2017)
8. Ai, H., Cheng, X.: Research on embedded access control security system and face recognition system. Measurement (2018)
9. Bluetooth transparent module HC-08 tutorial and simple application, CSDN (2021)
10. Neural Information Processing. Springer Science and Business Media LLC (2024)
11. Jia, T., Qiao, J., Guo, F.: Research and design of substation positioning system based on ultra-wide band. J. Phys. Conf. Ser. (2023)

12. Zhu, M., Shi, Y.: Application of ModBUS protocol communication. Microcomput. Inform. **2004**(06), 9–11 (2004)
13. Windows BLE programming net winform connecting Bluetooth 4.0, Blog Garden, 19 September 2018. https://www.cnblogs.com/webtojs/p/9675956.html
14. Xiao, Z., Wang, Y., Tian, B.: Research and application of ultra wideband positioning: review and prospect. J. Electron. **39**(01), 133–141 (2011)
15. Liu, S.: Design of intelligent multi robot formation obstacle avoidance control system. Autom. Instrument. **34**(01), 28–32 (2019)

A Configurable and Automated Testing Framework for Hardware Trojan Detection in FPGAs

Xiaodong Li, Song Chai$^{(\boxtimes)}$, Liwei Wang, and Hua Wang

Southwest Minzu University, Chengdu, China
s.tschai@swun.edu.cn

Abstract. With the widespread adoption of FPGA, the inevitable rise of security threats, including Hardware Trojan, poses significant risks. Detecting such trojan is crucial as they can lead to severe consequences. Consequently, various detection methods have been proposed. However, conventional approaches typically require extensive datasets for model training. To tackle this issue, this paper presents a customizable Hardware Trojan detection framework. This framework is designed to generate ample samples for FPGA hardware security testing, ensuring robust detection capabilities.

Keywords: FPGA · Automatic Testing · Hardware Trojan

1 Introduction

FPGA (Field Programmable Gate Array), through the utilization of specific hardware description languages, allows for flexible adjustments to its internal circuit structure [1]. Compared to dedicated or general-purpose chips, FPGA demonstrates advantages such as multi-threaded processing capabilities, efficient parallel computing mechanisms, and significantly shortened design cycles. Therefore, they have gained widespread application in the field of information systems [2]. For example, Xilinx's Virtex series is applied to high-performance real-time data processing, Kindex series is applied to control algorithms in hardware accelerated smart grids. Intel's Stratix series is applied to achieve real-time control systems, process sensor data, and monitor device status, Agilex series is applied to accelerate various computing tasks in servers, including artificial intelligence inference, data analysis, image processing.

With the increasing application of FPGA, it faces more and more security threats [3–6]. Hardware Trojan is one of them, which is a malicious design or modification implanted into FPGA chips to perform unauthorized operations or damage the system. FPGA being threatened by Hardware trojan can lead to various serious consequences, such as attackers being able to use Hardware trojan inserted into the FPGA to obtain sensitive data in the system, or remotely control infected FPGAs to perform various malicious operations, such as tampering with data, shutting down services, refusing services; Attackers can also cause FPGA system failures or crashes through Hardware

P. Siarry et al. (Eds.): WCNA 2023, LNEE 1361, pp. 314–322, 2025.
https://doi.org/10.1007/978-981-96-2409-6_31

trojan, resulting in system service interruptions, data loss, and even hardware device damage [7]. At present, Hardware Trojan detection methods is mainly divided into two types: netlist detection and real-time detection. Netlist detection achieves Hardware Trojan detection by analyzing the designed netlist. The disadvantage of this type of method is that it can only detect Trojans implanted during the design phase. Real-time detection is achieved by deploying sensors on the chip to sense the running status of the FPGA, detect abnormal running status, and determine whether Hardware Trojan have appeared. This method has a good detection rate, but it requires a high sample size of test data.

This article proposes a configurable hardware Trojan detection testing framework to address the issue of high sample size requirements for training models in deep learning detection methods in real-time detection. By employing configurable testing modes, the framework enables automated testing of target FPGA devices, thereby facilitating the collection of extensive sample data for FPGA hardware security testing. This framework provides essential conditions for the research and development of AI-based FPGA hardware security technologies.

2   Related Work

Thus far, two distinct detection methods have been proposed for identifying Hardware Trojans on FPGA: netlist analysis and real-time monitoring. The netlist analysis approach for Hardware Trojan detection has seen various methodologies proposed by different academic institutions. Notably, Nanjing University of Aeronautics and Astronautics introduced a method based on few-shot learning [8], while Xidian University proposed a gate-level netlist hardware Trojan detection and diagnosis method using support vector machines [9]. Additionally, Queen's University Belfast and Waseda University presented netlist detection methods based on machine learning techniques [10, 11]. However, netlist analysis is limited to detecting Trojan implanted during the design phase. If Trojan are inserted in subsequent stages, such as directly into the generated bitstream, netlist analysis becomes ineffective.

Real-time detection methods have been proposed by various institutions, including the use of online learning machine learning techniques by the University of Maryland to detect runtime anomalies [12], dynamic power-aware scheduling detection methods for real-time tasks by Krishnendu Guha [13], and a novel runtime monitoring technology for detecting Hardware Trojan on FPGA by Southwest Jiaotong University [14]. Both netlist and real-time detection methods, particularly those based on machine learning, require a substantial amount of data samples to support model training [15]. Therefore, this paper proposes a testing framework to acquire a large number of detection sample data, facilitating machine learning-based Hardware Trojan detection.

3   Design and Implementation of Testing Framework

3.1   System Architecture

The testing framework proposed in this paper consists of three main modules: the Test Scenario Configuration Module, the Bitstream Generation Module, and the Test Record Module, as depicted in Fig. 1. Initially, the Test Scenario Configuration Module is used to select parameters such as the type of Hardware Trojan, the target circuit type, the number of stages in the Ring oscillator (RO), and the layout and scale of the Ring oscillator network (RON). Subsequently, the Bitstream Generation Module automatically generates constraint files for the target circuit based on the configured test plan, followed by the automatic generation of corresponding bitstream files. Finally, the Test Record Module automates the process of downloading the generated bitstream files to the target FPGA device, initiating the instance, activating the Trojan, and recording and storing data from the RO.

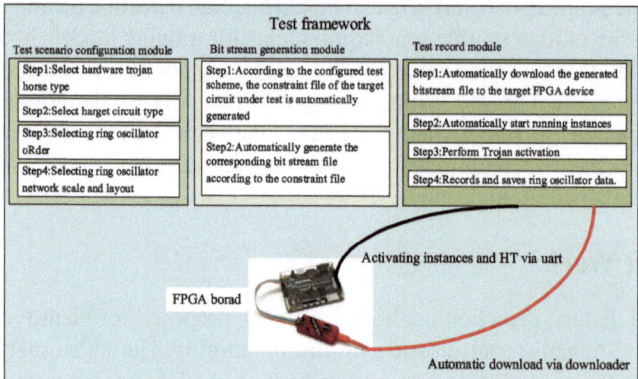

Fig. 1. Test framework diagram

3.2   Test Scenario Configuration Module

Based on the target FPGA device, the initial step involves configuring the type of trojan and selecting the desired trojan types for testing. Subsequently, configuring the target circuit involves selecting benchmark tests from Trust-Hub, as outlined in Table 1 [16]. Next, the number of stages is determined based on the configured target circuit to set the stages of the RO. The RO is a circuit structure formed by connecting an odd number of inverters end-to-end, as illustrated in Fig. 2. It operates without a stable state, with the output voltage oscillating between high and low levels. The oscillation frequency is influenced by the power supply voltage. If a hardware Trojan exists in the test circuit, whether activated or not, it will affect the voltage and consequently impact the oscillation frequency. Therefore, in this framework, RO are utilized as monitors. Following this, the scale of the RON is configured based on the target circuit, determining the size of the RON. Finally, the layout of the RON is configured.

3.3   Bitstream Generation Module

The functionalities implemented by this module are as follows: automatically generating constraint files for the target circuit based on the configured test plan, and subsequently generating corresponding bitstream files based on these constraints.

Table 1. Trust-Hub benchmark

Test circuit	HT	Trigger	Payload
AES	T500	Cheat code	Deny service
AES	T700	Cheat code	Leak data
AES	T1800	Cheat code	Degrade chip
RS232	T500	Time bomb	Deny service
RS232	T700	Cheat code	Deny service
BASICRSA	T300	Time bomb	Leak data

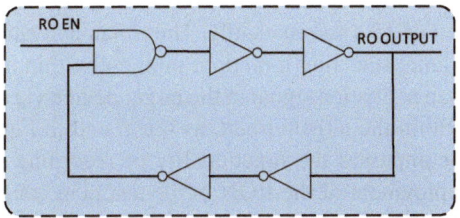

Fig. 2. Structure Diagram of Ring Oscillator

To automate the generation of constraint files for the target circuit, this module first partitions the FPGA test circuit logic blocks (Pblock). Utilizing the Xilinx ISE development environment and hardware description language, it writes the test circuit used for testing based on the configuration information of the test plan and the target FPGA device requirements. The manual layout is then conducted using Xilinx ISE's PlanAhead tool, where a Pblock is drawn. The size of the Pblock determines the FPGA resources available to the target circuit, its position defines where it resides within the FPGA, and the type of FPGA resources included within the Pblock constrains the resources accessible to the target circuit. Next, positional constraints for the RON are applied. This involves constraining the RON positions based on information obtained from the specified test plan regarding the number of stages and the scale of the oscillator network. Applying the same positional constraint method to each RO within the network ensures avoidance of potential conflicts during subsequent positional constraints for the target circuit. Subsequently, positional constraints for the target circuit are established. Initially, the total number of SLICE available in the target FPGA is determined using Xilinx ISE's PlanAhead tool. Areas where SLICE resources are unavailable are identified. Then, the SLICE resources required to deploy the target circuit onto the target FPGA are obtained. This step involves parsing the original positions of the Pblock slice, calculating the required SLICE count in the sample constraint file, ensuring that the SLICE count in the generated constraint file meets or exceeds this value. For the target circuit positional constraints, nine different layout patterns and four positional adjustments (jitter) are considered, resulting in 36 distinct layout constraint possibilities. Each layout pattern calculates start and end coordinates while accounting for potential jitter values, conducting boundary checks to ensure that Pblock do not exceed the SLICE range of the target FPGA. Finally, the 36 generated layout constraint files are automatically saved.

3.4 Test Record Module

The functionalities implemented by this module are as follows: automating the download of generated bitstream files to the target FPGA device, initiating the execution of instances automatically, activating the trojan, and recording and saving data from the RO.

Firstly, this module automates the loading of the bitstream file onto the target device by invoking Xilinx's iMPACT software. Subsequently, utilizing the serial port functionality within the framework, it autonomously sends instance activation signals to the target device. Upon receiving the activation signal, the target device initiates instance execution for a duration of 15 s automatically. The trojan activation functionality operates within this timeframe, selecting a random moment within a 10-s interval. At this moment, it sends a trojan activation signal to the target device via the serial port, thereby activating the trojan within the target circuit, as illustrated in Fig. 3 of the framework. Lastly, the module encompasses the functionality of recording and saving data from the RO. During the deployment of the RON in the test plan configuration module, the framework connects the RO to a counter. Utilizing the RO's oscillation frequency as the counter's clock, the corresponding counter value is transmitted to the framework via the serial port. Subsequently, the testing framework processes the collected counter values to derive the RO frequency. Once the frequency data generated during instance execution is collected, the testing framework automatically saves the data for future use.

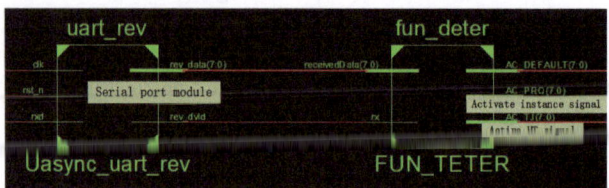

Fig. 3. Activation module diagram

4 Experiments and Discussion

4.1 Experimental Environment

This paper validates the effectiveness of a configurable Hardware Trojan detection testing framework on FPGA. The experiments were conducted using a computer as the development environment, equipped with Xilinx ISE Design Suite version 14.7 and other necessary software. The FPGA model used was Xilinx's SPARTAN-6 XC6SLX9. The benchmark circuit employed AES from Trust-hub, with two types of circuit trojans: T700 for information leakage and T500 for denial of service. The number of RO was configured as 16, and each set of test circuits was executed 50 times.

4.2 Test Framework Result Analysis

Through the configurable hardware trojan detection testing framework proposed in this paper, various datasets with different configurations can be automatically generated. To verify the feasibility of the testing framework, AES-T700 is selected as the target circuit for experimentation. The ring oscillator is configured with 5 stages and 16 oscillators in total, arranged in a 4X4 layout. Multiple test samples under different configurations are automatically generated. This section briefly enumerates the results of four different configurations, depicting layouts in the east, south, west, and north directions as shown in Fig. 4.

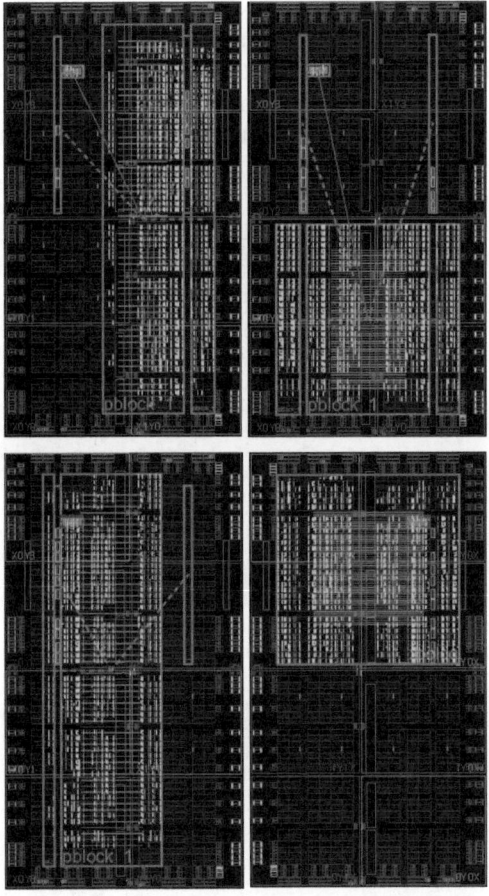

Fig. 4. The constraint graphs generated by this testing framework for different configurations. (purple box for target circuit constraint position, orange rectangle for ring oscillator) (Color figure online)

Furthermore, analyzing the collected test data yields the following results.

By observing Fig. 5a, it can be noted that there is a variation in the frequency of the RO before and after the triggering of the Hardware Trojan. Conclusion (1) can be drawn: Activation of the Hardware Trojan leads to a change in the operational state of the test circuit, impacting the frequency of the RO.

By observing Fig. 5b, it is evident that at the moment of triggering the Hardware Trojan, the impact on the RO [0–5] is most pronounced in the first plot, while in the second plot, the RO [0–9] and [12–15] are most affected. Conclusion (2) can be drawn: For the same type of Hardware Trojan, simultaneous activation results in varying impacts on different RO within the network, depending on the layout of the test circuit on the target device.

Observing Fig. 5c, it can be noted that in region ①, the RON in the second plot is more noticeably affected; in region ②, the RON in the first plot is more affected; and in region ③, when the Hardware Trojan is activated, both RON in the two plots are impacted to some extent, albeit different RO are affected differently. Conclusion (3) can be drawn: Regardless of whether the Hardware Trojans of different types are in an activated state or not, they exhibit varying effects on the test circuit, even with identical layouts and activation timings.

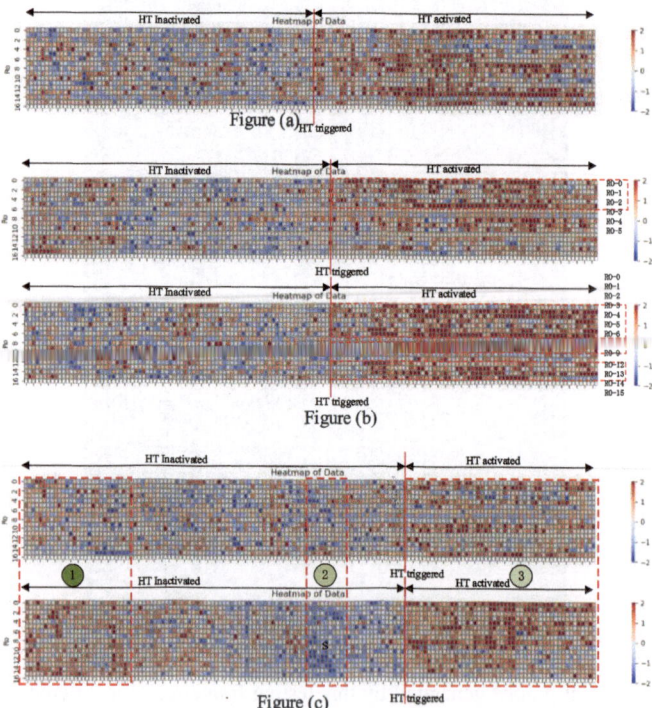

Fig. 5. Data analysis results

5 Conclusion

The paper introduces a configurable Hardware Trojan detection testing framework. Experimental results demonstrate that utilizing this testing framework enables the acquisition of a vast array of effective detection samples under various configurations, thereby providing ample samples for model training in Hardware Trojan detection algorithms.

Acknowledgment. This paper is supported by the Graduate Innovation Research Project of Southwest Minzu University (No. ZD2023663) and Sichuan Province Key Research and Development Projects (No. 2022YFG0334).

References

1. Compton, K., Hauck, S.: Reconfigurable computing: a survey of systems and software. ACM Comput. Surv. **34**, 2, 171–210 (2002)
2. Trimberger, S.M.: Three Ages of FPGAs: a retrospective on the first thirty years of FPGA technology. Proc. IEEE **103**(3), 318–331 (2015). https://doi.org/10.1109/JPROC.2015.2392104
3. Naveenkumar, R., Sivamangai, N.M., Napolean, A., Janani, V.: A survey on recent detection methods of the hardware Trojans. In: 2021 3rd International Conference on Signal Processing and Communication (ICPSC), Coimbatore, India (2021)
4. Hayashi, Y., Kawamura, S.: Survey of hardware Trojan threats and detection. In: 2020 International Symposium on Electromagnetic Compatibility - EMC EUROPE, Rome, Italy (2020)
5. Chen, Z., Guo, S., Wang, J., Li, Y., Lu, Z.: Toward FPGA security in IoT: a new detection technique for hardware Trojans. IEEE Internet Things J. **6**(4), 7061–7068 (2019). https://doi.org/10.1109/JIOT.2019.2914079
6. Hu, W., Chang, C.-H., Sengupta, A., Bhunia, S., Kastner, R., Li, H.: An overview of hardware security and trust: threats, countermeasures, and design tools. IEEE Trans. Comput. Aided Des. Integr. Circuits Syst. **40**(6), 1010–1038 (2021). https://doi.org/10.1109/TCAD.2020.3047976
7. Bharath Kumar, S., Prasanna, M., Bharath, P., Sheeba Joice, C.: FPGA Trojan attack impacts on architecture performance: a systematic approach revealed. In: 2023 8th International Conference on Communication and Electronics Systems (ICCES), Coimbatore, India (2023)
8. Lu, T., Zhou, F., Wu, N., Ge, F., Zhang, B.: Hardware Trojan detection method for gate-level netlists based on the idea of few-shot learning. In: 2021 IEEE 21st International Conference on Communication Technology (ICCT), Tianjin, China, pp. 301–305 (2021)
9. Du, M., Huang, Z., Chen, Y., Li, L., Wang, Q., Liu, J.: A HT detection and diagnosis method for gate-level netlists based on machine learning. In: 2021 IEEE 6th International Conference on Signal and Image Processing (ICSIP), Nanjing, China, pp. 1070–1074 (2021)
10. Yu, S., Gu, C., Liu, W., O'Neill, M.: Deep learning-based hardware Trojan detection with block-based netlist information extraction. IEEE Trans. Emerg. Topics Comput. **10**(4), 1837–1853 (2022). https://doi.org/10.1109/TETC.2021.3116484
11. Negishi, R., Togawa, N.: Evaluation of ensemble learning models for hardware-Trojan identification at gate-level netlists. In: 2024 IEEE International Conference on Consumer Electronics (ICCE), Las Vegas, NV, USA, pp. 1–6 (2024)
12. Kulkarni, A., Pino, Y., Mohsenin, T.: Adaptive real-time Trojan detection framework through machine learning. In: 2016 IEEE International Symposium on Hardware Oriented Security and Trust (HOST), McLean, VA, USA, pp. 120–123 (2016)

13. Guha, K., Majumder, A., Saha, D., et al.: Dynamic power-aware scheduling of real-time tasks for FPGA-based cyber physical systems against power draining hardware Trojan attacks. J. Supercomput. **76**, 8972–9009 (2020)
14. Cheng, J., Feng, Q., Li, C., et al.: Securing FPGAs in IoT: a new run-time monitoring technique against hardware Trojan. Wireless Netw. (2023)
15. Huang, Z., Wang, Q., Chen, Y., Jiang, X.: A survey on machine learning against hardware Trojan attacks: recent advances and challenges. IEEE Access **8**, 10796–10826 (2020). https://doi.org/10.1109/ACCESS.2020.2965016
16. Salmani, H., Tehranipoor, M.: Trust-hub: Chip-level Trojan benchmarks (2021). https://trust-hub.org/#/benchmarks/chip-level-trojan

Research on Visual Target Detection and Recognition of Shopping Robots Based on Improved YOLO Algorithm

Yufan Lu[✉]

Department of Electronic and Information Engineering, (Sussex Artificial Intelligence Institute),
Zhejiang Gongshang University, Hangzhou, China
2137010228@pop.zjgsu.edu.cn

Abstract. This research aims to improve the visual target detection and recognition capabilities of shopping robots in various sales environments by optimizing and improving the YOLO algorithm, in order to improve accuracy and real-time performance. The research method involves embedded spatial hierarchical sampling technology and it adapts to image processing of different sizes, uses a separate convolutional neural network structure to reduce computational complexity, and cultivates a more concise network model by refining the effective data of complex models. Experimental results show that the improved YOLO algorithm performs well in weak Its average accuracy has been significantly improved under light, medium light and strong light environments, especially in the detection of small items. A study shows that improved programming significantly improved the vision of shopping assistance robots. Recognition capabilities enable robots to provide more accurate and faster services in real shopping environments.

Keywords: YOLO algorithm · visual target detection · spatial pyramid pooling · depth-separable convolution

1 Introduction

In today's retail industry, the widespread deployment of shopping assistance robots has made consumers' shopping experience smoother, thereby improving the overall efficiency of business operations. The performance of these automated machines in business situations depends largely on the accuracy of their image recognition and the speed of their response. This project aims to develop an upgraded version of the YOLO algorithm to make it more in line with the application requirements of shopping robots, thereby improving the robot's work efficiency and consumers' shopping experience, and promoting the development of intelligent sales technology.

P. Siarry et al. (Eds.): WCNA 2023, LNEE 1361, pp. 323–332, 2025.
https://doi.org/10.1007/978-981-96-2409-6_32

2   Theoretical Overview

2.1   Basics of Computer Vision

In computer vision field, algorithms are able to not only detect and identify objects in images, but also understand the spatial relationships between these objects and their dynamics over time. Computer vision systems usually include steps such as image acquisition, pre-processing, feature extraction, detection/recognition task execution, and the use of post-processing techniques to optimize the results [3]. With the introduction of deep learning technology, many breakthroughs have been achieved in the field of computer vision, especially showing excellent capabilities in processing complex and highly unstructured data. Modern computer vision systems play a key role in self-driving cars, medical image analysis, automated surveillance, interactive gaming, intelligent video analysis, and many other fields, constantly pushing the boundaries of related technologies.

2.2   Development of Target Detection Technology

Object detection technology has experienced significant evolution from early simple algorithms to modern deep learning methods. Before deep learning became mainstream, object detection relied on hand-designed features and rule-based classifiers, such as the Viola-Jones detector and the HOG detector. These methods performed well for their era on object detection problems under some restricted conditions. However, these algorithms encounter performance bottlenecks in diverse and complex practical application scenarios, especially when dealing with scale changes, occlusions, and background noise in images [4]. With the breakthrough of AlexNet in the field of image recognition in 2012, target detection methods based on deep learning began to develop rapidly, changing the research and application landscape. Anchor-based methods such as Faster R-CNN significantly improve detection accuracy by generating region proposals and then using deep convolutional networks to refine classification and localization. At the same time, one-stage detection methods such as YOLO and SSD achieve faster processing speeds by simplifying the detection process, making real-time target detection possible [5]. In addition, the latest Anchor-free methods such as CenterNet locate targets by directly predicting key points, further reducing algorithm complexity and improving applicability [6].

2.3   Principle of YOLO Algorithm

The YOLO (You Only Look Once) algorithm is a revolutionary target detection method that converts the target detection problem into a single regression problem, directly mapping from image pixels to bounding box coordinates and class probabilities. The YOLO algorithm first divides the input image into an $S \times S$ grid, and predicts B bounding boxes and the confidence scores corresponding to these bounding boxes for each grid. At the same time, each grid is also responsible for predicting C conditional category probabilities (assuming There are C categories in the data set). Each bounding box consists of five predictions: four coordinate values and a confidence score that reflects the presence or absence of an object in the bounding box and the model's confidence in the accuracy of the box's location. YOLO's design allows it to look at the entire image when predicting, thereby better understanding the global contextual information in the image, unlike traditional region proposal-based methods, which often ignore global contextual information [7]. Figure 1 is the schematic diagram of the YOLO algorithm:

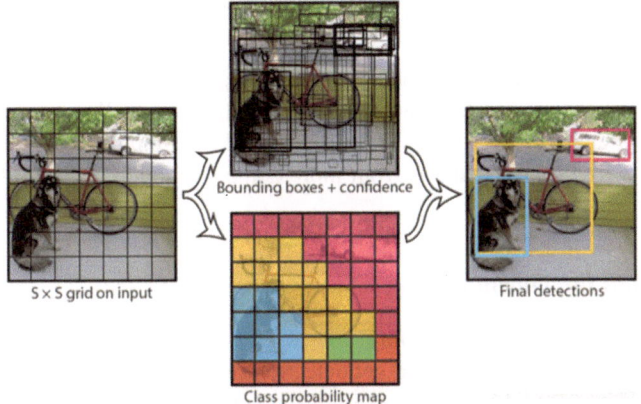

Fig. 1. Schematic diagram of YOLO algorithm

3 Shopping Robot Visual Target Detection and Recognition Based on Improved YOLO Algorithm

3.1 Design Concept of Improving YOLO Algorithm

The design concept of improving the YOLO algorithm mainly focuses on improving the accuracy and adaptability of the model, especially for small target recognition and occlusion problems [8]. First, based on the original YOLO model, the Spatial Pyramid Pooling (SPP) layer is introduced. This improvement enables the model to accept input images of any size while maintaining the scale invariance of the features. Specifically, the SPP layer can capture rich contextual information by implementing pooling kernels of different sizes ($1 \times 1, 2 \times 2, 4 \times 4$), thereby improving the model's ability to recognize targets of different sizes. The formula is expressed as follows:

$$\text{SPP}(x) = \left[\text{maxpool}_{1x1}(x), \text{maxpool}_{2x2}(x), \text{maxpool}_{4x4}(x)\right] \tag{1}$$

where x represents the input feature map. In addition, in order to accurately adjust the alignment between the predicted box and the real box, the improved algorithm uses the IoU loss function. This loss function directly optimizes the intersection-over-union ratio (IoU) between the predicted box and the real box [9]. The formula is as follows:

$$L_{\text{IoU}} = 1 - \frac{\text{pred} \cap \text{true}}{\text{predUtrue}} \tag{2}$$

Considering the real-time nature of the algorithm, a lightweight network architecture is adopted, and knowledge distillation technology is introduced to transfer knowledge through the teacher network (large high-precision model) to the student network (lightweight model). Assuming that the output of the teacher network is, and the output of the student network is, the loss function of knowledge distillation can be expressed as:

$$L_{KD} = \alpha \times L_{hard}(S(x), y) + (1 - \alpha) \times L_{soft}(S(x), T(x)) \tag{3}$$

where L_{hard} is the standard classification loss and L_{soft} is the soft label loss between the student network output and the teacher network output, typically using temperature - regulated cross-entropy. This design allows the model to maintain high-speed processing while also learning robust features from more complex models, effectively improving detection performance [10].

3.2 Algorithm Implementation

A) Data preprocessing and annotation
  In the shopping robot's visual target detection project, data preprocessing includes adjusting all images to 416x416 pixels to fit the network input. Furthermore, different visions are simulated by applying a series of data augmentation techniques such as image flipping (horizontal flipping probability 0.5), random cropping (up to 20% of the original image size) and color dithering (color adjustment factor between 0.6 and 1.4). Conditions to enhance the generalization ability of the model [11]. During the annotation process, each product image will be annotated with a bounding box and tagged with a specific category by a professional annotation team. For example, for an image of the "beverage" category, the annotation information is
  {"category":"drink","boundingbox": [$x_m in, y\_ min, x\_ max, y\_max$]},
  Among them, represents the coordinates of the upper left corner of the box, and represents the coordinates of the lower right corner of the box.

b) Modification of network architecture
  For the original YOLO network, depth-separable convolution is first introduced to reduce the amount of calculation, which is specifically reflected in replacing the traditional convolution layer with a depth-separable layer. For example, in a typical convolutional layer, if the input feature map size is $416 \times 416 \times 32$ and 256 3 × 3 convolution kernels are used for convolution, then the number of parameters of this layer is $32 \times 3 \times 3 \times 256 = 73{,}728$. After using depth-separable convolution, the number of parameters is calculated in two parts, depth convolution $32 \times 3 \times 3 \times 1 = 288$ and point-wise convolution $32 \times 1 \times 1 \times 256 = 8{,}192$, a total of $288 + 8{,}192 = 8{,}480$ parameters, significantly reducing model complexity. In addition, the Feature Pyramid Network (FPN) is integrated to enhance the detection ability of small objects. FPN effectively improves the representation ability of features through multi-scale feature fusion from the top to the bottom [13].

c) Training process and parameter optimization

During the training process, a combination of cross-entropy loss function and IoU loss function is used. The loss function is defined as follows:

$$L = \lambda_{\text{cls}} L_{\text{cls}} + \lambda_{\text{IoU}} L_{\text{IoU}} \tag{4}$$

Among them, λ_{cls} and λ_{IoU} are the weight coefficients used to balance the two losses. In practical applications, and $\lambda_{\text{cls}} = 1$ and $\lambda_{\text{IoU}} = 2.5$ may be set to focus on optimizing the accuracy of the bounding box. The Adam optimizer was used for parameter optimization, with the initial learning rate set to 0.001 and the learning rate reduced by 10% after every 20 epochs.

3.3 Systems Integrated into Shopping Robots

A) System architecture
The visual target detection system of the shopping robot is designed using a layered architecture, which mainly includes a data input layer, a processing layer and an interactive output layer. At the data input layer, the video stream is first pre-processed, including standardization, uniformly adjusting the image size to 416×416 pixels, and applying a Gaussian blur filter to reduce noise. The formula is expressed as:

$$I_{\text{out}} = \text{GaussianBlur}(I_{\text{in}}) \tag{5}$$

The core of the processing layer is the improved YOLO algorithm, which combines FPN and depth-separable convolution technology to enhance the model's ability to recognize objects of different scales. The interactive output layer displays processing results through a graphical user interface (GUI) and receives user instructions, including search requests or command adjustments [14].

b) Real-time target detection and tracking
Real-time target detection and tracking uses the improved YOLO network, and achieves the ability to process 30 frames per second by introducing batch processing and optimized algorithm processes. Specific improvements include adding a fast feature pyramid network based on YOLO to improve the accuracy of small target detection. The formula can be expressed as:

$$F_{\text{pyramid}} = \sum_{i=1}^{n} P_i * K_i \tag{6}$$

where P_i represents the feature map of the i-th layer, and K_i is the corresponding convolution kernel. Target tracking uses the Kalman filter algorithm, and the prediction model is expressed as a linear equation:

$$x_k = A x_{k-1} + B u_k + w \tag{7}$$

where x_k is the current state, A is the state transition matrix, u_k is the control input, B is the control input model, and w is the process noise. This method can effectively track dynamic targets in complex environments.

c) User interaction and feedback mechanism

Using the graphical interface of the touch screen, the software builds an interactive system that allows users to provide instant feedback and adjust instructions. For example, customers score the recognition accuracy through the evaluation system. This process is expressed in mathematical expressions. Within a certain range, the scores provided by netizens are integrated into a data set, the number of which is the sum of the scores. Artificial intelligence technology is used to analyze user feedback and adjust platform functions accordingly, such as adjusting thresholds or optimizing search algorithms, in order to enhance user pleasure and satisfaction [15]. In addition, the platform will analyze the user's purchase history and preferences, and use the recommendation algorithm to push personalized product recommendations to users, thereby enhancing the user's shopping experience.:

$$Score\$_{average} = \frac{\sum (\text{Scores})}{n} \tag{8}$$

where Scores is the list of ratings given by the user, and n is the number of ratings. The recommendation system may use collaborative filtering algorithm, the formula is:

$$s(u, i) = \frac{\sum_{v \in N} r_{v,i} \times \text{sim}(u, v)}{\sum_{v \in N} |\text{sim}(u, v)|} \tag{9}$$

where $s(u, i)$ represents user u's interest score for product i, is other user v's score for product i, and is the similarity between users u and v.

4   Experimental Testing and Result Analysis

4.1   Experimental Setup

This study focuses on an in-depth evaluation of the YOLO algorithm and observes its actual application effect in visual object detection and recognition in a shopping assistance robot that simulates a shopping mall environment. A series of tests were conducted under strictly controlled experimental conditions. This scenario includes a variety of lighting configurations, a variety of product displays, and various possible influences on consumers. It aims to simulate the changes in the scene in the real shopping process. In an experimental environment with three different lighting conditions: low light (100 lx), medium light (500 lx) and high light (1000 lx), the performance of the algorithm under various light intensities was analyzed. In the experiment, more than 50 different daily shopping products were selected as target detection objects, including large products that are easy to identify and small products that are easy to block. At least 20 pictures were taken of each product at different positions and angles, for a total of more than 1,000 test images. All images were pre-processed including resizing to 416 × 416 pixels to ensure consistency. To evaluate the detection accuracy, the image dataset was divided into a training set (80%) and a test set (20%) using a cross-validation method. In the training phase, adjust the optimization parameters, use the Adam optimizer, set the initial learning rate to 0.001, and set the learning rate decay to 10% every 10 epochs. The main evaluation index of detection accuracy is average precision (mAP), and this is compared with the performance of the traditional YOLO model to quantify the effect of improvement.

4.2 Test Process

During the execution phase of the experiment, the configuration of the data set was first set to ensure that the image quality and diversity met the test requirements. Each item was photographed under three different lighting conditions (100 lx, 500 lx, 1000 lx) in a light-controlled environment. To ensure that the detection algorithm can adapt to different visual environments, the image data is enhanced with Gaussian noise and color perturbation. The specific operations include randomly adjusting the brightness and contrast of the image with an adjustment factor within ±15%. Next, the enhanced image data is input into the training model. Trained using the improved YOLO algorithm, the model includes depthwise separable convolutional layers for feature extraction and a feature pyramid network for refined detection. During the training process, a series of batch size and learning rate adjustments were used, starting from the initial batch size of 32 and gradually reducing to 16 as the training progressed, and the learning rate started from 0.001, every 10 epochs Then it will be automatically reduced by 10% based on performance (Fig. 2).

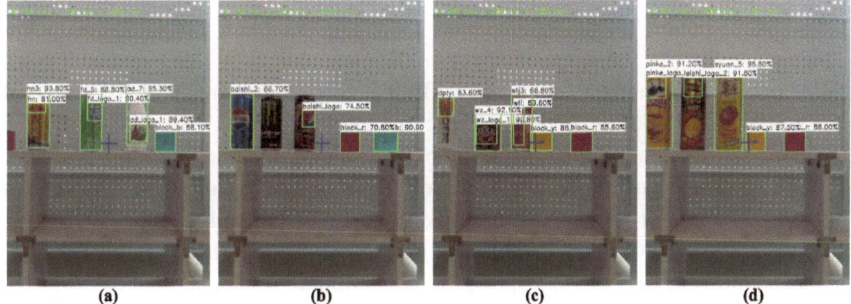

Fig. 2. Experimental results

4.3 Result Analysis

The detection accuracy and recall rate of each category of goods are also evaluated in detail. For example, for smaller items such as bottled drinks, the improved model's recall rate increased from 70% to 85%, while for larger items such as food packaging boxes, the recall rate increased from 80% to 90%. Details are shown in Table 1:

In addition, the performance of the algorithm under different lighting conditions was also analyzed, and Table 2 was obtained:

Table 1. TPerformance comparison of YOLO algorithm before and after improvement on different product categories

Product Category	mAP before improvement (%)	Improved mAP (%)	Recall rate before improvement (%)	Improved recall rate (%)
Bottled drinks	65	80	70	85
food packaging box	75	88	80	90
personal care products	60	75	65	80
Fresh fruits and vegetables	68	83	73	88

Table 2. Performance comparison under different lighting conditions

Lighting conditions	mAP before improvement (%)	Improved mAP (%)
Low light (100 lx)	55	70
Medium light (500 lx)	70	82
High light (1000 lx)	75	85

Research shows that the upgraded YOLO detection system, which incorporates spatial hierarchical pooling, independent convolutional neural networks, and knowledge extraction technology, has made leaps and bounds in accuracy, retrieval speed, and real-time processing capabilities. These improvements not only improve the algorithm's detection performance for small and difficult-to-identify objects, but also enhance the algorithm's stability under changing lighting conditions, thereby providing solid technical support for the practical application of shopping robots in retail environments.

5   Conclusion

This study aims at complex retail environments. By improving the upgraded version of the YOLO algorithm, it has significantly enhanced the recognition and discrimination of visual targets by retail assistance robots. Under diverse lighting conditions, the improved algorithm logic shows higher performance. The recognition accuracy and more comprehensive performance are improved, especially when dealing with small objects and those objects that are prone to occlusion. The integration of spatial pyramid pooling, depthwise separable convolution and knowledge distillation technology not only improves the accuracy of the algorithm while ensuring high computing efficiency and meeting the needs of real-time applications. In the future, we will focus on the application research of computing methods in more product categories, and actively explore the research of network optimization technology, aiming to Effectively enhance system stability and improve operating efficiency.

References

1. Shen, S.R., Zhang, X., Yan, W., et al.: An improved UAV target detection algorithm based on ASFF-YOLOv5s. Math. Biosci. Eng. MBE **20**(6), 10773–10789 (2023). https://doi.org/10.3934/mbe.2023478
2. Ying, Z., Lin, Z., Wu, Z., et al.: A modified-YOLOv5s model for detection of wire braided hose defects. Measurement **190**, 190 (2022). https://doi.org/10.1016/j.measurement.2021.110683
3. Zhang, Q., Wang, Y., Song, L.,et al. Using an improved YOLOv5s network for the automatic detection of silicon on wheat straw epidermis of micrographs. J. Field Robot. (2023). https://doi.org/10.1002/rob.22120
4. Chen, Z., Huang, C., Duan, L., et al.: Lightweight Surface Litter Detection Algorithm Based on Improved YOLOv5s **2023**(7), 1085–1102 (2023). https://doi.org/10.32604/cmc.2023.039451
5. Zhou, S., Zhao, J., Mei, S.Q., et al.: Research on improving YOLOv5s algorithm for fabric defect detection. Int. J. Cloth. Sci. Technol. **35**(1), 88–106 (2023). https://doi.org/10.1108/IJCST-11-2021-0165
6. Lin, H.: Multi-scale forest fire recognition model based on improved YOLOv5s. Forests **14** (2023). https://doi.org/10.3390/f14020315
7. Singh, K.J., Kapoor, D.S., Thakur, K., et al.: Computer-vision based object detection and recognition for service robot in indoor environment. Comput. Mater. Continua **72**(1) (2022)
8. Shahin, M., Chen, F.F., Hosseinzadeh, A., et al.: Robotics multi-modal recognition system via computer-based vision. Int. J. Adv. Manufac. Technol. 1–17 (2024)
9. Zhang, H., Song, R., Jiang, H.: Recent reviews on dynamic target detection based on vision. Recent Patents Eng. **17**(6), 120–133 (2023)
10. Li, T., Sun, M., He, Q., et al.: Tomato recognition and location algorithm based on improved YOLOv5. Comput. Electron. Agric. **208**, 107759 (2023)
11. Yang, G., Liu, X., Zhong, O., et al. : A vision-based fruit packaging robot. In: 2021 IEEE 6th International Conference on Computer and Communication Systems (ICCCS), pp. 297–302. IEEE (2021)
12. Rane, N.: YOLO and Faster R-CNN object detection for smart Industry 4.0 and Industry 5.0: applications, challenges, and opportunities. Available at SSRN 4624206 (2023)
13. Wu, D., Lv, S., Jiang, M., et al.: Using channel pruning-based YOLO v4 deep learning algorithm for the real-time and accurate detection of apple flowers in natural environments. Comput. Electron. Agric. **178**, 105742 (2020)
14. Hasibuan, N.N., Zarlis, M., Efendi, S.: Detection and tracking different type of cars with YOLO model combination and deep sort algorithm based on computer vision of traffic controlling. Sinkron: jurnal dan penelitian teknik informatika, **6**(1), 210–221 (2021)
15. Cao, Z., Liao, T., Song, W., et al.: Detecting the shuttlecock for a badminton robot: a YOLO based approach. Expert Syst. Appl. **164**, 113833 (2021)

A Review on the Large Language Model Augmented Knowledge Graph Question Answer: Task, Model, Advance and Outlook

Rongdong Yu, Dou Wang$^{(\boxtimes)}$, Xiaoyan Jia, Zhifeng Jiang, and Zhenwei Zhang

Block B, Zhejiang Energy Technology Innovation Center, Zhejiang Energy Digital Technology Co., Ltd., Hangzhou, China
{yurongdong,wangdou,jiaxiaoyan,jiangzhf,
zhangzhenw}@zjenergy.com.cn

Abstract. The development of Large Language Models (LLMs) has sparked a new wave in the domain of knowledge engineering. To gain deeper insights into the transformations catalyzed by LLMs to the field of Knowledge Graphs, this paper conducted a research around Knowledge Graph Question Answer (KGQA) task, and summarized the application forms of LLMs in KGQA task as well as the performance enhancement they brought. First, the paper provided a concise and comprehensive introduction to the KGQA task, focusing on its definition, methods, datasets, and evaluation metrics. Then the application of LLMs in KGQA task was discussed. Next, using three typical KGQA datasets as references, the accuracy improvement of LLM-augmented KGQA methods was visually demonstrated. From the comparison of results, it could be found that using LLM as a module in the general framework of KGQA methods could effectively improve the accuracy of the answers. Moreover, with the advancement of technology, LLMs were expected to be break away from the general framework of KGQA methods and directly answer the KGQA questions in zero-shot scenarios. Finally, this paper provides an outlook on the challenges that exist when LLMs applied to KGQA task.

Keywords: Knowledge Graph Question Answer · Large Language Models · Semantic Understanding · Enhance · Review

1 Introduction

Knowledge graph (KG) is a grid-like semantic knowledge base (KB) composed of nodes and edges. It could enable the organization, management and understanding of information by describing entities and their relationships in a structured way [1]. KG is usually represented in the form of directed graphs, where entities are represented by nodes in directed graphs, and relations are represented by edges (as shown in 0) (Fig. 1).

From the point of view of automated knowledge application, it was only the first step that converting a vast amount of unstructured knowledge in different forms into structured knowledge in the form of a KG. In order to reason and mine deeper information in the KG, the computers were required to understand the specific meanings of the entities and

© The Author(s) 2025
P. Siarry et al. (Eds.): WCNA 2023, LNEE 1361, pp. 333–347, 2025.
https://doi.org/10.1007/978-981-96-2409-6_33

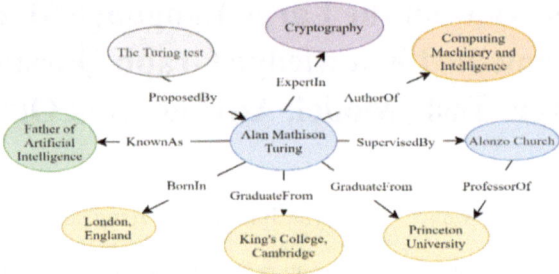

Fig. 1. An example of KG.

relationships in the KG. In recent years, LLMs have developed rapidly and demonstrated extraordinary capabilities in the field of natural language processing (NLP).

Currently, many researchers have started to apply LLMs to KG-related works, such as KG embedding, KG complementation, KG construction, KG to text generation, KG question answer (KGQA), and so on. And there were related review articles. However, most of the review articles focused on the process of KG construction [2]. In order to analyze the application potential of LLMs in KG in a more targeted way, this paper investigated the current research status of LLMs in KGQA task, and discussed the application history and future development trend of LLMs in the field of KGQA. The primary contributions of this paper include the following:

(1) Regarding KGQA task, this paper provided a complete and comprehensive summary. From the problem definition of KGQA, it was systematically introduced that the two main problem types, common datasets, LLM-augmented KGQA methods, and common evaluation metrics for KGQA methods.
(2) This paper intuitively demonstrated the enhancement that LLM technology brought to KGQA task by presenting the changes in accuracy rates on three typical KGQA datasets.
(3) This paper provides an outlook on the opportunities and challenges of LLM in KGQA task based on the existing techniques and methods.

2   Related Work

2.1   Knowledge Graph

The initial spark of KG emerged in the field of machine translation. As early as 1956, Richens introduced the concept of Semantic Net and constructed the connection between thing and its attributes and relations in the form of arrows pointing while exploring the operational logic of machine translation [3] (Fig. 2).

In the 1960s, Quillian and Collins proposed the Semantic Network in a series of studies aimed at machine translation [4]. Semantic Network is essentially a highly interconnected network constructed from nodes and relationships between nodes that expressed human knowledge constructs in a network format. In 1976, the medical diagnostic expert system MYCIN [5] was designed. In 1977, the concept of Knowledge Engineering (KE)

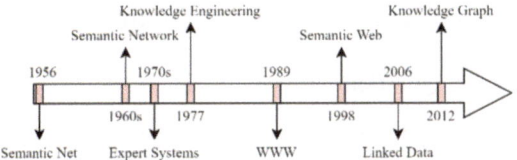

Fig. 2. The development history of KG.

was introduced by Feigenbaum [6]. In 1989, the World Wide Web (WWW) was created by Berners-Lee, which realized the mesh linking of documents [7]. On the basic of WWW, in 1998, Berners-Lee further proposed the concept of Semantic Web, which combined WWW and artificial intelligence to achieve the computer's understanding and reasoning of information [8]. In 2006, Berners-Lee proposed the concept of Linked Data and proposed four principles [9]. In 2012, Google released a search engine based on KG [10].

2.2 Large Language Models

LLMs usually referred to as language models pre-trained on large-scale corpora consisting hundreds of billions of parameters. Since the release of ChatGPT [11] in 2022, LLMs have swept through the fields of NLP in a devastating manner. They even pushed forward the change of computer vision field (Fig. 3).

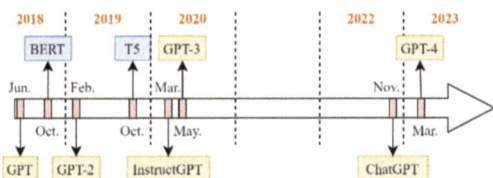

Fig. 3. The development history of LLM related to KGQA.

In 2017, Reference [12] proposed the Transformer architecture. Transformer was initially used to solve the machine translation problem, but it quickly sweep the entire NLP field due to its powerful feature extraction ability and parallel computing ability. In 2018, OpenAI proposed a self-supervised pre-trained language model, which had since opened the era of GPT (Generative Pre-Trained Transformer) in the field of NLP [13]. In the same year, Google proposed BERT (Bidirectional Encoder Representation from Transformer), which implemented a bidirectional self-supervised pre-trained language model [14]. In February 2019, OpenAI proposed GPT-2, which could perform various language tasks without any explicit supervision [15]. In May 2020, OpenAI proposed GPT-3 [16]. In March 2023, OpenAI proposed the large-scale multimodal model GPT-4, which added multimodal inputs such as speech, image, and video to text inputs, further advancing the Artificial General Intelligence (AGI) [17].

3 LLM-Augmented Knowledge Graph Question Answer

3.1 Knowledge Graph Question Answer

KG Question Answer (KGQA) aims to retrieve relevant facts to retrieve answers to natural language questions based on structured knowledge stored in the KG [18]. According to the type of question, KGQA could be divided into simple and complex questions.

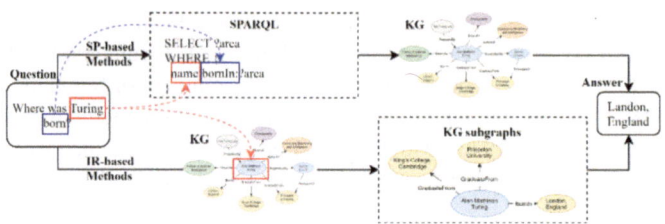

Fig. 4. Comparison of two kinds of KGQA methods

In general, KGQA methods include SP-based methods and IR-based methods [19] (as shown in Fig. 4). SP methods usually transform natural language questions into descriptive language could be executed in KGs, and retrieve the answer to the question from the KGs through query language execution [19]. IR methods, on the other hand, find the KG subgraphs related to the question using the information extracted from the question, and then use techniques such as neural networks to rank the candidate facts in the subgraphs to find matching answer.

Simple Questions

Simple questions usually mean that only knowledge of one of the triples in KG is required to answer the question. For instance, the natural language question "Where was Turing born?" is answered by KGQA as "Landon, England", which is retrieved by the entity-relationship pair < Alan Mathison Turing, born_in > in the KG. In this paper, KGQA based on Simple questions referred to as S-KGQA. In S-KGQA, the answer of question could be determined when the relation and predicate of the triples were found, so the SP method could be directly reduced to a classification problem [20].

In general, the S-KGQA methods comprise four modules: Mention Detection, Entity Disambiguation, Relation Detection and Answer Query. In some studies, Mention Detection and Entity Disambiguation were merged into Entity Linking [21].

Mention Detection means finding the mentions that represent entities in a natural language question, which often also known as the named entity recognition task. The common methods included RNN (Recurrent Neural Network) [22], CNN (Convolutional Neural Network) [21], and so on. With the development of LLM technology, models such as BERT [23] were gradually applied to the Mention Detection process.

Entity Disambiguation means retrieving candidate entities based on mentions and ranking candidate entities based on relevance to the question. The common methods included Fuzzy String Matching [22], TF-IDF Scores [24], CNN [21], and so on.

Relation Detection means finding the correct relation from the problem that could correspond to KGs. For SP-based KGQSA methods, Relation Detection is usually a classification problem that uses models such as RNN [25], LSTM [22] to encode the problem sequence and classify the problem sequences according to the candidate relations in KGs. For IR-based KGQSA methods, the candidate relations in KGs and the schematic representation of the problem are encoded, and then the candidate relations are sorted by relevance to the problem. Common encoding methods included CNN [21], LSTM [26],, BERT [27], and so on.

Answer Query means retrieve answers from KG based on candidate entities and candidate relationships. The weighted scores of candidate entities and candidate relationships were used as the scores of candidate answers.

Complex Questions

Complex Questions based KGQA also known as C-KGQA [28]. Multiple subject entities and multiple master relationships in KG would be involved in the process of C-KGQA [29] and the solution process often faced challenges such as multi-hop reasoning、constrained relations、numerical operations, and so on [28]. Figure 5 was an example for C-KGQA. Consider a complex problem. First, the entity "Annie Ernaux" need to be found according to the subject entity "Nobel Prize winner in Literature" with "2022" as the constraint. Second, the relationship "author_of" need to be found according to the question's mention "novel". Third, several novels need to be found according to the entity "Annie Ernaux" and the relationship "author_of". Fourth, the publication date of each novel need to be searched. Finally, the publication date need to be sort chronologically in order to determine "the first" as the answer to the question.

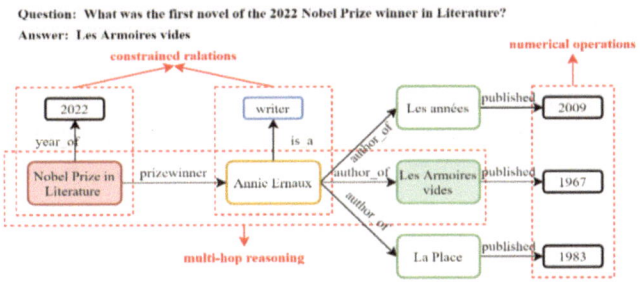

Fig. 5. Example of KGQA and its solution process for a complex problem. Multi-hop reasoning, constrained relations, and numerical operations are indicated by dashed boxes.

Methodologically, C-KGQA could still be categorized into SP and IR. As the problem involves multiple relationships and entities, the researchers tried to decompose the complex problem. Reference [30] decomposed the problem into a set of sub-problems through a template and obtained the final answer by combining the answers to the sub-problems. Reference [29] used computation plan to decompose the query graph into partial queries. Then the candidates for each partial queries were generated sequentially and the optimal answer was found by semantic matching.

In addition to the form of problem decomposition, some studies have added entities and relationships on candidate paths layer by layer through an iterative approach. Reference [31] found candidate paths to the problems through staged entity and relationship retrieval. Reference [32] designed the greedy search method, which began with the topic entity, and retrieved the optimal relationship and the optimal entity sequentially until the optimal path and the optimal answer were found.

3.2 KGQA Dataset

Simple Questions Datasets

WebQuestions [33]. A QA dataset containing 5810 (question, answer) pairs built based on the Freebase knowledge base. The proportion of simple questions is at 84%.

WebQuestionSP [34]. A QA dataset based on Freebase knowledge base. The dataset contains full semantic parses in SPARQL queries for 4737 questions, with partial annotations provided for the remaining 1073 questions.

30M Factoid Questions [35]. A Freebase-based QA dataset automatically constructed by the model. It contains 30M QA pairs, and the quality of the questions is comparable to that constructed by humans.

FreebaseQA [36]. A QA dataset consisting of about 54K matches. The dataset uses the TriviaQA dataset as the primary source, and also includes questions crawled from the knowledge quiz sites KnowQuiz, QuizBalls, and QuizZone.

Complex Questions Datasets

ComplexQuestions [37]. A dataset constructed specifically for complex problems, consists of 2100 multi-constrained QA pair from 3 sources.

LC-QuAD [38]. A DBpedia-based dataset of complex questions with 18% of simple 1-hop questions. The dataset contains 5000 pairs of problems and their corresponding SPARQL queries.

ComplexWebQuestions [39]. A complex question dataset built on the WebQuestionsSP dataset, including question pairs and their corresponding SPARQL queries.

LC-QuAD 2.0 [40]. A complex problem dataset using Wikidata and DBpedia datasets. It contains 30000 problems, and also provides corresponding SPARQL queries.

Domain Datasets

MetaQA [41]. MetaQA is a substantial multi-hop KGQA benchmark for the film domain. It provides a KG consisting of 135k triples, 43k entities and 9 relations. The dataset comprises over 400000 questions divided into 1-hop, 2-hop and 3-hop based on the number of hops experienced in answering them. Each question is annotated with header entities, answers and the class of entities involved in the reasoning path.

XAI4Wind [42]. XAI4Wind includes wind turbine operation and maintenance KG and a corresponding QA dataset. The KG contains 537 nodes and 1059 relationships (9 different types), incorporating diverse types of heterogeneous information such as alarm overview and so on. Based on the above KG, 2361 unique QC pairs (natural language question-Cypher query) were firstly constructed, and 73105 QC pairs were obtained as QA dataset after expansion.

3.3 LLM-Augmented KGQA

With the development of LLMs techniques, they are beginning to be applied to the KGQA task, and the following provides a concise overview of the application of LLMs techniques in different methods.

S-KGQA

In Reference [23], a fine-tuned LLM was utilized for both entity span prediction and relation prediction. The experimental results showed that LLM-based prediction model significantly outperformed BiLSTM-based baseline. Reference [43] used LLM-CRF for mention detection and used LLM-Softmax for entity disambiguation and predicate mapping. The experimental results demonstrated the performance improvement of LLM bringing to the KGQA problem.

Reference [27] proposed a relationally aware attention model based on LLM. The model toke question-fact pairs as model inputs and captured the interaction between questions and candidate facts. Finally, the answer of the question was retrieved by calculating the semantic similarity between them. Reference [44] proposed DEKCOR (Descriptive Knowledge for Commonsense Question Answering) model, which input the question, the options of the question, the detected relationships and the external entities descriptions into the LLM model to calculate the correlation coefficients between the question and the options thereby obtaining the answer.

C-KGQA

Reference [45] integrated a logic programming language into LLM to solve KGQA task. Firstly, the study converted the triples in the dataset into first-order logic predicates. Secondly, LLM was used to generate problem-specific logical forms. Thirdly, the tokens of the logical form were replaced with the corresponding entities in KG to generate Prolog queries. Finally, the queries were executed to retrieve the answers and the logical paths in KG. Reference [46] proposed a LLM and GNN (Graph Neural Networks) based joint reasoning model ReLMKG, which used LLM to extract the implicit knowledge at the text level and used GNN to extract the explicit knowledge of the KG. The model used LLM to generate the encoding of questions and textual paths, which guided the reasoning module to propagate and aggregate messages on KG.

Reference [47] divided the complex QA task into three phases: question decomposition, constraint extraction and question inference. In the question decomposition part, the BTAM model was designed to decompose complex problems and the LLM was used an encoder to encode the semantic information of the question. Reference [48] used LLM to generate candidate answers and extract entities from the initial questions. Then the subgraphs containing the extracted entities were constructed according to Wikidata and the candidate answers were generated. Finally, the subgraphs and original subgraphs were linearized using the Transformer Encoder models and sorted to find the answer.

3.4 KGQA Evaluation Metrics

The evaluation metrics of KGQA methods include accuracy, precision, F-score, Hits@K, and so on.

Accuracy represents the proportion of correctly answered questions out of the total number of questions, and the calculation formula is as follow:

$$accuracy = |A \cap \hat{A}|/|Q| \tag{1}$$

where Q is the set of questions, A is the correct answers of the questions, $\hat{A}$ is the predicted answers of the questions.

Precision represents the ratio of correct answers to the total number of predicted answers, and calculated using the following formula:

$$precision = |A_q \cap \hat{A}_q|/|\hat{A}_q| \tag{2}$$

where A_q is the correct answers set of the question q, $\hat{A}_q$ is the predicted answers of the question q.

Recall represents the ratio of correct answers out of the total number of answers for a single question. The calculation formula is as follow:

$$recall = |A_q \cap \hat{A}_q|/|A_q| \tag{3}$$

F-score represents the weighted harmonic mean of precision and recall, which could accommodate both of them without being affected by unbalanced samples. The calculation formula is as follow:

$$F = \frac{\left(\alpha^2 + 1\right) \times precision \times recall}{\alpha^2 \times precision + recall} \tag{4}$$

where α is the hyper-parameter. When $\alpha = 1$, it is the most common metric F1 score.

Hits@K represents the ratio of questions that contain the correct answer in the first K answers. When $K = 1$, hits@1 is equivalent to accuracy.

3.5 Technological Development

In order to better demonstrate the development of KGQA methods, this paper described the different methods and a comparison of the test accuracy of these methods on WebQuestionsSP, MetaQA and QALD-9.

PullNet used a graph convolutional neural network to construct subgraphs related to the question and another Graph Neural Network (GNN) to retrieve the answer from the subgraphs [49].

ReLMKG used the BERT model to encode the question and the text path obtained from the retrieval, and a graph neural network to encode the KG to find the answer by means of joint learning [46]. From the experimental results, there is a more obvious improvement in accuracy (as shown in TABLE I.).

Reference [50] used LLMs to directly answer questions in the KGQA dataset and evaluated the output accuracy of LLMs by designing an answer assessment methodology. ChatGPT and GPT-4 demonstrated superior performance in zero-shot situations.

VRN, the baseline method for MetaQA dataset, used an end-to-end probabilistic model for multi-hop problem solving [41].

Reference [45] used the T5 model to convert natural language questions into Prolog queries, achieving high response accuracy even when trained with only a small amount of data. The results in Table 2. Table were the model training with 1000 randomly sampled samples from the training set and there were no error responses on MetaQA dataset.

EDGQA consisted of question decomposition and query generation phases. The question decomposition phase generated an Entity Description Graph (EDG) that describes the natural language question. The query generation phase generated and combined subqueries using EDG to get the answer [51] (Tables 1, 2 and 3).

Table 1. Performance on WebQuestionsSP dataset

Method	Architecture	hits@1
PullNet [49]	LSTM + GNN	68.10
ReLMKG [46]	BERT + GNN	74.10
FLAN-T5 [50]	T5	59.87
GPT-3 [50]	GPT-3	67.68
ChatGPT [50]	ChatGPT	83.70
GPT-4 [50]	GPT-4	90.45

Table. 2. Performance comparison of different methods for KGQA on MetaQA dataset (Hits@1)

Method	Architecture	1-hop	2-hop	3-hop
VRN [41]	Probabilistic Modeling	97.50	89.90	62.50
PullNet [49]	LSTM + GNN	97.00	99.90	91.40
UNIKGQA [52]	PLM + GNN	98.00	99.90	99.90
T5-small + Prolog [45]	T5	100.00	100.00	100.00

KGQAN used the BERT model to translate questions into triples, searched the triples in KG to find possible matches, and scored the answers based on existing utilisation of generic word embedding models [53].

Table 3. Performance on QALD-9 dataset

Method	Architecture	F1
EDGQA [51]	NLP Parser	32.00
KGQAN [53]	BERT	44.07
FLAN-T5 [50]	T5	30.17
GPT-3 [50]	GPT-3	38.54
ChatGPT [50]	ChatGPT	45.71
GPT-4 [50]	GPT-4	57.20

3.6 The Advantages of LLMs Applied to KGQA

Prior to the applications of language models in KGQA task, it typically used LSTM, CNN and GNN models to encode problems and graphs for answering. The IR methods of KGQA has begun to use the similarity between candidate answers and questions for answer reasoning. However, that similarity reality inherently lacks semantic matching without language models. Even though the similarity score is high, it does not necessarily mean that the two are semantically the same, and thus such KGQA methods have a bottleneck in answering fuzzy questions with deep semantics. For KGQA methods that require the use of a query language for retrieval in KG, any simple entity or relation mismatch would result in a wrong answer. However, the practical application of the KGQA system inevitably involves problems of spelling or grammatical errors. That kind of mistake is difficult to handle with LSTM, CNN or GNN based methods. From this perspective, if the KGQA methods could relate to the context of the questions and understand the questions, candidate entities and relationships at the semantic level, it can mitigate the errors caused by input errors to a certain extent.

With the development of language models, PLMs (Pre-trained Language Models) began to replace LSTM and CNN modules in the encoding and parsing modules of KGQA methods. PLM provides a semantic-based unified encoding approach capable of mapping unstructured text and structured KG information into a cohesive semantic space. The use of PLMs modules could go some way towards solving the problem of semantic mismatches that existed in previous methods and provide better robustness in the face of erroneous inputs. The results of related studies also showed that the KGQA method using PLMs module has higher accuracy [53].

Then after that, the real LLMs were born. Initially, LLM remained as a module in the generic KGQA methodology. However, in fact, LLMs had learnt sufficient generic knowledge from a large amount of natural language during the training process and had a certain degree of reasoning ability. So researchers had started to try to move away from KGQA generic framework and directly used LLMs to answer questions in the KGQA dataset. It has been shown that LLMs, have exceeded the zero-point capabilities of traditional deep learning and knowledge representation models on some KGQA datasets.

4   Future

In the precious sections, the application of LLM techniques in KGQA task was discussed. However, there are still many problems and challenges in the application of LLM techniques.

The primary issue regarding the use of LLMs in KGQA may be in terms of evaluation metrics. If LLMs are only used as a module in the KGQA methods, the evaluation process is the same as the original KGQA methods. However, if separated from the basic framework of the KGQA methods, the accuracy of the LLMs' responses may be difficult to reliably evaluate. The correctness of ChatGPT responses has even been evaluated manually in existing studies.

The knowledge embedded in LLMs is mainly general domain knowledge, which may make it difficult to achieve the desired results for KGQA task in specialized domains. Therefore, fine-tuning of LLMs is required for some specialized tasks. In addition, timely updating of LLMs is likewise required to respond to knowledge updates; however, many state-of-the-art LLMs are closed-source models, which poses a hindrance to model iteration and updating for professional mission needs. On the other hand, LLMs are very large in size and number of parameters, and there are resource and efficiency bottlenecks in the training and deployment of LLMs.

Although LLMs have been applied to the KGQA task, the essence is that LLMs understand the text of questions and KGs rather than the KG itself. Therefore, how to make LLMs truly understand the huge and complex structured information contained in the KGs still needs to be deeply researched.

In addition, knowledge graphs are now not only limited to textual data, but expanded to multimodal data such as text, image, audio, and video. Therefore, how to exploit the value of LLMs in multimodal KGs also deserves more in-depth research.

Acknowledgement. This work was supported in part by the project "Development of Fault Warning and Safety Protection Model for Integrated Energy Service Stations," funded by Grant No. 13000002022139003 from Zhejiang Energy Digital Technology Co., Ltd. The authors express their gratitude for the financial support provided by this project.

References

1. Tian, L., Zhou, X., Wu, Y.-P., Zhou, W.-T., Zhang, J.-H., Zhang, T.-S.: Knowledge graph and knowledge reasoning: a systematic review. J. Electron. Sci. Technol. **20**(2), 1–19 2022/06/01/ 2022. https://doi.org/10.1016/j.jnlest.2022.100159
2. Pan, J.Z., Razniewski, S., Kalo, J.-C., Singhania, S., Chen, J., Dietze, S., et al.: Large Language Models and Knowledge Graphs: Opportunities and Challenges. ArXiv, vol. abs/2308.06374, pp. 1–30 (2023)
3. Richens, R.H.: Preprogramming for mechanical translation. Mech. Transl. **3**(1), 20–25 (1956)
4. Collins, A.M., Quillian, M.R.: Retrieval time from semantic memory. J. Verbal Learn. Verbal Behav. **8**(2), 240–247, (1969), 04/01/ 1969. https://doi.org/10.1016/S0022-5371(69)80069-1
5. Van Melle, W.: MYCIN: a knowledge-based consultation program for infectious disease diagnosis. Int. J. Man-Mach. Stud. **10**(3), 313–322 (1978). 05/01/1978. https://doi.org/10.1016/S0020-7373(78)80049-2

6. Felgenbaum, E.A.: The art of artificial intelligence: themes and case studies of knowledge engineering. Presented at the International Joint Conference on Artificial Intelligence, Cambridge, USA (1977)
7. Berners-Lee, T., Cailliau, R., Groff, J.-F., Pollermann, B.J.I.R.: World-wide web: the information universe. Electron. Networking **2**(1), 461–471 (1992)
8. Berners-Lee, T., Hendler, J., Lassila, O.: The semantic web. Sci. Am. **284**(5) (2001)
9. Berners-Lee, T.: Linked data. https://www.w3.org/DesignIssues/LinkedData.html. Accessed 1–12 2024
10. Singhal, A.: Introducing the knowledge graph: Things, not strings. https://www.blog.google/products/search/introducingknowledge-graph-things-not/. Accessed
11. OpenAI. ChatGPT: Get Instant Answers, Find Creative Inspiration, Learn Something New. https://openai.com/chatgpt. Accessed
12. Vaswani, A., Shazeer, N., Parmar, N., Uszkoreit, J., Jones, L., Gomez, A.N., et al.: Attention is all you need. Presented at the International Conference on Neural Information Processing Systems, Long Beach, California, USA (2017)
13. Radford, A., Narasimhan, K.: improving language understanding by generative pre-training (2018)
14. Devlin, J., Chang, M.-W., Lee, K., Toutanova, K.: BERT: pre-training of Deep Bidirectional Transformers for Language Understanding, ArXiv, vol. abs/1810.04805, pp. 1–16 (2018)
15. Radford, A., Wu, J., Child, R., Luan, D., Amodei, D., Sutskever, I.: Language Models are Unsupervised Multitask Learners (2019)
16. Brown, T.B., Mann, B., Ryder, N., Subbiah, M., Kaplan, J., Dhariwal, P., et al.: Language Models are Few-Shot Learners, ArXiv, vol. abs/2005.14165, pp. 1–75 (2020)
17. OpenAI, GPT-4 Technical Report, ArXiv, vol. abs/2303.08774, pp. 1–99 (2023)
18. Pan, S., Luo, L., Wang, Y., Chen, C., Wang, J., Wu, X.: Unifying large language models and knowledge graphs: a roadmap. IEEE Trans. Knowl. Data Eng. Early Access, 1–20 (2024). https://doi.org/10.1109/TKDE.2024.3352100
19. Wu, P., Zhang, X., Feng, Z.: A survey of question answering over knowledge base. In: Zhu, X., Qin, B., Zhu, X., Liu, M., Qian, L. (eds.) China Conference on Knowledge Graph and Semantic Computing, Singapore. Springer Singapore, pp. 86–97 (2019)
20. Hu, N., Wu, Y., Qi, G., Min, D., Chen, J., Pan, J.Z., et al.: An empirical study of pre-trained language models in simple knowledge graph question answering. World Wide Web **26**(1), 2855–2886 (2023). https://doi.org/10.1007/s11280-023-01166-y
21. Yin, W., Yu, M., Xiang, B., Zhou, B., Schütze, H.: Simple question answering by attentive convolutional neural network. In: International Conference on Computational Linguistics, Osaka, Japan, 2016: The COLING 2016 Organizing Committee, pp. 1746–1756 (2016)
22. Mohammed, S., Shi, P., Lin, J.: Strong baselines for simple question answering over knowledge graphs with and without neural networks. In: Conference of the North American Chapter of the Association for Computational Linguistics: Human Language Technologies, New Orleans, Louisiana, 2018: Association for Computational Linguistics, pp. 291–296. https://doi.org/10.18653/v1/N18-2047
23. Lukovnikov, D., Fischer, A., Lehmann, J.: Pretrained transformers for simple question answering over knowledge graphs. In: Ghidini, C., et al. (eds.) International Semantic Web Conference, pp. 470–486. Springer, Cham (2019)
24. Ture, F., Jojic, O.: No need to pay attention: simple recurrent neural networks work! In: Conference on Empirical Methods in Natural Language Processing, Copenhagen, Denmark, 2017: Association for Computational Linguistics, pp. 2866–2872 (2017)
25. Petrochuk, M., Zettlemoyer, L.: SimpleQuestions nearly solved: a new upperbound and baseline approach. In: Conference on Empirical Methods in Natural Language Processing, Brussels, Belgium, 2018: Association for Computational Linguistics, pp. 554–558 (2018)

26. Sidiropoulos, G., Voskarides, N., Kanoulas, E.: Knowledge graph simple question answering for unseen domains. In: Automated Knowledge Base Construction, Online (2020)

27. Luo, D., Su, J., Yu, S.: A BERT-based approach with relation-aware attention for knowledge base question answering. In: International Joint Conference on Neural Networks (IJCNN), Glasgow, UK, 19–24 July 2020 2020, pp. 1–8. IEEE. https://doi.org/10.1109/IJCNN48605.2020.9207186

28. Song, Y., Li, W., Dai, G., Shang, X.: Advancements in complex knowledge graph question answering: a survey. Electronics **12**(21), 1–16 (2023). https://www.mdpi.com/2079-9292/12/21/4395

29. Bhutani, N., Zheng, X., Jagadish, H.V.: Learning to answer complex questions over knowledge bases with query composition. Presented at the ACM International Conference on Information and Knowledge Management, Beijing, China (2019). https://doi.org/10.1145/3357384.3358033

30. Zheng, W., Yu, J.X., Zou, L., Cheng, H.: Question answering over knowledge graphs: question understanding via template decomposition. Proc. VLDB Endowment **11**(11), 1373–1386 (2018). https://doi.org/10.14778/3236187.3236192

31. Yih, W.-T., Chang, M.-W., He, X., Gao, J.: Semantic parsing via staged query graph generation: question answering with knowledge base. In: Annual Meeting of the Association for Computational Linguistics, Beijing, China, 2015: Association for Computational Linguistics, pp. 1321–1331 (2015)

32. Chen, Z.-Y., Chang, C.-H., Chen, Y.-P., Nayak, J., Ku, L.-W.: UHop: an unrestricted-hop relation extraction framework for knowledge-based question answering. In: Conference of the North American Chapter of the Association for Computational Linguistics: Human Language Technologies, Minneapolis, Minnesota, pp. 345–356. Association for Computational Linguistics (2019)

33. Berant, J., Chou, A., Frostig, R., Liang, P.: Semantic parsing on freebase from question-answer pairs. In: Conference on Empirical Methods in Natural Language Processing, Seattle, Washington, USA. Association for Computational Linguistics, pp. 1533–1544 (2013)

34. Yih, W.-t., Richardson, M., Meek, C., Chang, M.-W., Suh, J.: The value of semantic parse labeling for knowledge base question answering. In: Annual Meeting of the Association for Computational Linguistics, Berlin, Germany. Association for Computational Linguistics, pp. 201–206 (2016)

35. Serban, I.V., García-Durán, A., Gulcehre, C., Ahn, S., Chandar, S., Courville, A., et al.: Generating factoid questions with recurrent neural networks: The 30M factoid question-answer corpus. In: Annual Meeting of the Association for Computational Linguistics, Berlin, Germany. Association for Computational Linguistics, pp. 588–598 (2016)

36. Jiang, K., Wu, D., Jiang, H.: FreebaseQA: a new factoid QA data set matching trivia-style question-answer pairs with freebase. In: Conference of the North American Chapter of the Association for Computational Linguistics, Minneapolis, Minnesota, pp. 318–323. Association for Computational Linguistics (2019)

37. Bao, J., Duan, N., Yan, Z., Zhou, M., Zhao, T.: Constraint-based question answering with knowledge graph. In: International Conference on Computational Linguistics, Osaka, Japan, 2016, no. 2503–2514: The COLING 2016 Organizing Committee

38. Trivedi, P., Maheshwari, G., Dubey, M., Lehmann, J.: LC-QuAD: a corpus for complex question answering over knowledge graphs. In: International Semantic Web Conference, Vienna, Austria, d'Amato, C., et al. (eds.) 2017// 2017: Springer International Publishing, pp. 210–218

39. Talmor, A., Berant, J.: The web as a knowledge-base for answering complex questions. In: Conference of the North American Chapter of the Association for Computational Linguistics, New Orleans, Louisiana, 2018. Association for Computational Linguistics, pp. 641–651 (2018)

40. Dubey, M., Banerjee, D., Abdelkawi, A., Lehmann, J.: LC-QuAD 2.0: a large dataset for complex question answering over Wikidata and DBpedia. In: Ghidini, C., et al. (eds.) International Semantic Web Conference, Auckland, New Zealand, pp. 69–78. Springer (2019)

41. Zhang, Y., Dai, H., Kozareva, Z., Smola, A.J., Song, L.: Variational reasoning for question answering with knowledge graph. presented at the AAAI Conference on Artificial Intelligence, New Orleans, Louisiana, USA (2018)

42. Chatterjee, J., Dethlefs, N.: Automated question-answering for interactive decision support in operations & maintenance of wind Turbines. IEEE Access **10**, 84710–84737 (2022). https://doi.org/10.1109/ACCESS.2022.3197167

43. Liu, A., Huang, Z., Lu, H., Wang, X., Yuan, C.: BB-KBQA: BERT-Based Knowledge Base Question Answering. Presented at the Chinese Computational Linguistics: 18th China National Conference, Kunming, China, October 18–20, 2019 (2019). https://doi.org/10.1007/978-3-030-32381-3_7

44. Xu, Y., Zhu, C., Xu, R., Liu, Y., Zeng, M., Huang, X.: Fusing context into knowledge graph for commonsense question answering. In: International Joint Conference on Natural Language Processing (IJCNLP), pp. 1201–1207. Association for Computational Linguistics (2021)

45. Madani, N., Srihari, R.K., Joseph, K.: Domain specific question answering over knowledge graphs using logical programming and large language models. ArXiv, vol. abs/2303.02206, pp. 1–6 (2023)

46. Cao, X., Liu, Y.: ReLMKG: reasoning with pre-trained language models and knowledge graphs for complex question answering. Appl. Intell. **53**(10), 12032–12046 (2023). https://doi.org/10.1007/s10489-022-04123-w

47. Yang, L., Guo, H., Dai, Y., Chen, W.: A method for complex question-answering over knowledge graph. Appl. Sci. **13**(8), 1–23 (2023). https://www.mdpi.com/2076-3417/13/8/5055

48. Salnikov, M., Le, H., Rajput, P., Nikishina, I., Braslavski, P., Malykh, V. et al.: Large language models meet knowledge graphs to answer factoid questions. ArXiv, vol. abs/2310.02166 (2023)

49. Sun, H., Bedrax-Weiss, T., Cohen, W.: PullNet: open domain question answering with iterative retrieval on knowledge bases and text. In: Conference on Empirical Methods in Natural Language Processing, Hong Kong, China. Association for Computational Linguistics, pp. 2380–2390 (2019)

50. Tan, Y., Min, D., Li, Y., Li, W., Hu, N., Chen, Y., et al.: Can ChatGPT replace traditional KBQA models? an in-depth analysis of the question answering performance of the GPT LLM family. In: Payne, T.R., et al. (eds.) International Semantic Web Conference, pp. 348–367. Springer, Cham (2003)

51. Hu, X., Shu, Y., Huang, X., Qu, Y.: EDG-based question decomposition for complex question answering over knowledge bases. presented at the International Semantic Web Conference (2021). https://doi.org/10.1007/978-3-030-88361-4_8

52. Jiang, J., Zhou, K., Zhao, X., Wen, J.-R.: UniKGQA: unified retrieval and reasoning for solving multi-hop question answering over knowledge graph. In: International Conference on Learning Representations (2023)

53. Omar, R., Dhall, I., Kalnis, P., Mansour, E.: A universal question-answering platform for knowledge graphs. Proc. ACM Manage. Data **1**(1), 1–25 (2023). https://doi.org/10.1145/3588911

An Offline Learning-Based Anti-interference Communication Scheme for UAV Networks

Tao Tang[1,2,3], Runhui Zhao[1,2,3], Hong Wen[1,2,3]([✉]), Xuewei Feng[1,2,3], Weihong Shi[1,2,3], and Yulin Peng[1,2,3]

[1] School of Aeronautics and Astronautics, University of Electronic Science and Technology of China, Chengdu 611731, People's Republic of China
{taotang,shiweihong,yulin}@std.uestc.edu.cn,
uestcrunhui@gmail.com, sunlike@uestc.edu.cn
[2] Aircraft Swarm Intelligent Sensing and Cooperative Control Key Laboratory of Sichuan Province, Chengdu 611731, People's Republic of China
[3] Sichuan Intelligent IoT Communication Technology Engineering Research Center, UESTC, Chengdu, China

Abstract. Unmanned Aerial Vehicles (UAV) face challenges from advanced interference technologies, making them susceptible to malicious node attacks, data interception, and tampering. Traditional anti-interference decisions have limitations, as they cannot adaptively adjust to changes in interference signals. Moreover, anti-interference communication models based on Deep Reinforcement Learning (DRL) require prolonged interactions with the environment, demanding high requirements for anti-interference environments. This paper investigates an offline anti-interference decision based on Decision-Transformers, which can quickly and stably acquire practical anti-interference decision models. Simulation experiments have verified the effectiveness of this algorithm in making anti-interference decisions under AWGN and fading channel conditions. Furthermore, this offline approach can achieve the expected reward targets with fewer training iterations.

Keywords: UAV · Communication · Anti-interference decisions · Deep reinforcement learning · Decision-Transformers

1 Introduction

Unmanned Aerial Vehicles (UAVs), with their low cost, rapid deployment capabilities, and high maneuverability, are extensively used in both civilian and military domains. However, UAV terminals, which communicate through wireless networks, are vulnerable to malicious node attacks, data interception, and tampering. Literature [1] introduced and verified the anti-jamming effectiveness of artificial noise in MIMO communication systems, thereby highlighting a significant threat to UAV communication security. Furthermore, the advent of software-defined radio technology has substantially lowered the barrier for attacks.

In response to these security threats, traditional anti-jamming techniques, such as frequency hopping [2], spread spectrum [3], time slot changes, power changes, and

P. Siarry et al. (Eds.): WCNA 2023, LNEE 1361, pp. 348–354, 2025.
https://doi.org/10.1007/978-981-96-2409-6_34

communication rate changes [4, 5], along with anti-jamming decision libraries, exhibit certain limitations and are only effective against specific types of interference. Literature [6] demonstrates the application of ML algorithms in the communication anti-jamming field, showcasing exceptional adaptability and the capability to discern interference signals. Literature [7, 8] proposed a relay communication anti-jamming algorithm based on Q-learning that could circumvent malicious interference through channel hopping. Qi [9] developed an improved hopping strategy based on Q-learning, providing a solution for broadband communication systems under interference attacks when the transmitter and receiver have incomplete understanding of the interference patterns. In literature [10], the authors engineered a Deep Reinforcement Learning (DRL) scheme to process signal waterfall charts, achieving finite states and optimal anti-jamming strategies. Literature [11] merged Recurrent Neural Networks (RNN) with Deep Q-Networks (DQN), effectively boosting the system's anti-jamming performance.

The Decision Transformer (DT), as an innovative offline reinforcement learning approach, converts reinforcement learning challenges into sequential prediction problems [12–14], expediting the learning process. Moreover, [15] presented a rapid anti-jamming communication scheme based on domain knowledge reuse, which accelerates learning by exploiting the similarities between state-action pairs.

Although adaptive decision schemes based on reinforcement learning and deep reinforcement learning have addressed some drawbacks of traditional solutions, Deep Reinforcement Learning (DRL) in complex environments is hindered by the high-dimensional state space issue, necessitating prolonged interaction with the environment for stable training, which is untenable for UAV terminals. This study focuses on fully leveraging the computational capabilities of ground control stations to construct a three-tier architecture comprising ground control stations, control terminals, and UAV terminals. It explores a time series prediction model for offline training of DRL-based anti-jamming decision trajectories, aiming to enhance the anti-jamming capability of UAV communication links.

2 Communication Anti-jamming System Model Based on UAV

In the anti-jamming architecture of Unmanned Aerial Vehicles (UAVs), a three-tier structure plays a crucial role, comprising the UAV terminal, control terminal, and ground station as shown in Fig. 1. The UAV terminal is responsible for collecting data and executing control commands. Serving as an intermediary layer, the control terminal performs lightweight data processing and computation, and issues control instructions. The ground station, on the other hand, undertakes tasks related to large-scale data processing, storage, and advanced analysis, providing the necessary computational power to support the decision-making process.

2.1 Reinforcement Learning for Anti-jamming Decision-Making

The anti-jamming decision-making process is divided into three stages: Initially, the control terminal sends pilot data to the UAV terminal at fixed time slots, and the UAV terminal sends the same data back to the control terminal. The control terminal then

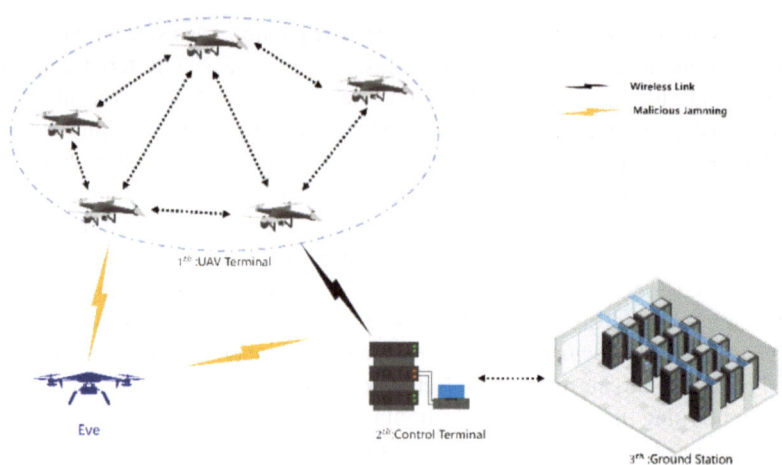

Fig. 1. UAV Communication Link Environment

analyzes the error rate or packet loss rate based on the returned data. If the communication status is abnormal, it instructs the terminal to reduce power and initiate interference detection. In the second stage, the ground station extracts the characteristics of the interference signal from the uplink data, uses machine learning to identify the interference, and notifies the terminal through the feedback link. Finally, based on the interference information, the ground station updates the transmission strategy and guides the UAV terminal to re-upload the data to optimize communication quality.

The entire anti-jamming decision-making process can be abstracted as a Markov Decision Process (MDP), which includes a state space, action space, immediate rewards, and action functions to implement intelligent anti-jamming decisions. The action space defines all possible actions that can be taken before the next transmission time slot upon receiving feedback from the main station. These actions include changing the transmission signal pattern, symbol rate, coding method, Signal-to-Interference-plus-Noise Ratio (SINR), and anti-jamming method. The design of the action space aims to ensure the stability of the DRL algorithm, allowing the system to select the most appropriate anti-jamming strategy based on the current state and action when encountering interference. The reward space is the feedback of the current state and behavior of the Agent. Define our reward function using the following five indicators: communication quality, communication rate, SINR, signal processing time, and signal bandwidth Unlike other indicators, agent can only receive positive rewards when the communication quality and communication rate reach a certain threshold.

The Deep Q-Network (DQN) uses a deep neural network to approximate the Q-function, aiming to choose the optimal anti-interference communication strategy for UAV. Using an experience replay buffer to record every transformation in the current environment, including the current state $s = s_t$, action taken $a = a_t$, real-time reward $r(s, a, s')$, and the next state $s' = s_{t+1}$.

In DQN, the use of a target Q-network is intended to minimize the estimation error of the Q values, bringing the learned $Q(s, a)$ closer to the optimal Q function. By randomly sampling a batch of data from the experience replay pool and using these samples to iteratively optimize the model until the main network converges. During forward propagation, the network calculates the network output z_i layer by layer, from the input layer to the output layer, through linear transformations involving connection weights $W(c)$ and biases $\Theta(c)$, followed by an activation function ϕ. The backward propagation process adjusts the network parameters based on the error of the output units to optimize network performance.

Through continuous training and optimization of the DQN, UAVs can gradually learn the optimal anti-jamming communication strategy under specific environmental conditions. However, when faced with large state spaces, the extensive number of states and actions may lead to situations where historical experiences cannot cover new states. This can result in the agent being unable to learn new strategies during training, or even facing the risk of training collapse.

2.2 Offline Anti-jamming Decision Based on the Decision Transformer

The Decision Transformer (DT) integrates sequence models with reinforcement learning (RL) techniques, aimed at solving decision problems by directly predicting the next action from sequences of states, actions, and cumulative rewards. DT constructs a sequence model for decision-making based on sequences of states s_t,actions a_t, and rewards r_t, The state model includes the interference method $j_{i,n\tau}$, interference power $p_{j,i,n\tau}$, signal power transmitted by node $p_{r,i,n\tau}$, transmission signal bandwidth $b_{i,n\tau}$, signal processing time $t_{i,n\tau}$, and communication quality $q_{i,n\tau}$, Actions involve adjusting the node's transmission signal pattern $g_{i,(n+1)\tau}$, symbol rate $f_{b,I,(n+1)\tau}$, coding method $c_{i,(n+1)\tau}$, SINR $sinr_{r,i,(n+1)\tau}$, and anti-jamming method $a_{i,(n+1)\tau}$ The reward model comprises five parts: communication quality, communication rate, SINR, anti-jamming processing time, and signal bandwidth.

The core of the decision-making process is to use historical state data: s_{t-1} and s_t to predict the next optimal action a_{t+1}. The sequence s_t can be represented as:

$$S_t = \{(s_1, a_1, r_1), (s_2, a_2, r_2), \ldots, (s_t, a_t, r_t)\} \tag{1}$$

The cumulative reward at time step t, G_t can be represented by the following formula:

$$G_t = \sum_{k=0}^{T-t} \gamma^k r_{t+k} \tag{2}$$

where γ is the discount factor, T is the upper limit of the time range, and r_{t+k} is the immediate reward obtained at time step $t + k$. The Decision Transformer (DT) takes the cumulative reward G_t as one of its input parameters, allowing the model to directly utilize the accumulated reward information to guide decision-making, in order to predict the best action to be taken in the current state. In this way, DT can effectively capture key information in RL tasks. When the model receives the current state s_t and the target cumulative reward G, it outputs a prediction for the next action $\hat{a}_{t+1}$:

$$\hat{a}_{t+1} = DT_\theta(S_{t-1}, s_t, G_t) \tag{3}$$

The model is trained by minimizing the difference between the predicted action and the actual action. The loss function can be represented as:

$$\min_{\theta} \mathbb{E}_{(s,a,r)\sim\mathcal{D}}\left[\mathcal{L}(DT_{\theta}(S_{t-1}, s_t, G_t), a)\right] \qquad (4)$$

DT_{θ} is the Decision Transformer model parameterized by θ, $\mathcal{D}$ is the dataset, and $\mathcal{L}$ is the loss function. The training process of DT involves optimizing its parameters θ to minimize the prediction error across the entire training set. Typically, this is achieved through gradient descent or its variants. The training optimization process under learning rate αcan be represented as:

$$\theta \leftarrow \theta - \alpha\nabla_{\theta}\mathbb{E}_{(s,a,r)\mathcal{D}}[\mathcal{L}(DT_{\theta}(s, G_t), a)] \qquad (5)$$

The Decision Transformer relies on sequences of state-action-reward obtained from historical experiences. Different DRL algorithms affect the distribution of states, actions, and rewards in the training sequence, thereby influencing the training outcome of the DT model to some extent. Thus, we can construct the entire system framework: the first part is the control terminal, which collects or records the historical decision data of each node (agent) and uploads it to the ground station. The second part, the ground station, organizes and filters the agents' historical decision trajectories to create a training dataset. The third part is the trajectory data, used for offline training of the DT model, resulting in a generalized anti-jamming decision model. The model parameters are shared with each node for anti-jamming decision inference.

The Decision Transformer relies on sequences of state-action-reward obtained from historical experiences. Different DRL algorithms affect the distribution of states, actions, and rewards in the training sequence, thereby influencing the training outcome of the DT model to some extent. Thus, we can construct the entire system framework: the first part is the control terminal, which collects or records the historical decision data of each node (agent) and uploads it to the ground station. The second part, the ground station, organizes and filters the agents' historical decision trajectories to create a training dataset. The third part is the trajectory data, used for offline training of the DT model, resulting in a generalized anti-jamming decision model. The model parameters are shared with each node for anti-jamming decision inference.

3 System Simulation and Results Modeling

Custom reinforcement learning environments were developed using Python 3.9 and the gym package. The deep learning networks were constructed with the Pytorch1.12 framework and deployed with CUDA 11.6.

Figure 2(a) shows the reward of DQN under fading channels. At lower training epochs, the cumulative reward is also low, but as the number of epochs increases, the reward value continually improves. Figure 2(b) presents the cumulative rewards of DT under different targets in fading channels. The red graph in Fig. 2(c) represents the reward of DQN in an AWGN channel, while Fig. 2(d) shows the cumulative rewards of DT under different targets in an AWGN channel. In the same channel conditions, DQN's reward performance is superior to that of DT; however, DQN requires a longer training time. In the DT algorithm, selecting a higher target yields better results.

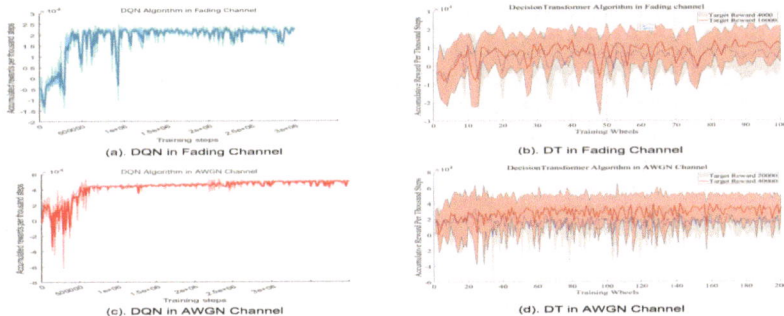

Fig. 2. The training and evaluation of DQN and DT in AWGN and fading channel environments. Peroration

4 Conclusions

Under the three-layer architecture of UAV communication, relying on the ground main station-control end-terminal scheme, the communication anti-jamming decision-making scheme based on DT overcomes the problems of slow convergence of traditional RL and DRL online training and long interaction time between nodes. In the AWGN channel, the offline decision-making scheme can achieve the same performance as the DRL model with fewer iterations required. Meanwhile, under fading channel conditions, our new DT model performs well with reasonable training data selected, achieving better anti-jamming effects.

Funding Statement. This work is supported by the National Natural Science Foundation of China (NSFC) under Grants U23B2021 and 62201132.

References

1. Yu, Z., Song, H., Wen, H., Liu, Y., Li, W.: A hardware simulation platform of artificial noise-assisted MIMO communication system based on LabVIEW-USRP. In: 2023 IEEE 34th Annual International Symposium on Personal, Indoor and Mobile Radio Communications (PIMRC), Toronto, ON, Canada, pp. 1–6 (2023). https://doi.org/10.1109/PIMRC56721.2023.10294015
2. Qi, J., Zhang, H., Qi, X., Peng, M.: Deep reinforcement learning based hopping strategy for wideband anti-jamming wireless communications. IEEE Trans. Vehicular Technol., October 2023
3. Li, X., Chen, J., Ling, X., Wu, T.: Deep reinforcement learning-based anti-jamming algorithm using dual action network. IEEE Trans. Wireless Commun. **22**(7), 4625–4637 (2023)
4. Pelechrinis, K., Broustis, I., Krish-namurthy, S.V., Gkantsidis, C.: A measurement-driven anti-jamming system for 802.11 networks. IEEE/ACM Trans. Netw. **19**(4), 1208–1222 (2011). https://doi.org/10.1109/TNET.2011.2106139
5. Liao, R.F., et al.: Security enhancement for mobile edge computing through physical layer authentication. IEEE Access **7**, 116390–116401 (2019)
6. Yao, F., Jia, L.: A collaborative multi-agent reinforcement learning anti-jamming algorithm in wireless networks. IEEE Wireless Commun. Lett. **8**(4), 1024–1027 (2019). https://doi.org/10.1109/LWC.2019.2904486

7. Zhang, Z., Wu, Q., Zhang, B., Peng, J.: Intelligent anti-jamming relay communication system based on reinforcement learning. In: Proc. 2nd Int. Conf. Commun. Eng. Technol. (ICCET), Nagoya, Japan, 2019, pp. 52–56 (2019). https://doi.org/10.1109/ICCET.2019.8726916

8. Elleuch, I., Pourranjbar, A., Kaddoum, G.: A novel distributed multi-agent reinforcement learning algorithm against jamming attacks. IEEE Commun. Lett. **25**(10), 3204–3208 (2021). https://doi.org/10.1109/LCOMM.2021.3097290

9. Qi, J., Zhang, H., Qi, X., Peng, M.: Deep reinforcement learning based hopping strategy for wideband anti-jamming wireless communications. IEEE Trans. Veh. Technol. https://doi.org/10.1109/TVT.2023.3324387

10. Liu, X., Xu, Y., Jia, L., Wu, Q., Anpalagan, A.: Anti-jamming communications using spectrum waterfall: a deep rein-forcement learning approach. IEEE Commun. Lett. **22**(5), 998–1001 (2018). https://doi.org/10.1109/LCOMM.2018.2815018

11. Chang, X., Li, Y., Zhao, Y., Du, Y., Liu, D.: An improved anti-jamming method based on deep reinforcement learning and feature engineering. IEEE Access **10**, 69992–70000 (2022). https://doi.org/10.1109/ACCESS.2022.3187030

12. Chen, L., et al.: Decision transformer: reinforcement learning via sequence modeling. Adv. Neural. Inf. Process. Syst. **34**, 15084–15097 (2021)

13. Lee, K.H., et al.: Multi-game decision transformers. Adv. Neural Inf. Process. Syst. **35**, 27921–27936 (2022)

14. Zhou, Q., Niu, Y., Xiang, P., Li, Y.: Intra-domain knowledge reuse assisted reinforcement learning for fast anti-jamming communication. IEEE Trans. Inf. Forensics Secur. **18**, 4707–4720 (2023). https://doi.org/10.1109/TIFS.2023.3284611

15. Li, Z., Lu, Y., Li, X., Wang, Z., Qiao, W., Liu, Y.: UAV networks against multiple maneuvering smart jamming with knowledge-based reinforcement learning. In: IEEE Internet Things J. **8**(15), 12289–12310 (2021). https://doi.org/10.1109/JIOT.2021.3062659.S. Yang, J. Li and B. He, "A novel interference suppression method in spread

A Solution to User Fairness in NOMA Heterogeneous Networks Based on Channel Selection and Power Optimization

Dongpo Zhang[1](✉), Chunlei Chen[2], Zhaorong Wang[1], and Xiujuan An[1]

[1] The 36th Research Institute of China Electronic Technology Corporation, Jiaxing, Zhejiang, China
65866151@qq.com

[2] Department of Electronic and Communication Engineering, East China University of Science and Technology, Shanghai, China

Abstract. In the optimization process aimed at maximizing the overall transmission rate in 5G heterogeneous non-orthogonal multiple access networks, the issue of unequal resource allocation arises. This phenomenon directly results in significantly lower minimum rates for channels with poor transmission conditions. Under such circumstances, the service quality of affected users suffers negative impacts, leading to unbalanced service experiences among users. To address this issue, this paper proposes a strategy that decouples the overall optimization problem and combines optimal channel pairing selection with user power allocation optimization to ensure the maximization of the minimum user rate and address fairness issues among users. Firstly, this paper elaborates on the decoupling method for the overall optimization problem. Subsequently, it details the implementation of optimal user pairing selection and user power allocation optimization. Simulation verifications are conducted using system parameters that closely resemble actual scenarios. The simulation results demonstrate that the algorithm proposed in this paper significantly improves the minimum user rate without increasing system complexity, effectively addressing fairness issues among users to a certain extent.

Keywords: 5G NOMA Heterogeneous Networks · Unequal Resource Allocation · User Power Allocation Optimization · Fairness Issues

1 Introduction

With the explosive growth of 5G mobile communication devices in recent years, the demand for data volume has also increased, putting pressure on spectrum resources. In order to address the shortage of spectrum resources, the industry has proposed Non-Orthogonal Multiple Access (NOMA) technology [1–3]. NOMA utilizes encoding overlay technology at the transmitter to superimpose channels of different users, and employs Successive Interference Cancellation (SIC) technology at the receiver to decode received signals [4, 5], thereby allowing multiple users to reuse the same frequency band resources, greatly alleviating the dilemma of spectrum scarcity. Heterogeneous networks deploy small base stations (SBS) around traditional macro base stations

© The Author(s) 2025
P. Siarry et al. (Eds.): WCNA 2023, LNEE 1361, pp. 355–365, 2025.
https://doi.org/10.1007/978-981-96-2409-6_35

(MBS) to improve channel gain, enhance communication quality in user-dense areas or other special environments, and expand coverage. Due to the reduced distance between users and base stations, transmission is more effective, resulting in a better user experience [6–8]. Non-Orthogonal Access Heterogeneous Networks combine heterogeneous networks with NOMA technology to achieve more efficient resource utilization, better data transmission performance, and improved user experience.

Currently, research on Non-Orthogonal Access Heterogeneous Networks focuses primarily on increasing the system's total rate, i.e., maximizing the system's total rate under given power conditions [9–13]. However, optimizing the total rate may lead the system to allocate resources automatically to channels with better conditions, thus compressing the available resources for channels with poorer conditions. This results in a very low minimum speed for some users, slow data transmission speeds, and sacrifices fairness among users to some extent, leading to poor service quality for certain users and a subpar user experience. This paper aims to address user fairness while considering the total system rate, by studying optimal user-base station pairing and power allocation methods to improve user fairness.

2 System Model and Optimization Problem

2.1 Heterogeneous Network System Model of Non-orthogonal Multiple Access

This paper considers the downlink scenario of heterogeneous networks, as illustrated in Fig. 1, where the MBS is located at the center of the cell, with each MBS serving M users. Additionally, there are S SBSs distributed around the MBS, with each SBS serving U users. In this paper, Macro-base User Equipment (MUE) still accesses in an orthogonal multiple access manner, while Small-base User Equipment (SUE) accesses using NOMA, with each SBS occupying one channel.

Non-Orthogonal Access Heterogeneous Networks differ from single-base station non-orthogonal access networks in that, due to the presence of multiple base stations in heterogeneous networks, users not only experience serial interference from encoding overlay but also interference from other base stations, known as inter-tier interference. In some scenarios, inter-tier interference can lead to significant performance degradation.

In this paper, to fully utilize the spectrum, one MUE and one SBS are multiplexed onto the same subchannel. Therefore, the number of MUEs is the same as the number of SBSs, i.e., M = S. The received signal of MUE is represented by Eq. (1):

$$y_m(t) = h_{m,m}\sqrt{P_m}s_m(t) + \sum_{j=1}^{S} h_{m,j}\left(\sum_{i=1}^{U} \sqrt{P_{si}}s_i(t)\right) + z(t) \tag{1}$$

In the equation, y_m represents the actual received signal of MUE, $h_{m,m}$ is the channel gain corresponding to the user, P_m is the corresponding transmit power, $s_m(t)$ denotes the transmit signal of the user, $h_{m,j}$ represents the channel gain from SBS j to MUE m, $P_{s,i}$ is the transmit power of the i-th SBS, $s_i(t)$ denotes the transmit signal of the i-th SBS, and $z(t)$ denotes additive white Gaussian noise with mean 0 and variance σ^2. From the above definitions, it is evident that the first term on the right-hand side of Eq. (1) represents the ideal received signal under noise-free and interference-free conditions,

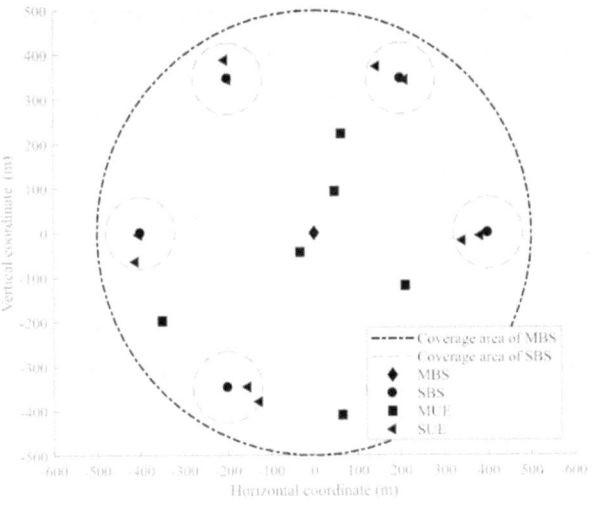

Fig. 1. Heterogeneous network system model with non-orthogonal access

the second term represents the inter-tier interference from all SBSs, and the third term represents the channel additive white noise.

$$y_{si}(t) = h_{si,si}\sqrt{P_{mi}}s_{mi}(t) + h_{si,si}\sum_{\substack{j=1 \\ j \neq i}}^{S} \sqrt{P_{mj}}s_{mj}(t) + h_{m,si}\sum_{m=1}^{M} \sqrt{P_m}s_m(t) + z(t)$$

(2)

In the equation, $y_{si}(t)$ represents the actual received signal of SUE, $h_{si,si}$ is the channel gain between the user and the corresponding SBS, P_{mi} is the transmit power of the corresponding SBS, $s_{mi}(t)$ denotes the transmit signal of the corresponding SBS, P_{mj} is the transmit power of the j-th SBS, $s_{mj}(t)$ denotes the transmit signal of the j-th SBS, $h_{m,si}$ represents the channel gain between SUE and the m-th MBS, where the meanings of P_m, $s_m(t)$, and $z(t)$ are the same as in Eq. (1). The first term on the right-hand side of Eq. (2) represents the ideal received signal of SUE under noise-free and interference-free conditions, the second term represents the intra-cell interference, the third term represents the inter-tier interference from MBSs, and the fourth term represents the channel additive white noise.

Based on Eq. (1) and Eq. (2), the achievable rates of MUE and the i-th SUE can be expressed as:

$$R_m = Blog_2\left(1 + \frac{P_m h_m}{\sum_{m=1}^{S} \alpha_{s,m} h_{mj} \sum_{i=1}^{U} P_{si} + \sigma^2}\right)$$

(3)

$$R_{si} = Blog_2\left(1 + \frac{P_{si} h_{si,si}}{\sum_{m=1}^{M} \alpha_{s,m} P_m h_{m,si} + h_{si,si} \sum_{i=1}^{U-1} P_{si} + \sigma^2}\right)$$

(4)

In the equations, R_m represents the achievable rate of MUE, R_{si} represents the achievable rate of the i-th SUE, B denotes the subchannel bandwidth, and $\alpha_{s,m}$ represents the power allocation factor, which is a binary number. MUE considers signals from SBSs as interference, while SUE not only treats signals from MBSs as interference but also, after decoding and eliminating signals from users with higher power using SIC technology, treats signals from other SBSs with lower power as interference.

2.2 System Optimization Problem

This paper aims to balance the total system rate and user fairness, thus optimizing objectives include maximizing the total system transmission rate and maximizing the minimum user rate. The system optimization equation is as follows:

$$\max_{S,U} R$$

$$R_m \geq R_{min}$$

$$R_m \geq R_{min}$$

$$\sum_{m=1}^{M} P_m \leq P_{max}^M$$

$$\sum_{i=1}^{U} P_{si} \leq P_{max}^s \tag{5}$$

$$\sum_{m=1}^{M} \alpha_{s,m} \leq 1$$

$$\sum_{s=1}^{S} \alpha_{s,m} \leq 1$$

In the above equation, R represents the total system rate, R_{min} represents the minimum user rate. P_{max}^M denotes the maximum total power of the MBS, and P_{max}^S represents the maximum total power of the small stations.

Based on the optimization equation above, it is observed that the optimization problem is a mixed integer nonlinear programming problem, making it unable to directly obtain a global optimal solution. Therefore, to find feasible solutions, this paper decouples the system optimization problem into two subproblems and utilizes corresponding algorithms to optimize each subproblem separately. The aim is to find an approximate optimal solution in practical applications.

3 Channel Selection and Power Allocation Optimization

3.1 Decoupling Method for System Optimization Problem

As mentioned in the previous section, solving the system optimization problem represented by Eq. (5) directly to obtain a global optimal solution is challenging. Considering that the optimization objectives are to maximize the system rate and maximize the minimum user rate, the main factors affecting these two optimization objectives are the optimization selection of channels and the optimization allocation of corresponding power.

Therefore, the system-wide optimization problem is decoupled into two subproblems: channel optimization selection and power optimization allocation.

The decoupling method mainly consists of two steps. Firstly, a channel gain ratio matrix is generated based on the channel gains. Then, utilizing the Hungarian optimization algorithm with the channel gain ratio matrix as input, a channel selection matrix is generated to determine the pairing of MUEs and SBSs. Subsequently, power allocation optimization is conducted using the binary search method to obtain the final results of user pairing and power allocation. The decoupling method for the system optimization problem is illustrated in Fig. 2.

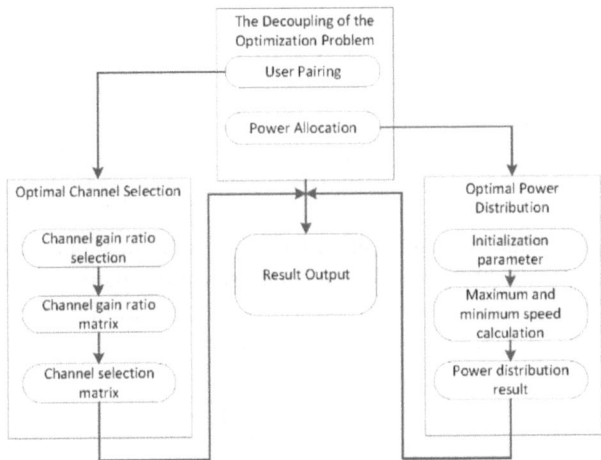

Fig. 2. System optimization problem decoupling

3.2 Channel Selection Optimization

Due to the inter-tier interference between MBS and SBS in the optimization Eq. (5), channel selection should not only aim to maximize channel gains but also strive to minimize interference as much as possible. Although users have multiple base stations to choose from, and each base station can serve multiple users, once a channel is selected, a user can only be paired with one base station at a time, meaning that channel selection between users and base stations corresponds one-to-one. How to achieve channel selection between users and base stations, as defined by the problem of objective allocation [14-17], is evidently a typical objective allocation problem, which can be optimized using the Hungarian algorithm. Based on the aforementioned analysis, selecting the ratio of channel gain coefficients serves as the input for the Hungarian algorithm:

$$H_{ra} = \frac{h_{m,s1} + h_{s,m}}{h_{m,m}} \tag{6}$$

The user's channel gain coefficient ratio is the sum of the interference from the MBS to the SBS channel with the highest gain and the interference from the SBS to the MUE, divided by the gain of the user's corresponding MBS. The advantage of selecting the ratio of channel gain coefficients lies in the fact that if the interference from MUE to SUE is significant, then the value of $h_{m,s1}/h_{m,m}$ will be relatively large. Similarly, if the interference from the SBS to the MUE is significant, then the value of $h_{s,m}/h_{m,m}$ will be relatively large. Using Eq. (6) as input for the Hungarian algorithm, considering the impact of both types of interference, we can obtain results with minimized overall interference.

Furthermore, expanding upon this, we can derive the matrix of channel gain coefficient ratios:

$$
\begin{bmatrix}
\frac{h_{1,11}}{h_{1,1}} + \frac{h_{1,1}}{h_{1,1}} & \frac{h_{1,21}}{h_{1,1}} + \frac{h_{2,1}}{h_{1,1}} & \cdots & \frac{h_{1,S1}}{h_{1,1}} + \frac{h_{S,1}}{h_{1,1}} \\
\frac{h_{2,11}}{h_{2,2}} + \frac{h_{1,2}}{h_{2,2}} & \frac{h_{2,21}}{h_{2,2}} + \frac{h_{2,2}}{h_{2,2}} & \cdots & \frac{h_{2,S1}}{h_{2,2}} + \frac{h_{S,2}}{h_{2,2}} \\
\vdots & & \ddots & \vdots \\
\vdots & & \ddots & \vdots \\
\frac{h_{M,11}}{h_{M,M}} + \frac{h_{1,M}}{h_{M,M}} & \cdots & \cdots & \frac{h_{M,S1}}{h_{M,M}} + \frac{h_{S,M}}{h_{M,M}}
\end{bmatrix}
\tag{7}
$$

By applying the Hungarian algorithm to solve the aforementioned matrix, we obtain the channel selection matrix. The final channel selection matrix is derived from the identity matrix through row-column exchanges. Assuming the dimension of the channel coefficient matrix is 4×4, a possible optimized channel selection matrix is as follows:

$$
C = \begin{bmatrix}
0 & 0 & 0 & 1 \\
0 & 0 & 0 & 0 \\
0 & 0 & 1 & 0 \\
1 & 0 & 0 & 0
\end{bmatrix}
\tag{8}
$$

In Eq. (8), rows represent MUEs, columns represent SBSs, and each row and each column have only one number as 1, with the rest being 0. This denotes that the i-th MUE is paired with the j-th base station, achieving optimal channel selection.

3.3 Power Distribution Optimization

Due to the presence of inter-tier interference, the system's power allocation optimization problem can be solved by obtaining equations among the channel matrix, power allocation matrix, and signal-to-interference-plus-noise ratio (SINR). The equations are as follows:

$$
\boldsymbol{H}^{(U+1)\times(U+1)} \times \boldsymbol{P}^{(U+1)\times 1} = \boldsymbol{SIR}^{(U+1)\times 1}
\tag{9}
$$

In the equations, H represents the channel matrix, P represents the power allocation matrix, SIR represents the signal-to-interference-plus-noise ratio (SINR) matrix, and U represents the number of SUEs.

The computation of the system's power allocation matrix can be achieved through solving a system of equations or by matrix inversion. This paper adopts the approach of matrix inversion. Taking the example of SBS user $U = 3$, where there is 1 MUE and 3 SUEs, the expansion of H, P and SIR is as follows:

$$H^{4\times4} = \begin{bmatrix} -h_{m,m} & h_{s,m} \times \gamma_m & h_{s,m} \times \gamma_m & h_{s,m} \times \gamma_m \\ h_{m,s1} \times \gamma_{s1} & -h_{s1,s1} & 0 & 0 \\ h_{m,s2} \times \gamma_{s2} & h_{s2,s2} \times \gamma_{s2} & -h_{s2,s2} & 0 \\ h_{m,s3} \times \gamma_{s3} & h_{s3,s3} \times \gamma_{s3} & h_{s3,s3} \times \gamma_{s3} & h_{s3,s3} \times \gamma_{s3} \end{bmatrix} \tag{10}$$

$$P = [P_m, P_{s1}, P_{s2}, P_{s3}]^T \tag{11}$$

$$SIR = \left[\gamma_m \times \sigma_m^2, \gamma_{s1} \times \sigma_{s1}^2, \gamma_{s2} \times \sigma_{s2}^2, \gamma_{s3} \times \sigma_{s3}^2\right]^T \tag{12}$$

In the equations above, P_m represents the power allocation value for MUE, P_{s1}, P_{s2}, and P_{s3} represent the power allocation values for the three users of the small base station. $\sigma_m^2, \sigma_{s1}^2, \sigma_{s2}^2$, and σ_{s3}^2 represent the thermal noise of the MBS and the three corresponding SBSs, γ_m represents the SINR of MUE, and γ_{s1}, γ_{s2}, and γ_{s3} represent the SINRs of the three SUEs:

$$\gamma_m = \frac{P_m h_m}{\sum_{m=1}^{S} \alpha_{s,m} h_{m,j} \sum_{i=1}^{U} P_{si} + \sigma^2} \tag{13}$$

$$\gamma_{si} = \frac{P_{si} h_{si,si}}{\sum_{m=1}^{M} \alpha_{s,m} P_m h_{m,si} + h_{si,si} \sum_{i=1}^{U-1} P_{si} + \sigma^2}, i = 1,2,3\ldots\ldots S \tag{14}$$

The power allocation matrix can be obtained using the following formula:

$$P = H^{-1} \times SIR \tag{15}$$

In practical applications, the binary search algorithm is employed to obtain the power allocation matrix. Since maximizing the minimum value is the optimization goal, we have $\gamma_m = \gamma_{s1} = \gamma_{s2} = \gamma_{s3}$, and we set its value equal to g. Suppose the maximum power values for MBS and SBS are P_{max}^M and P_{max}^S, respectively, and the corresponding maximum SINR and minimum SINR are γ_{max} and γ_{min} respectively. Based on the actual environment, we set the error threshold to ε. The algorithm proceeds as shown in Fig. 3, obtaining the final power allocation result and the corresponding SINR γ_{mid} at that time. The algorithm terminates when the condition $|\gamma_{max} - \gamma_{min}| \leq \varepsilon$ is met.

4 Simulation Result

To validate the effectiveness of the algorithm, simulations were conducted using MAT-LAB. The simulated system scenario is as follows: the MBS is located at the center of the cell with a coverage radius rm of 500 m, while the SBSs are uniformly distributed around the MBS with a coverage radius rs of 40 m. The minimum distance between the MUE and the MBS is 50 m, and the minimum distance between the SUE and the SBS is 2 m. The system bandwidth B is 5 MHz, and the noise power spectral density is -174 dBm.

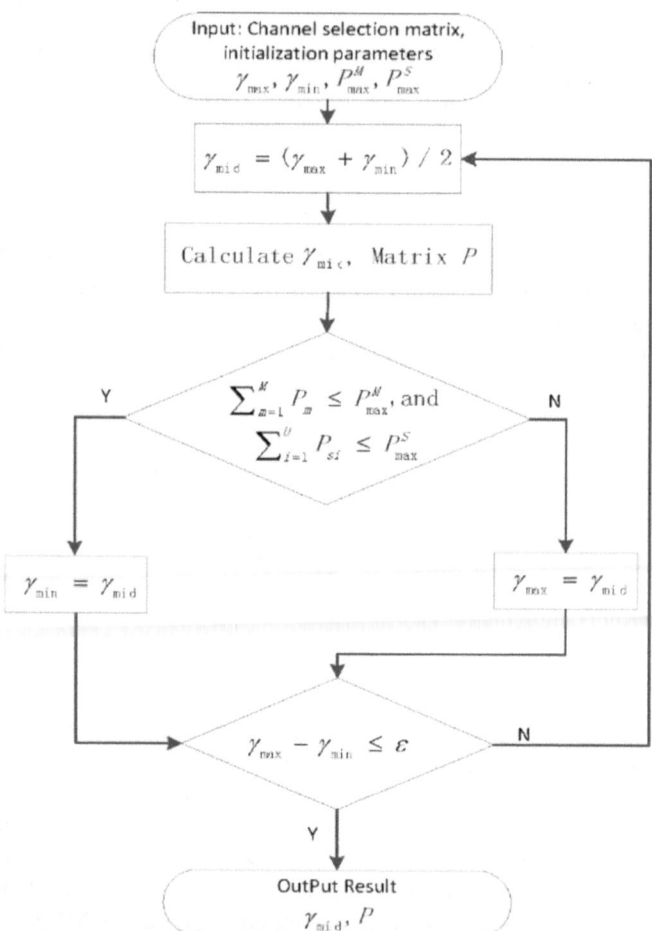

Fig. 3. Power distribution matrix solution algorithm flow

The specific simulation parameters are shown in Table 1.

Figure 4 presents a comparison of the minimum user rates between the random channel allocation method and the proposed channel allocation method in this paper, based

Table 1. Non-orthogonal access heterogeneous network system simulation parameters

Parameter	Value	Parameter	Value
R_M (m)	500	R_s (m)	40
P_{max}^M (dBm)	46	P_{max}^s (dBm)	23
σ^2 (dBm)	-74	B(MHz)	5

on 1000 Monte Carlo simulations, for different numbers of SBSs. It can be observed that under the same conditions, the minimum user rate obtained using the proposed channel allocation method is at least 0.5 Mbit/s higher than that obtained using the random channel allocation method, and this advantage becomes more pronounced with an increasing number of SBSs. It is also evident that the minimum user rate decreases with an increase in the number of SBSs. This is because with more SBSs and MUEs, the mutual interference becomes more severe, and the channel coefficients of users also deteriorate, thereby limiting the minimum user rate. Therefore, to ensure that the minimum user rate is at an acceptable level, the number of MUEs and SBSs should be kept within a certain range and should not be increased indefinitely.

Figure 5 compares the minimum user rates of the system under different SBS power conditions when using the proposed optimized power allocation method and the traditional Fractional Transmit Power Allocation (FTPA) method [18]. From the simulation results, it can be observed that under different transmission powers, the proposed optimized power allocation method outperforms the FTPA method in terms of minimum user rates.

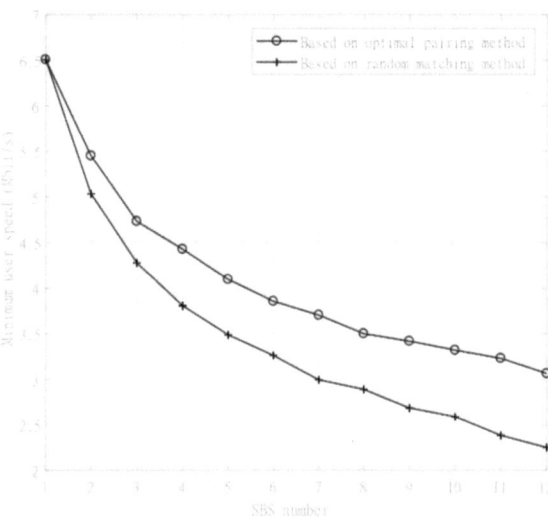

Fig. 4. The minimum speed varies with the number of SBSs

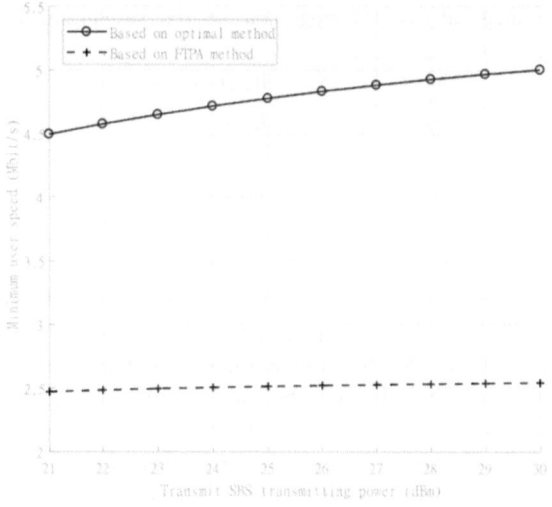

Fig. 5. The minimum speed varies with the transmit power of SBSs

5 Conclusion

This paper investigates the fairness issue in non-orthogonal heterogeneous network systems. Under the condition of maximizing the total system rate, with the goal of maximizing the minimum user rate, a method based on optimizing channel selection and power allocation is proposed to improve user fairness. Corresponding optimization algorithms are adopted to solve the problems of channel optimization selection and power allocation. Numerical simulation results verify the effectiveness of the proposed algorithm, demonstrating that under the same conditions, it can achieve a larger minimum user rate compared to traditional methods.

References

1. Ge, R., Li, G., et al.: Survey on power domain non-orthogonal multiple access technology in satellite communication networks. Mob. Commun. **43**(5), 33–39 (2019)
2. Guo, F., Lu, H., Jiang, X., Zhang, M., Wu, J., Chen, C.W.: QoS-aware user grouping strategy for downlink multi-cell NOMA systems. IEEE Trans. Wireless Commun. **20**(12), 7871–7887 (2021)
3. Zhang, C., Wang, H.: Research on energy-efficient transmission scheme for IRS-assisted NOMA systems. J. Chongqing Univ. Posts Telecommun. (Natural Sci. Edition) **35**(01), 16–22 (2023)
4. Shuai, H., Guo, K., et al.: On the performance of non-orthogonal multiple access integrated satellite-terrestrial networks in imperfect constraints. J. Electron. Inf. Technol. **45**(02), 16–22 (2023)
5. Yu, Z., Hou, J.: Optimization technology of downlink interference coordination in heterogeneous network based on non-orthogonal multiple access. Sci. Technol. Eng. **23**(01), 236–245 (2023)

6. Yi, X.: A multi-user schedule method for non-orthogonal multiple access in 5G heterogeneous network. J. Northwestern Polytech. Univ. **37**(02), 337–343 (2019)
7. Zhang, H., Zhang, Z., Long, K.: Resource allocation in NOMA heterogeneous network based on MEC. J. Commun. **41**(04), 27–33 (2019)
8. Cheng, Y., Tian, H., Liu, Z.: Collaborative optimization of joint user association and power control in NOMA heterogeneous network. Comput. Sci. **48**(03), 269–274 (2020)
9. Manglayev, T., Kizilirmak, R.C., Kho, Y.H., et al.: AI based power allocation for NOMA. Wireless Personal Commun. **124**, 253–3261 (2022)
10. Xiao, H., Ren, C., Nie, Z., Li, M.: Beamforming algorithm for multi-base station cooperation based on linearly-decrease inertia weight particle swarm optimization. J. Univ. Electron. Sci. Technol. China **44**(05), 663–667 (2015)
11. Rezvani, S., Jorswieck, E.A., Yamchi, N.M., Javan, M.R.: Optimal SIC ordering and power allocation in downlink multi-cell NOMA systems. IEEE Trans. Wireless Commun. **21**(6), 3553–3569 (2022)
12. Lee, H., Park, J., Lee, S.H., Lee, I.: Message-passing based user association and bandwidth allocation in HetNets with wireless backhaul. IEEE Trans. Wireless Commun. **22**(1), 704–717 (2023)
13. Zhang, S., Kang, G.: Energy efficient power allocation with NOMA in downlink heterogeneous networks. J. Electron. Inf. Technol. **42**(11), 2656–2663 (2020)
14. Lopes, P.A.C., Yadav, S.S., Ilic, A.: Fast block distributed CUDA implementation of the Hungarian algorithm. J. Parallel Distribut. Comput. **130**, 50–62 (2019)
15. Shopov, V.K., Markova, V.D.: Application of Hungarian algorithm for assignment problem. In: 2021 International Conference on Information Technologies (InfoTech), Varna, Bulgaria, pp. 1–4 (2021)
16. Liu, Y., Tong, M.: An application of Hungarian algorithm to the multi-target assignment. Fire Control Command Control **27**(4), 31–34 (2002)
17. Chopra, S., Notarstefano, G., Rice, M., Egerstedt, M.: A distributed version of the Hungarian method for multirobot assignment. IEEE Trans. Rob. **33**(4), 932–947 (2017)
18. Islam, S.M.R., Zeng, M., Dobre, O.A., Kwak, K.-S.: Resource allocation for downlink NOMA systems: key techniques and open issues. IEEE Wirel. Commun. **25**(2), 40–47 (2018)

Digital Twin for Power Load Forecasting

Zhijun Wang, Riyu Cong, Ruihong Wang, and Zhihui Wang[✉]

School of Electronic Information Engineering, Inner Mongolia University, Hohhot, China
wzhbit2007@163.com

Abstract. In this work, a novel Digital Twin model using attention mechanism integrated with LSTM to forecast the future power load of a specific user is developed. The power load prediction research is done in detail by taking into account important factors such as temperature, humidity, and the price of electricity. Therefore, LSTM networks are adopted for deep learning of the historical power load data, while the attention mechanism is used to assign weights to the significance of various factors that affect the power load and make better predictions of the future power load. The results of the presented experiment show the improved prediction accuracy and stability of the model in comparison with the existing power load prediction models. The present study also introduces a new and effective method for the power load forecasting.

Keywords: Digital Twin · LSTM · Attention Mechanism · Electricity power Load Forecasting

1 Introduction

Electricity in the contemporary world has become one of the most vital energy sources and a vital element of the infrastructure that plays an essential role in maintaining the stability of the socio-economic system. Hence, forecasting of electricity power load [1] has always been a big task for electric power systems particularly under varying environment and economic conditions. In order to improve the effciency of work in infrastructure, make electric power system to more stablity and meet the requirements that people's increasing demands for electric powers, the accuracy of electric power load forecasting is very important.

At present, there are many limitations in study of electricity power load forecasting in the field of time sequence analysis, it including some destabilizing factors that may effect the electricity power load such as rambling fluctuation of data and observation error [2]. Now we need to new methods for sloving the problems. Some models like Holt-Winters [3, 4] and the Auto Regressive Integrated Moving Average (ARIMA) [5, 6] are used to time sequence analysis, but if you want to an exact forecasting result by the such models, nut only you need enormous data, but also make the data more smoother ahead of time. So, the electricity power load forecasting used several technologies of forecasting based on machine learning, the principal ones are Support Vector Machine(SVM) [7, 8] and neural networks [9, 10]. Among them, the Long Short-Term Memory Network (LSTM)

P. Siarry et al. (Eds.): WCNA 2023, LNEE 1361, pp. 366–374, 2025.
https://doi.org/10.1007/978-981-96-2409-6_36

and Gated Recurrent Unit (GRU) are two different variants of Recurrent Neural Network (RNN), and both have exhibited a high level of predictability in time series forecasting. The LSTM network uses the curve sequence of the past electricity consumption [11] as a sequence to predict the electricity consumption at the next time step. Multi-layer GRU has been implemented for developing models for forecasting electricity consumption [12]. However, these two methods do not possess the right level of precision and do not generalize the features of the training data well enough to provide the best predictive results.

To overcome this problem, this paper presents a predictive digital twin (DT) [13] model that employs LSTM networks with attention mechanism [14]. This model is used to forecast the electricity load for the subsequent day provided with data. The structure of this paper is as follows: Sect. 2 discusses the definition of DT and the choice of attention mechanism, Sect. 3 describes the algorithm, Sect. 4 describes the experimental results and the analysis, and Sect. 5 concludes this paper and suggests the future work.

2 Background Knowledge

2.1 Definition of Digital Twin

Digital Twin: Digital twin is the term that is used to refer to the virtual replica of a real-life object or system. DT is an innovative technology for digital change and wise enhancement, which supports functions like the observation and forecasting of the real world based on data and physical models. DT is based on the accurate modeling of physical entities. DT [15] can be divided into the following six stages based on the depth of functionality:

(1) **Independent DT.** It is the realistic simulation of an environment that is usually created through the use of three dimensional graphics. 3D modeling can be considered as the basis for an independent DT.

(2) **Descriptive DT.** When a data pipeline is established to feed data into the independent DT, a visualization platform is generated. This platform also enables tracking of the power consumption status of each of the Point of Interest (POI) for the users.

(3) **Diagnostic DT.** After the data has been inputted into the descriptive DT, the diagnostic DT is capable of diagnosing the user.

(4) **Predictive DT.** The user's future power consumption can be predicted by the predictive DT through the collection of data such as temperature, humidity, electricity prices, etc.

(5) **Normative DT.** It can give the user risk analysis and suggestions according to the context of their environment.

(6) **Autonomous DT.** It can also self-explore factors that affect power load by employing unmanned aerial vehicles (drones).

This article is particularly centered on the predictive DT, which seeks to make precise estimations of the power load of users based on the past records.

2.2 The Selection of Attention Mechanism

The goal of this paper is to enhance LSTM networks to realize predictive DT.

LSTM is one of the types of Recurrent Neural Network (RNN) [16]. It can model long-term dependencies in sequences and is useful in processing sequential data, a problem that traditional RNNs face when processing long sequences because of the gradient explosion or gradient vanishing. Nevertheless, LSTM does not pay attention to the content of the input sequence, but depends on the changing of the hidden states and memory states to handle the sequence. In the case of sequential data, LSTM only captures the characteristics of the data using long-term dependencies in the sequence. In order to help LSTM pay attention to serial content at the same time, the attention mechanism is added to LSTM.

The thought of attention mechanism was evolved based on study that research on phenomena of attention in neuroscience. It be applied to filed of machine learning and deep learning widely, In the course of calculating the probability of attention's distribution, each words in sentences will be assigned a probability [17]. From this, two types of attention mechanism have been developed which are the soft attention mechanism and the hard attention mechanism. On one hand, the soft attention mechanism makes it possible for the model to weight all the input elements and thus, the model can attend to all the elements in the sequence. This enables the model can according to the importance and relevance of input sequence to processing data adaptively. On the other hand, the hard attention mechanism is a discrete attention mechanism that chooses a particular element from the sequence randomly. Hard attention mechanism often involves using a discrete sampling technique such as using random variables or probability distribution to select a position or element. Because of the discrete nature of hard attention, it often necessitates intricate optimization procedures during the training phase. Since gradients can be computed directly instead of using a random process and since the soft attention mechanism can be easily incorporated with the prediction algorithms, we choose the soft attention mechanism.

3 Algorithm

The following are the main components of the model as illustrated in the architectural design shown in Fig. 1; input-output layer, attention mechanism as illustrated in Fig. 2, LSTM network, and weight optimization layer. The input data comprises six attributes that can affect the power consumption of a particular electricity user and the output data refers to the power load of the particular user in the subsequent time period.

The weight a_t^k of the input sequence is calculated based on the previous hidden state h_{t-1} and cell state c_{t-1} in LSTM, and the computed $\widetilde{X}_t$ is then fed back into the LSTM unit. In the process of electricity load forecasting, data preprocessing must be performed first to meet the input requirements of the attention mechanism. Given the power consumption sequence Seq $= \{s_1, s_2, s_3, \ldots, s_N\}$, we divide it into a training sequence Seq$_{train}$ $= \{s_1, s_2, s_3, \ldots, s_M\}$ and a testing sequence Seq$_{test}$ $= \{s_{M+1}, s_{M+2}, s_{M+3}, \ldots, s_N\}$, where the subscript N represents the total length of the data sequence, M is the length of the training

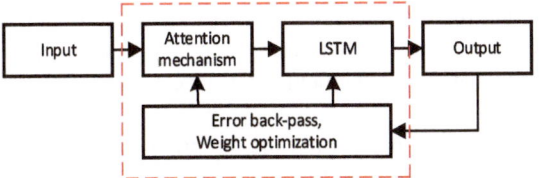

Fig. 1. Averall architecture of the model

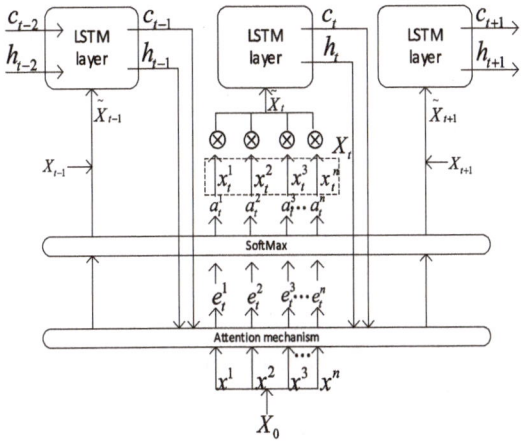

Fig. 2. Attention mechanism

sequence. At this point, the sequence is divided into n sub-sequences, and the value of n is calculated using the following formula:

$$n = [\frac{M - T}{k}] + 1 \tag{1}$$

where T is the length of each subsequence, which also represents the number of LSTM units, and k is the step size for shifting the data each time it is segmented.

The set of subsequences is denoted as $X_0 = Seq_{train} = \{x^1, x^2, x^3, ...x^n\}$, where x^2 represents the second subsequence, x^3 is the third subsequence, and so on. X_t^1 is an element of $X_t = \{x_t^1 x_t^2, x_t^3, ... x_t^n\}$, the attention mechanism is constructed by referencing the previous hidden state h_{t-1} and cell state c_{t-1} in the LSTM unit through input X_0, with $t = \{1,2,3...T\}$ representing the value of the initial subsequence at time t.where:

$$e_t^k = V_e^T \tanh(W_e[h_{t-1}; c_{t-1}] + u_e x_t^k + B_e) \tag{2}$$

$$a_t^k = \frac{\exp(e_t^k)}{\sum_{i=1}^n \exp(e_t^i)} \tag{3}$$

where V_e, W_e, U_e are parameters to be learned, B_e is the bias term, and a_t^k, the attention weights $k = \{1,2,3,...,n\}$ quantify the importance of the input consumption sequence at time t. The SoftMax function is used on k = {1,2,3,...,n} to normalize the attention weights so that they sum up to 1. The training segments are weighted according to Eqs. (2)–(3) giving higher weights to segments that have a larger impact on the prediction performance.

In electricity power load forecasting, the attention mechanism in the model gives more weight to the time periods with high peak power and a strong impact of the environment than the time periods with low values. This attention mechanism can be trained as a feed-forward network and can be trained together with other parts of the LSTM. By utilizing these attention weights, the sequence can dynamically extract relevant information: By utilizing these attention weights, the sequence can dynamically extract relevant information:

$$\widetilde{X}_t = (a_t^1 x_t^1, a_t^2 x_t^2, a_t^3 x_t^3, ..., a_t^n x_t^n)^T \tag{4}$$

4 Experiment

4.1 Dataset

The dataset contains six variables of a particular electricity user from 1 January 2006 to 1 January 2010, which is for a period of five years. These are the dry bulb temperature, dew point temperature, wet bulb temperature, humidity, price of electricity, and the electricity power load. This is done every half an hour, which means that the number of samples taken per day is 48, and the total number of data points in the dataset is 87648.

For the experiment, we trained the model using an NVIDIA GeForce RTX 4070 as the training device. In this work, we developed a deep learning framework using keras 2. 7. 0 version. The value of batch size was set to 30 and the value of time step was set to 48. The feature dimension was 6. In this work, we applied the sliding window technique, which means that we forecasted the power load of the 49th time step using the features of the previous 48 time steps (data from the previous day). We chose the Adam optimizer with the learning rate set to 0.01. In order to assess the model's predictability, loss function, R2, MAE, RMSE, and MAPE were employed. We compared the experimental results using three different approaches: a back propagation neural network (BP neural network), a Long Short-Term Memory neural network, and LSTM with attention mechanism.

4.2 Comparison of Loss Functions

From the loss values of the three experiments as shown in the Figs. 3,4,5, it is clear that the LSTM with attention mechanism converges faster and has the lowest loss. The following Fig. 6 presents the LSTM model integrated with attention mechanism for the prediction and it is evident that the predicted values are almost similar to the actual values with least error.

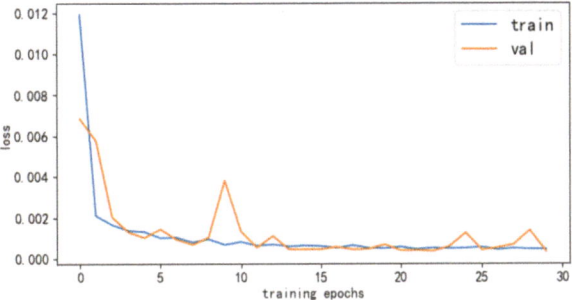

Fig. 3. Loss value of BP

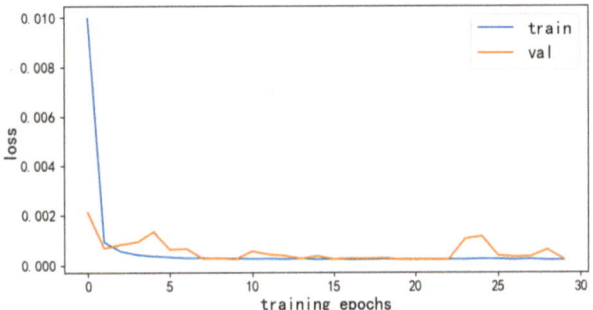

Fig. 4. Loss value of LSTM

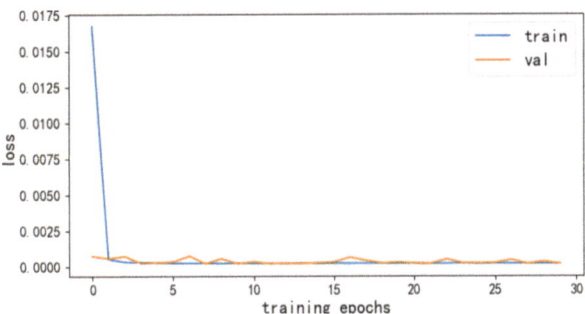

Fig. 5. Loss value of LSTM combined with attention mechanism

4.3   The Comparison of Model Evaluation Metrics

We compare the model accuracy using the four indicators in Table 1. A higher R^2 value indicates a more accurate model, while smaller values for the other evaluation metrics also indicate higher accuracy. Table 1 illustrates this comparison:

From the comparison made in the above table, it can be observed that the performance of BP is moderate while LSTM has a better value for accuracy. Comparing the results of all models, LSTM with attention mechanism has the highest R2 value closest to 1 and

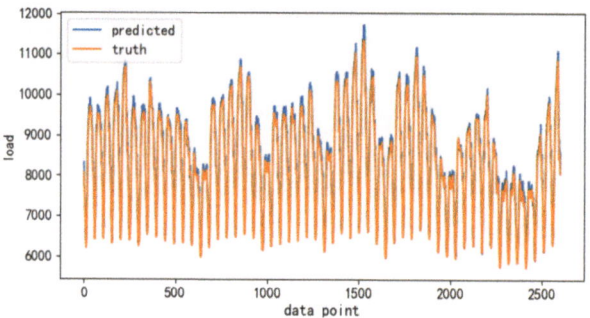

Fig. 6. Prediction results

Table 1. Model comparison

model/evaluation metrics	R^2	MAE	RMSE	MAPE
BP	0.98749	109.5	139.8	0.0133
LSTM	0.99263	84.0	107.2	0.0103
LSTM combined with attention mechanism	0.99267	83.6	106.9	0.0102

the lowest MAE, RMSE, and MAPE values, which means the best prediction accuracy. This indicates that the kind of attention mechanism that we selected is quite appropriate.

5   Conclusion

The incorporation of attention mechanism and LSTM in the predictive DT for power load forecasting has enhanced the load prediction accuracy. This model takes the input of the power load for the previous 48 time steps and predicts the power load for the next time step. Not only does it improve the accuracy of LSTM in load forecasting, but it also provides accurate predictions when environmental factors are present. Thus, the power system can be operated more efficiently based on the optimized operations and resource allocation with the help of this predictive DT model.

We also plan to expand the other parts of the DT in the future to enhance the students' learning experience. The independent DT can be created using the help of modeling tools such as Unity3D. When the data is populated, it will be possible to create a visualization dashboard to track the users' electricity usage in real-time. Furthermore, by integrating the Python programming language with the networking capabilities of Unity, we are able to produce real-time predictions and, therefore, set up a normative DT.

Acknowledgments. This research is supported by Inner Mongolia Natural Science Foundation Project (No. 2022MS06021).

References

1. Lindberg, K.B., Seljom, P., Madsen, H., Fischer, D., Korpås, M.: Long-term electricity load forecasting: current and future trends. Utilities Policy **58**, 102–119 (2019)
2. Colak, I., Sagiroglu, S., Fulli, G., Yesilbudak, M., Covrig, C.F.: A survey on the critical issues in smart grid technologies. Renew. Sustain. Energy Rev. **54**, 396–405 (2016)
3. Ventura, L.M.B., de Oliveira Pinto, F., Soares, L.M., Luna, A.S., Gioda, A.: Forecast of daily PM 2.5 concentrations applying artificial neural networks and holt-winters models. Air Qual. Atmos. Health **12**, 317–325 (2019)
4. Stability of multiple seasonal holt-winters models applied to hourly electricity demand in Spain
5. Rizkya, I., Syahputri, K., Sari, R. M., Siregar, I., Utaminingrum, J.: Autoregressive integrated moving average (ARIMA) model of forecast demand in distribution Centre. In IOP Conference Series: Materials Science and Engineering (Vol. 598, no. 1, p. 012071). IOP Publishing, August 2019
6. Rahayu, W.S., Juwono, P.T., Soetopo, W.: Discharge prediction of Amprong river using the ARIMA (autoregressive integrated moving average) model. In: IOP Conference Series: Earth and Environmental Science (Vol. 437, No. 1, p. 012032). IOP Publishing, February 2020
7. Dai, X., Sheng, K., Shu, F.: Ship power load forecasting based on PSO-SVM. Math. Biosci. Eng. **19**, 4547–4567 (2022)
8. Ahmad, W., et al.: Towards short term electricity load forecasting using improved support vector machine and extreme learning machine. Energies **13**(11), 2907 (2020)
9. Li, X., Wu, Y., Tang, C., Fu, Y., Zhang, L.: Improving generalization of convolutional neural network through contrastive augmentation. Knowl.-Based Syst. **272**, 110543 (2023)
10. Duan, F., Chapeau-Blondeau, F., Abbott, D.: Optimized injection of noise in activation functions to improve generalization of neural networks. Chaos Solitons Fractals **178**, 114363 (2024)
11. Jin, Y., Guo, H., Wang, J., Song, A.: A hybrid system based on LSTM for short-term power load forecasting. Energies **13**(23), 6241 (2020)
12. Ke, K., Hongbin, S., Chengkang, Z., Brown, C.: Short-term electrical load forecasting method based on stacked auto-encoding and GRU neural network. Evol. Intel. **12**, 385–394 (2019)
13. Jia, W., Wang, W., Zhang, Z.: From simple digital twin to complex digital twin part i: a novel modeling method for multi-scale and multi-scenario digital twin. Adv. Eng. Inform. **53**, 101706 (2022)
14. Niu, Z., Zhong, G., Yu, H.: A review on the attention mechanism of deep learning. Neurocomputing **452**, 48–62 (2021)
15. Stadtmann, F., et al.: Digital twins in wind energy: Emerging technologies and industry-informed future directions. IEEE Access (2023)
16. Schmidt, R.M.: Recurrent neural networks (RNNs): A gentle introduction and overview. arXiv preprint arXiv:1912.05911 (2019)
17. Bahdanau, D., Cho, K., Bengio, Y.: Neural machine translation by jointly learning to align and translate. arXiv preprint arXiv:1409.0473 (2014)

Malicious Node Detection and Robust Federated Aggregation Algorithm Based on TD3

Fan Sun, Hong Wen[✉], Wenjing Hou, Huanhuan Song, Yongfeng Wang,
Ruixiang Yao, and Dibao Yan

Institute of Aeronautics and Astronautics, University of Electronic Science and Technology of
China, Chengdu, China
{sunlike,hwj,hhs_communi,wyf_kt}@uestc.edu.cn,
Yizhi@std.uestc.edu.cn

Abstract. Malicious client nodes in federated learning can corrupt the global
model of central server by modifying dataset or submitted model, making it less
accurate or distorted. A TD3-based malicious node detection and robust federated aggregation algorithm is proposed. The algorithm excludes malicious clients
based on model distance, and uses reinforcement learning to assign model weight
to client nodes to reduce the influence of malicious nodes gradually. The experimental results show that the proposed algorithm is effective against data and model
modification attacks and is superior to traditional aggregation algorithms such as
FedAvg, Krum and MKrum.

Keywords: federated learning · node detection · robust aggregation · TD3
algorithm

1 Introduction

Traditional machine learning algorithms may lead to uncontrolled data flow and sensitive
data leakage problems, because it requires users to upload source data to a cloud server
with high computing power for centralized training. Federated learning can train the
global optimal model in the scenario of multi-party data source aggregation, transfer
the data storage and model training stage of machine learning to local users, and only
interact with the central server to update the model and effectively guarantees the privacy
security of users [1].

Although federated learning solves the problem of data islanding, the sharing of
federated learning model parameters will bring new opportunities to attackers when the
model is aggregated. For example, federated learning may be affected by data poisoning
attack. Data poisoning means that an attacker adds malicious data or tampered data to the
training process to make the trained model conform to the expectation, so as to destroy
the model or tamper with the model results. Malicious clients may deliberately disrupt
the learning process of the model through data poisoning, thus affecting the performance
of the entire model [2, 3].

© The Author(s) 2025
P. Siarry et al. (Eds.): WCNA 2023, LNEE 1361, pp. 375–383, 2025.
https://doi.org/10.1007/978-981-96-2409-6_37

Current research on federated learning mainly focus on adversarial training[4], model pruning[5], robust aggregation algorithms[6], regularization[7], and node detection[8]. These studies attempt to enhance the safety of federated learning through different methods.

In this paper, a malicious node detection and robust federated learning strategy based on TD3 reinforcement learning is proposed. Malicious client nodes are excluded based on model distance, and the model weights of each client node are dynamically adjusted to reduce the influence of malicious clients on the global model. The effectiveness of the proposed algorithm against data and model modification attacks is verified by experiments on label flipping attack and gradient ascent attack. Meanwhile, the proposed algorithm is compared with traditional aggregation algorithms such as FedAvg, Krum and MKrum. Experiments show that the proposed robust aggregation algorithm is better than traditional methods in terms of model accuracy and loss.

2 System Model

2.1 Label Flipping Attacks

Label flipping attacks represent a security threat in federated learning, where malicious participants intentionally modify the labels of their data to disrupt or manipulate the model's learning process. Label flipping attacks act directly at the data level by changing the labels of the data. The aim of this attack is to introduce incorrect learning signals, causing the model to make erroneous predictions or judgments.

In label flipping attacks, a malicious client m intentionally changes the label y_m in its dataset (x_m, y_m) to an incorrect label y'_m. The client still calculates the loss function and gradients following the normal process, but these gradients are now based on incorrect labels, thus misleading the overall learning direction of the model.

Suppose the loss function $L(\theta; x, y)$ reflects the degree of alignment between the model parameters θ and the data point (x, y). In label flipping attacks, the gradient computed by the malicious client is no longer $\nabla_\theta L(\theta; x_m, y_m)$ but $\nabla_\theta L(\theta; x_m, y'_m)$. This means that although the correct gradient calculation method is used, the input label is incorrect, thereby generating gradients based on a wrong target.

The parameter update formula, when malicious clients participate, becomes:

$$\theta \leftarrow \theta - \eta \cdot \frac{1}{K}\left(\sum_{k=1, k \neq m}^{K} \nabla_\theta L(\theta; x_k, y_k) + \nabla_\theta L(\theta; x_m, y'_m)\right) \tag{1}$$

In this scenario, $\nabla_\theta L(\theta; x_m, y'_m)$ is the gradient calculated based on the incorrect label y'_m This approach can lead the model to learn incorrect information on a global scale, especially when there is a significant difference between the incorrect and correct labels. Consequently, the model's performance and generalization capabilities may be severely compromised.

2.2 Gradient Ascent Attacks

Gradient ascent attacks in federated learning are a form of attack where malicious participants use the gradient information they send to disrupt or manipulate the entire learning process. In gradient ascent attacks, attackers typically attempt to modify their gradient contributions to enhance the model's performance on their specific targets or data, which usually negatively impacts the performance of all other honest participants. The general mathematical formulation of this attack is as follows:

Let θ be the model parameters and $L(\theta; x, y)$ be the loss function, where x and y represent the data and labels, respectively. In the normal federated learning process, each client k calculates the gradient of the loss function $\nabla_\theta L_k(\theta)$ with respect to its data, and sends it to the central server. The server then averages these gradients and updates the model parameters:

$$\theta \leftarrow \theta - \eta \cdot \frac{1}{K} \sum_{k=1}^{K} \nabla_\theta L_k(\theta) \tag{2}$$

wherein, η is the learning rate, and K is the number of participating clients.

In the case of gradient ascent attacks, assume that one or more clients are malicious. A malicious client m might send incorrect gradients $\nabla_\theta L'_m(\theta)$ to maximize their impact, which may be crafted to push the model parameters in a direction that enhances the attacker's performance on specific tasks. Malicious gradients could have a higher magnitude or differ in specific directions from the true gradients. Under such circumstances, the formula for parameter updates becomes:

$$\theta \leftarrow \theta - \eta \cdot \left(\frac{1}{K} \left(\sum_{k=1, k \neq m}^{K} \nabla_\theta L_k(\theta) + \nabla_\theta L'_m(\theta) \right) \right) \tag{33}$$

In this scenario, $\nabla_\theta L'_m(\theta)$ represents the maliciously altered gradient, which can be crafted using various strategies such as amplifying the magnitude of the gradient or altering its direction to mislead the model's learning process.

3 Algorithm Design

We proposes an algorithm based on the TD3 (Twin Delayed Deep Deterministic policy gradient) for malicious node detection and robust federated aggregation. Initially, some malicious nodes are excluded based on model distance, and then a deep reinforcement learning agent optimizes the entire training process. The agent uses the weights generated by the action network to precisely adjust the weight of each client's model update, effectively preventing malicious clients from damaging the global model training.

The procedure of our algorithm is as follows:

(1) Setup the basic federated learning environment, initialize parameters such as selecting datasets, training models, setting up a trusted model replay pool and so on.

(2) Set up the reinforcement learning environment, initialize two critic networks Q_{θ_1} and Q_{θ_2}, and their corresponding target networks $Q_{\theta_1'}$ and $Q_{\theta_2'}$. Initialize an actor network μ_{ϕ} and its corresponding target actor network $\mu_{\phi'}$ Initialize the experience replay pool to store experience tuples (s, a, r, s', d), where s is the state, a is the action, r is the reward, s' is the next state, and d indicates whether it is a terminal state.

(3) The server distributes the current model $G(t)$ to all clients, who perform local model training and then upload their updates to the aggregation server. The server uses Euclidean distance $dis_{i,j}^{euc}$ and cosine distance $dis_{i,j}^{cos}$ to calculate model update distances $dis_{i,j}^{choice}$, storing updates that fall below the distance standard in the trusted model replay pool. $dis_{i,j}^{euc}$ Helps to exclude malicious clients who change the weight of model parameters to increase their influence, while $dis_{i,j}^{cos}$ helps to exclude malicious clients with directional differences from benign clients.

(4) The agent performs actions in the environment according to the current policy $\mu_{\phi}(s)$, adding exploration noise for action smoothing. Based on the weight factors generated by the action network, the agent selects the current trusted model from the replay pool for global aggregation, producing a new round of global model $G(t+1)$. Update the reward r based on the current model's accuracy and also update the state s'. Store the transition tuple (s, a, r, s', d) in the experience replay pool.

(5) The agent extracts samples from the replay pool to update the Actor and Critic networks. For each sampled (s, a, r, s', d), use the target actor network $\mu_{\phi'}(s')$ and action noise to generate the next action a', and use the target critic networks $Q_{\theta_1'}$ and $Q_{\theta_2'}$ to calculate the target accuracy, update the critic networks Q_{θ_1} and Q_{θ_2} based on the target accuracy and minimizing the loss function, and Update the Actor Network μ_{ϕ}. Optimize ϕ by maximizing $Q_{\theta_1}(s, \mu_{\phi}(s))$. Use a soft update strategy to update the parameters of the target networks.

(6) Repeat steps (3)-(5).

4 Experimental Results Analysis

4.1 Basic Experimental Setup

The experiment employs the FMNIST dataset for image classification training, using a CNN model for image processing and recognition. The setup includes 100 client nodes participating, with clients chosen proportionally to participate in cooperative training. The number of global federated aggregation rounds is set to 20, with local training rounds set at 10. We controls the degree of attacks by modifying the proportion of malicious clients among all participating training clients.Below are the simulation experiment results and analysis of label flipping attacks and model replacement attacks under different degrees of attack and aggregation algorithms.

4.2 Label Flipping Attacks

Figure 1 shows the effects of label flipping attacks at different attack intensities. When there are no attacks, the model's accuracy remains at its highest (close to 90%), indicating good model performance under normal conditions. As the proportion of label flipping attacks increases (10%, 30%, 50%), there is a significant drop in model accuracy. The lowest accuracy occurs at a 50% attack ratio, illustrating the most destructive impact of the attack at this level. Under no attack conditions, the loss gradually decreases and stabilizes, indicating that the model can effectively learn under normal training. An increase in attack ratio leads to a significant increase in loss, especially at the 50% attack ratio, indicating the model's unstable performance under extreme attacks.

Figure 2 demonstrates the effectiveness of the proposed algorithm in defending against label flipping attacks. Compared to Fig. 1, the algorithm maintains a high accuracy rate (over 85%) across all attack ratios, almost equivalent to the no-attack scenario, showing that the algorithm effectively defends against label flipping attacks. The loss also quickly decreases and remains at a low level in all cases, further confirming the stability and efficiency advantages of the proposed algorithm.

Figure 3 shows the effectiveness of different aggregation rules in defending against label flipping attacks at a 30% attack level. The three traditional aggregation algorithms demonstrate some resilience against label flipping attacks, but the proposed algorithm (Ours) shows significant advantages in terms of accuracy and loss compared to FedAvg, Krum, and MKrum.

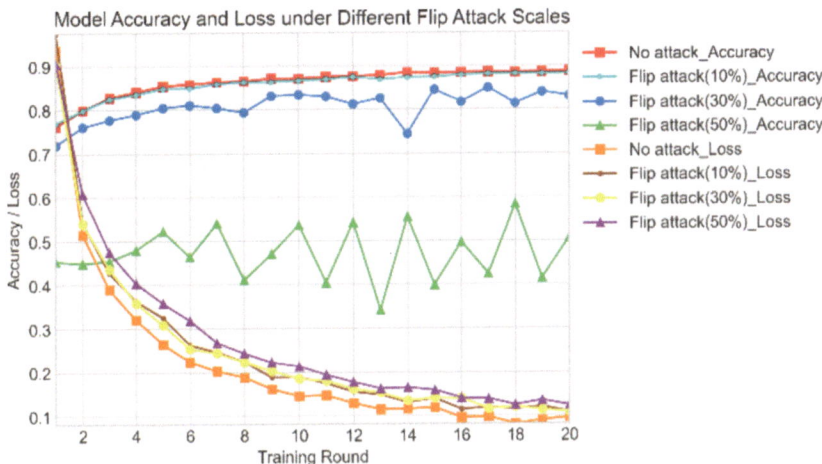

Fig. 1. Effects of label flipping attacks under different attack scales

4.3 Gradient Ascent Attacks

We investigates distributed multiple gradient ascent attacks in experiments. Distributed multiple ascent attacks occur in the 5th, 10th, 15th, and 19th rounds.

Figure 4 demonstrates the performance of the proposed algorithm in defending against distributed multiple gradient ascent attacks. The graph shows the model's response to distributed attacks at various intensities (10%, 20%, 30%) and the performance of the proposed algorithm in resisting these attacks. As the proportion of gradient

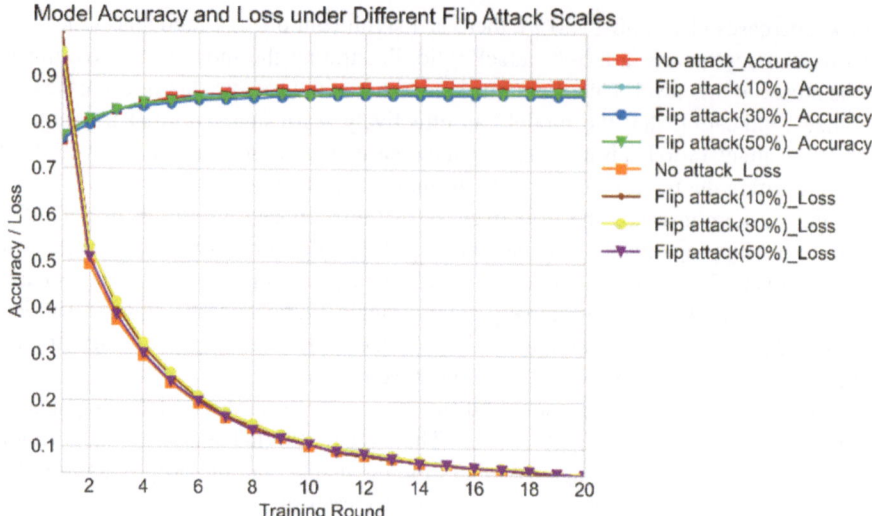

Fig. 2. Effectiveness of our algorithm in defending against label flipping attacks

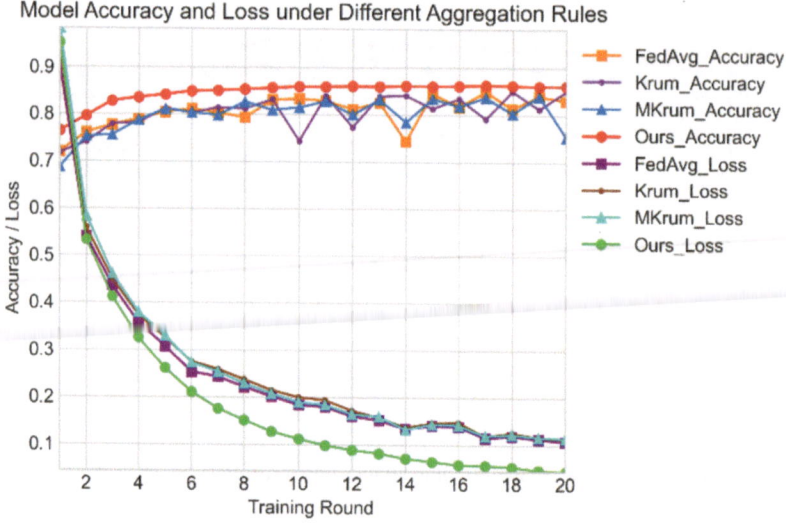

Fig. 3. Effectiveness of different aggregation rules in defending against label flipping attacks

ascent attacks increases, the model's accuracy significantly decreases, and loss fluctuation intensifies. However, the proposed algorithm is only slightly affected at a 30% attack intensity, and the model performs almost as well as when unattacked at lower attack intensities.

Figure 5 shows the effectiveness of different aggregation rules in defending against gradient ascent attacks at a 20% attack intensity. Compared to other aggregation rules,

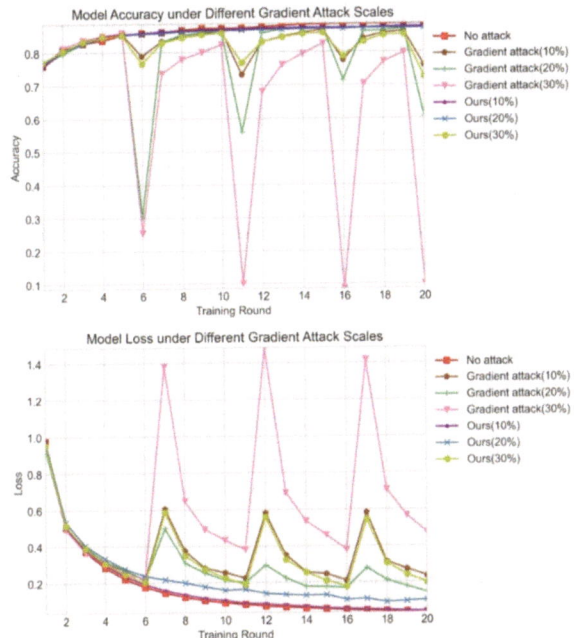

Fig. 4. Effectiveness of our algorithm in defending against multiple gradient ascent attack

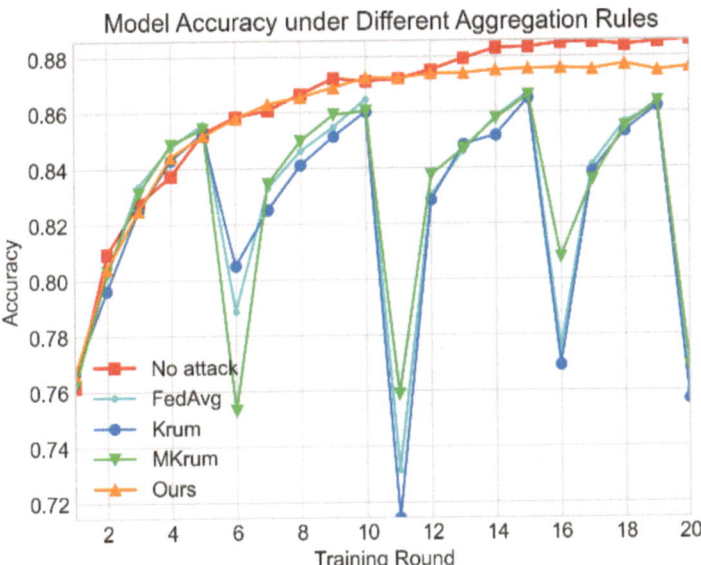

Fig. 5. Effectiveness of different aggregation rules in defending against gradient ascent attacks

the proposed algorithm exhibits higher accuracy under attack and a stable upward accuracy trend, with the final model performance close to that of the unattacked state. The other three traditional aggregation algorithms experience significant fluctuations in performance when facing gradient ascent attacks, all being adversely affected and leading to considerable training errors.

5 Conclusion

This paper introduces a new federated learning defense algorithm that utilizes TD3 reinforcement learning to detect malicious nodes and design robust federated learning strategies. This algorithm dynamically adapts and resists various potential security attacks such as label flipping attack and gradient ascent attack. Not only can it effectively identify and defend against malicious attacks, but it also enhances the overall security defenses of the model while maintaining efficient learning and model accuracy. Through experimental validation in simulated attack scenarios, this paper demonstrates the effective defense capabilities of the algorithm against the aforementioned types of attacks and proves that this algorithm outperforms traditional aggregation algorithms such as FedAvg, Krum, and Multi-Krum.

Acknowledgement. This work is supported by NSFC under Grant 61901089.

References

1. Kumar, K.N., Mohan, C.K., Cenkeramaddi, L.R.: The impact of adversarial attacks on federated learning: a survey. IEEE Trans. Pattern Anal. Mach. Intell. **46**(5), 2672–2691 (2024)
2. Bhagoji, A.N., Chakraborty, S., Mittal, P., et al.: Analyzing federated learning through an adversarial lens. In: International Conference on Machine Learning, pp. 634–643. PMLR (2019)
3. Liu, Z., Liu, Z., Yang, X.: Poisoning attack based on data feature selection in federated learning In: 2023 13th International Conference on Cloud Computing, Data Science & Engineering (Confluence), Noida, India, pp. 106–110 (2023)
4. Shah, D., Dube, P., Chakraborty, S., Verma, A.: Adversarial training in communication constrained federated learning (2021)
5. Jiang, Y., et al.: Model pruning enables efficient federated learning on edge devices. IEEE Trans. Neural Netw. Learn. Syst. **34**(12), 10374–10386 (2023)
6. Blanchard, P., El Mhamdis, E.M., Guerraoui, R., Stainer, J.: Machine learning with adversaries: byzantine tolerant gradient descent. In: Advances in Neural Information Processing Systems, pp. 118–128 (2017)
7. Jiang, X., Sun, S., Wang, Y., Liu, M.: Towards federated learning against noisy labels via local self-regularization. In: Proceedings of the 31st ACM International Conference on Information & Knowledge Management, pp. 862–873 (2022)
8. Rodríguez-Barroso, N., Marténez-Cámara, E., Luzón, M.V., Herrera, F.: Dynamic defense against byzantine poisoning attacks in federated learning. Futur. Gener. Comput. Syst.Comput. Syst. **133**, 1–9 (2022)

Getting Effect Prediction Model of CVTPD-PSL Technology Based on Artificial Neural Network

Caizhen Zhang[(✉)], Zongzhi Li, and Zaixing Wang

School of Electronic and Information Engineering, Lanzhou Jiaotong University,
Lanzhou 730070, China
zhangcaizhen@mail.lzjtu.cn

Abstract. In view of technology characteristic of Continuously Variable Phosphorous Getting Process Using a Porous Silicon Layer (PSL-CVTPDG), a prediction model based on Artificial Neural Network(ANN) is put forward for predicting effect of PSL-CVTPDG process. Establish, train, test and verify as long as simulation of the prediction model were finished by means of ANN function in MATLAB. Experimental results show that the prediction and actually measured values are very close to, the output follows the tracks of the expectation value very well, which reflects that the ANN is an effective method for predicting the gettering effect of porous silicon materials.

Keywords: porous silicon layer (PSL) · getting · Artificial Neural Network (ANN) · prediction model

1 Introduction

Continuously Variable Phosphorous Getting Process Using Porous Silicon Layer(PSL-CVTPDG) [1] can improve the performance index of electrical performance based on SOG-Si photovoltaic cell. But at present, the preparation of PSL must be involved in hydrofluoric acid. In the preparation process, the human body will inhale some poisonous gases that can not be excreted, which will damage the body of operating personnel through a long-term operation. Meanwhile, the gettering effect can not be measured until the gettering process is completely to the end, however, gettering process often need sample modulation, measurement, parameters modification, and then modulated repeatedly. As a result, it will increase the test cost and the amount of inhalation of hydrofluoric acid for operating personnel. Therefore, if establishing a more accurate model for predicting the gettering effect based on the experimental data have been obtained, it has a very important practical significance to optimize the process parameters of PSL-CVTPDG and save the cost of experiment.

There are many parameters affecting PSL-CVTPDG gettering effect, and the relationship between the affecting parameters and gettering effect is complex nonlinear and strong coupling. Therefore, using conventional method to establish the model of prediction on quality of gettering effect and analyzing the influence of technical parameters on gettering effect are of great difficulties. The Neural Network(NN) is a theory based on

P. Siarry et al. (Eds.): WCNA 2023, LNEE 1361, pp. 384–391, 2025.
https://doi.org/10.1007/978-981-96-2409-6_38

the structure and function of human brain. It has adaptive, parallelism, fault tolerance and strong non-linear processing ability, which can approximate a nonlinear function with arbitrary precision [2–6]. So, according to the characteristics of PSL-CVTPDG process, a prediction model based on Artificial Neural Network(ANN) is put forward for predicting the effect of PSL-CVTPDG process. The simulation results show that for the PSL-CVTPDG gettering effect, a nonlinear function with multiple inputs and multiple outputs that determined by a series of process parameters, using BPNN method with nonlinear approximation ability can obtain ideal gettering effect prediction.

2   Design of the ANN Prediction Model

In view of technical characteristics of PSL-CVTPDG, the four process parameters, i.e., the high temperature gettering temperature T_H, high temperature gettering time t_H, low temperature gettering temperature T_L and low temperature gettering time t_L have great influence on the gettering effect and are easy to be detected, which were selected as variables of input layer in ANN. The minority carrier lifetime change rate $\Delta\tau$ was defined as variable of output layer in ANN.

It has been approved that the tri-layer ANN possessing the S shape and linear transfer functions respectively in the hidden and output layer, respectively, can compel any non-linear function by any precision [6–8]. Therefore, an ANN with a single hidden layer given in Fig. 1 was chosen to construct the prediction model.

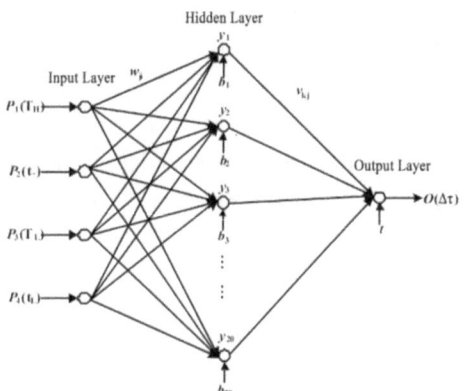

Fig. 1. Structure of ANN

By repeated calculation using the empirical formula [8]:

$$n_i = \sqrt{n+m} + 6 \tag{1}$$

the optimum number of hidden layer nodes n_i was determined as 20. Where $P_i(i = 1 \sim 4)$, y_j $(j = 1 \sim 20)$ and $o_k(k = 1)$ are neurons in the input, hidden and output layers, respectively. W_{ij} is the corresponding weights between the neuron i in the input layer and the neuron j in the hidden layer, $b_1 \sim b_{20}$ are the threshold values of the neuron j in the hidden layer, v_{kj} represents the corresponding weights between the neuron j in the hidden layer and the neuron k in the output layer, and t is the threshold of output neuron k.

3   Training and Prediction of the ANN Model

36 Sets representative data including the results of the orthogonal test [1] were extracted to construct the sample database shown in Table 1.

Table 1. Sample database based on the experimental results

Sample	1	2	3	4	5	6	7	8	9	10	11	12
$T_H(°C)$	1100	1000	900	1100	1000	900	1100	1000	900	1100	1100	1100
$t_H(min)$	90	90	90	60	60	60	30	30	30	60	45	30
$T_L(°C)$	750	800	700	700	750	800	800	700	750	700	700	700
$t_L(min)$	60	90	30	90	30	60	30	60	90	30	45	60
$\Delta\tau(\mu s)$	6.5	2.23	4.3	4.31	4.01	4.37	1.2	3.64	8.14	0.28	2.28	1.55
Sample	13	14	15	16	17	18	19	20	21	22	23	24
$T_H(°C)$	1000	1000	900	900	900	800	800	800	1100	1100	1100	1000
$t_H(min)$	60	45	60	45	30	60	45	30	30	45	60	45
$T_L(°C)$	700	700	700	700	700	700	700	700	750	750	750	750
$t_L(min)$	30	45	30	45	60	30	45	60	60	45	30	45
$\Delta\tau(\mu s)$	5.76	6.34	10.77	10.06	8.97	8.9	8.65	6.75	1.38	0.56	0.12	3.86
Sample	25	26	27	28	29	30	31	32	33	34	35	36
$T_H(°C)$	1000	900	900	900	900	1000	1100	1100	1100	1100	1000	1000
$t_H(min)$	30	60	60	45	30	60	60	30	45	60	60	45
$t_L(°C)$	750	750	750	750	750	700	700	800	800	800	800	800
$t_L(min)$	60	30	60	45	60	60	60	60	45	30	30	45
$\Delta\tau(\mu s)$	3.73	10.56	11.26	10.2	9.88	4.02	4.23	1.12	1.03	0.00	2.68	2.75

Figure 2 shows the flow chart of ANN prediction model. The input parameters T_H, t_H, T_L and t_L were collectively referred to as the system input vector X, $\Delta\tau$ was named as the system output vector Z.

Since each component of the system input X along with the system output Z of the prediction model have different dimensions, normally considering limitation in the numerical range of excitation functions and preventing the small numerical information being overwhelmed by the large numerical information, X and Z vectors cannot be directly used as the network input and output during the ANN mapping, which should be normalized [6] using the formula (2) and (3) to keep them within the [0,1] range.

$$P = AX + B \tag{2}$$

$$O = CZ + D \tag{3}$$

where A,B,C and D can be calculated by the formula (4), (5), (6) and (7),which are normalized transformation and deviation matrixes for input and output layers of the ANN, respectively.

$$A = \frac{1}{X_{max} - X_{min}} \tag{4}$$

$$B = \frac{-X_{min}}{X_{max} - X_{min}} \tag{5}$$

$$C = \frac{1}{Z_{max} - Z_{min}} \tag{6}$$

$$D = \frac{-Z_{min}}{Z_{max} - Z_{min}} \tag{7}$$

The normalized vectors P and O were used as the ANN input data and target data for training the configured ANN shown in part I. One fourth of the normalized sample database given in Table 1 is used for verification, one quarter for testing, and the other for training the network.

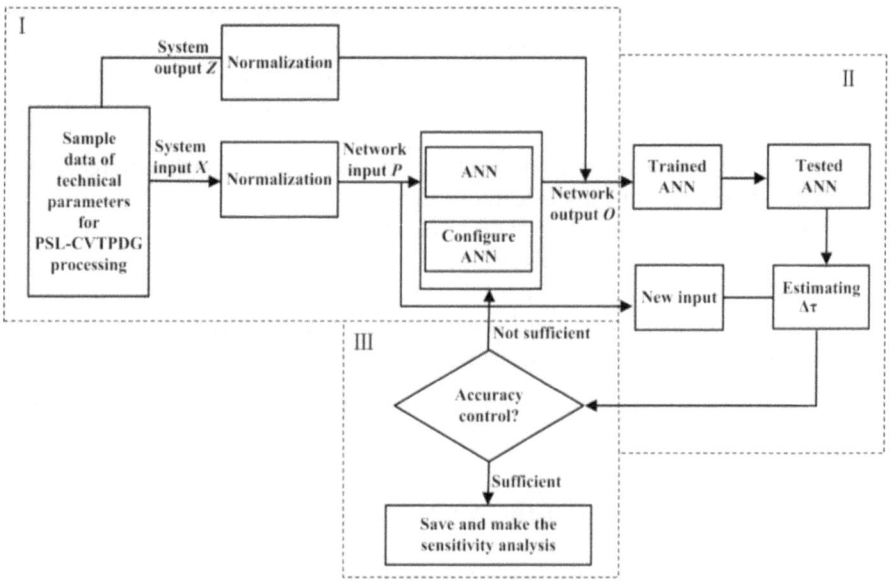

Fig. 2. Flow chart of ANN prediction model

The training steps of the prediction model based on ANN for PSL-CVTPDG process were following:

(1) Initialize weights and thresholds of the constructed ANN.
(2) Calculate the output of the ANN.

> For the constructed tri-layer ANN, the Sigmoid and linear transfer functions were adopted by the hidden and output neurons respectively, so the mapping between network output O and network input P is:

$$O_j = \frac{\exp(e_j) - \exp(-e_j)}{\exp(e_j) + \exp(-e_j)} \quad e_j = P_i w_{ji} + b_j \quad i = 1, 2 \ldots .4; \tag{8}$$

$$O_k = e_k \quad e_k = O_j v_{kj} + t_k \quad k = 1; \; j = 1, 2 \ldots \ldots .20 \tag{9}$$

(3) Calculate the training error between the output of the ANN and the experimental data to make a judgment whether the error is below an acceptable level. If the error does not satisfy the requirement, compare the current training error with the last training error, correct weights and thresholds according to the learning algorithm and turn to the step(2). Otherwise, the weights and thresholds between the connecting neurons then were determined, as shown in part II, Fig. 2, the trained ANN and the testing along with the verifying sample data were employed for testing and verifying the initial parameters to observe how close values were obtained to experimental effective lifetime change rate of the minority carriers. In the later stage of part II, Fig. 2, the new parameter set supplied in Table 2 was generated to perform the simulations for predicting new $\Delta\tau$ and the trained ANN can be used to make the parameters sensitivity analysis. In part III, Fig. 2, the most promising parameter giving the higher $\Delta\tau$ and the results of the sensitivity analysis were applied in a new experiment to validate the accuracy of the predicting model. If the accuracy was not sufficient, the ANN was reconfigured by turning back to part I, Fig. 2.

> The weights w_{ij}, v_{kj} and the thresholds b_j, t_k above are trained by the established ANN.

(4) Compare the current training error with the last training error, and then correct weight and threshold according to the learning algorithm.

> Make a judgment that whether the error satisfies a predetermined precision index. If it is satisfied, the training ends and the weights and thresholds between the connection units have been determined. Then extracting the sample data from the sample bank for testing and verification, to achive network test and verification, and output the prediction result; Otherwise, switch to step (2).

After setting the training parameters, the train function in Matlab can be called to complete the training of ANN.

4   Test and Verification of the ANN Model

In order to test the validity of the model further, 12 groups of new data given in Table 2 were selected to simulate the ANN.

The simulation results are shown in Fig. 3. Finally, all the training, testing, validation and simulating data were collected together and a linear regression analysis was carried out on the output of the ANN and the corresponding expected output vectors. The results are shown in Fig. 4.

Table 2. New parameters set for simulation

Sample	1	2	3	4	5	6	7	8	9	10	11	12
$T_H(°C)$	800	900	900	900	900	900	900	1000	1000	1000	1100	1100
$t_H(min)$	15	15	30	45	60	30	60	15	30	60	15	30
$T_L(°C)$	700	700	700	700	700	750	800	700	700	750	700	800
$t_L(min)$	30	45	60	45	30	90	60	30	60	30	90	30
$\Delta\tau(\mu s)$	0.25	2.23	8.89	10.32	10.58	8.14	4.37	4.31	3.64	4.01	4.31	1.22

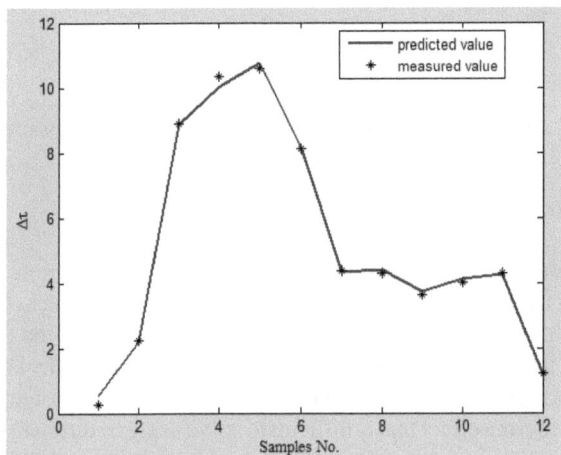

Fig. 3. Simulation result of the gettering effect

Figure 3 indicates that the predicting value is close to the measured value, and the mean relative error range of $\Delta\tau$ is 0.88%~1.86%. Figure 4 also shows that the output value track the desired value better, and the corresponding value of R exceeds 0.98. Thus, the proposed prediction model based on ANN for predicting gettering effect of PSL-CVTPDG process has higher prediction precision for the sample input, which is in accordance with the demand of the actual industrial process. That is to say, the model can accurately reflect the change trend of minority carrier lifetime after gettering.

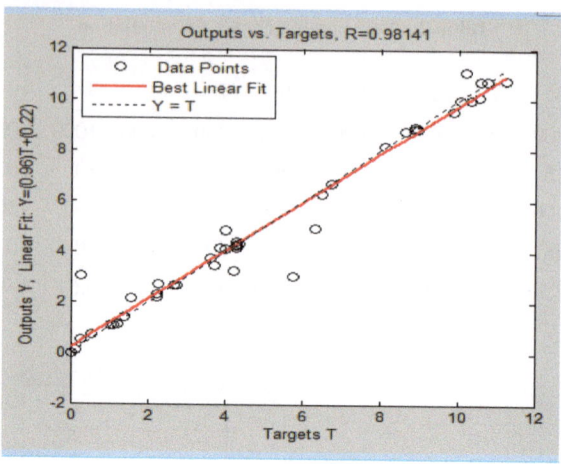

Fig. 4. Result of the linear regression model for PSL-CVTPGD process prediction analysis

5   Conclusion

In view of technology characteristic of PSL-CVTPDG,a prediction model based on Artifical Neural Network(ANN) is put forward for predicting effect of PSL-CVTPDG process. Establish, train, test and verify as long as simulation of the prediction model were finished by means of ANN function. The experimental results show that the prediction model has higher prediction accuracy, and can effectively solve the prediction of PSL-CVTPDG gettering effect that is difficult to solve using traditional mechanism modeling, which can provide some guidance for the optimization control of PSL-CVTPDG gettering process.

Acknowledgments. This work was supported by China Scholarship Council 2023 Western Region Talent Trainging Special Project (202308020193), the Natural Science Foundation of Gansu Province(23JRRA844), the Teaching Reform Project of Lanzhou Jiaotong University (JGY202301).

References

1. Zhang, C.Z., Liu, S., Wang, Y.S., Chen, Y.G.: Performance improvements of solar-grade crystalline silicon by continuously variable temperature phosphorous gettering process using a porous silicon layer. Mater. Sci. Semiconductor Process. **13**(3), 209–213 (2010)
2. Chen, W.C., Lee, A.H.I., Deng, W.J., Liu, K.Y.: The implementation of neural network for semiconductor PECVD process. Expert Syst. Appl. **32**, 1148–1153 (2007)
3. Yang, C.H., Deconinck, G., Gui, W.H., Li, Y.G.: An optimal power-dispatching system using neural network for the electrochemical process of zinc depending on varying prices of electricity. IEEE Trans. Neural Netw. **13**(1), 229–236 (2002)
4. Diego, L., Mark, H.W., Simon, Y.F., Hector, A.M.: A speed and accuracy test of backpropagation and RBF neural networks for small-signal models of active devices. Eng. Appl. Artif. Intell. **19**, 883–890 (2006)

5. Sagar, P., Afzal, H.K., Ashish, K., et al.: Sustainable cement replacement using waste eggshells: a review on mechanical properties of eggshell concrete and strength prediction using artificial neural network. Case Stud. Constr. Mater. **18**, e02160 (2023)
6. Elias, D.R.L., Marlon, M.S., Carlos, H.L., et al.: Nonlinear receding-horizon filter approximation with neural networks for fast state of charge estimation of lithium-ion batteries. J. Energy Storage **68**, 107677 (2023)
7. Kasenee, T., Rapat, P., Vitchaya, S., et al.: Convolutional neural network based artificial intelligent to perform real-time sematic segmentation on gastric intestinal metaplasia. Gastrointestinal Endoscopy **96**(6), AB228 (2022)
8. Shoffi, I.S., Riyanarto, S., Joko, S.: Estimating gas concentration using artificial neural network for electronic nose. Procedia Comput. Sci. **124**, 181–188 (2017)

Author Index

© The Editor(s) (if applicable) and The Author(s) 2025
P. Siarry et al. (Eds.): WCNA 2023, LNEE 1361, pp. 393–394, 2025.
https://doi.org/10.1007/978-981-96-2409-6